**Engineering Materials**

This series provides topical information on innovative, structural and functional materials and composites with applications in optical, electrical, mechanical, civil, aeronautical, medical, bio- and nano-engineering. The individual volumes are complete, comprehensive monographs covering the structure, properties, manufacturing process and applications of these materials. This multidisciplinary series is devoted to professionals, students and all those interested in the latest developments in the Materials Science field, that look for a carefully selected collection of high quality review articles on their respective field of expertise.

**Indexed at Compendex (2021) and Scopus (2022)**

Alessio Malandruccolo

# Superaustenitic Stainless Steels

## A Comprehensive Overview

Springer

Alessio Malandruccolo
Department of Industrial Engineering
University of Trento
Trento, Italy

ISSN 1612-1317 ISSN 1868-1212 (electronic)
Engineering Materials
ISBN 978-3-031-68743-3 ISBN 978-3-031-68744-0 (eBook)
https://doi.org/10.1007/978-3-031-68744-0



This Springer imprint is published by the registered company Springer Nature Switzerland AG
The registered company address is: Gewerbestrasse 11, 6330 Cham, Switzerland

*To all those who have taught me something,*
*to my teachers and mentors.*

# Foreword

I have been working in the world of stainless steels for several decades, basically since I finished my military service after completing my engineering studies at the Politecnico di Milano (Italy).

Education is certainly an important basis for being able to have a global vision and to create a *forma mentis* capable of rationally absorb other specific notions, which are necessary to become specialized in a given field and to have a proactive approach to the various problems that arise from time to time during an individual's working career. Equally important, however, is the path that is followed in the practice. This is the so-called *field experiences* that must properly shape the mind, make it flexible and capable of storing, in an organic manner, the greatest number of notions to be quickly recalled at the appropriate time.

Of course, as in the educational path, opportunities and good fortune of meeting people who can transmit their previous experience, which should always be treasured, also count a lot in the working field. Crucial, for me, was the guidance of a professional I found at the beginning of my career who was able to skillfully convey basic concepts to me.

In all these years spent in the same field, I have had the opportunity to study and touch the applications of stainless steels in the most diverse sectors: chemical and petrochemical, food, construction and infrastructure, automotive, power generation, just to mention the most known of them. All these sectors are characterized by different issues about the use of stainless steel, both in terms of physical and mechanical performance and in terms of ability to resist corrosive phenomena.

In the field of stainless steel and superalloys' production, there has been an increase in the types produced in recent decades. This is a direct consequence of the progress in technological innovations in the various industrial sectors and a result of end users' demands for increasingly higher performance and better characteristics.

There is certainly no shortage of bibliographical references that give even very detailed technical information on stainless steels, although percentage-wise, in terms of production, this family represents only a small slice in the general panorama of steels. However, when searching for news on "niche" materials, such as superaustenitics, superduplex, it is indeed possible to find several references in the literature, but still present not in an organic way and with very fragmented information. For this reason, professionals involved in design or "decision makers" struggle quite a bit to find news useful for their work and are very often forced to create knowledge on their own skin in a path that is not always studded with success.

So, when the author told me that he was considering writing a text dedicated to superaustenitic stainless steels, I greatly appreciated the initiative and encouraged him in starting the work and carrying it out as soon as possible, precisely because I realized the gaps that existed in the technical and scientific support for people working in the field.

Therefore, I was sure that this text was a necessary building block of the culture on a very particular type of steel. I was also sure of the quality of the result, knowing first of all the professional expertise of the author, gained initially through experience conducted in the field of stainless steel and nickel alloys production and later refined in academia. Therefore, I had perceived all the necessary elements, mentioned at the beginning, to obtain a successful text in terms of content. Indeed, for optimal results, it is essential not only to fully master a subject and to have experienced it on one's own skin, but also to have the skills to tell it in an easy and comprehensive way to an extremely diverse audience.

I had immediately glimpsed the combination of the author's high professionalism and communication skills when I first met him in 2018 during a seminar organized for the stainless steel industry. In fact, looking at the table of contents of the proposed topics and following some of the steps at the beginning of the writing of this text, I had full confirmation of my initial feelings.

All aspects of these "premium" stainless steels have been comprehensively addressed, where the boundary with nickel alloys is very blurred and in some cases there are application fields where performance can be overlapped. Moreover, precisely in lieu of the distinctive nature of these steels, they have been compared with more well-known types of stainless steels in order to provide an effective comparison. Therefore, the professional does not have to go searching for technical news among specific papers presented at dedicated conferences or in specialized texts that deal with only part of the issues: here he can find a handbook with everything he needs for a 360-degree view. The author has also cleverly enriched the text with detailed theoretical information to support the explanation of phenomena that can be found in the industrial practice in which he has spent a substantial part of his professional career. In addition to this, there are many insights for personal study organized in order to keep interest and curiosity alive in the reader.

I see considerable international potential in this text and hope it can become a true guide, constituting one of the main reference points in the world panorama of scientific literature related to the “stainless” world.

Fausto Capelli
Technical Advisor of Centro Inox
and Director of the “Inossidabile”
Magazine
Milan, Italy

# Acknowledgements

I must mention here and express my gratitude for the help and support received by the Springer team, especially to Marina Forlizzi, Barbara Amorese, and Kavitha Palanisamy.

Despite the purely technical nature of this work, there is a photograph within Chap. 1 that truly fascinated me from the first time I saw it many years ago. Hence, thanks are certainly owed to my old friend Cristian Pavan for contributing to this book by giving me that photograph taken during one of his travels: the Eiffel Tower featured in Chap. 1.

This text could not have discussed superaustenitic stainless steels in a comprehensive way without presenting even a micrograph of them in order to appreciate their microstructure, as can be done by technicians who deal with analysis on these materials every day. Since one of the goals pursued by this text is to provide tools, knowledge and insights to understand superaustenitic stainless steels, the presence of pictures of the microstructure is essential. Accordingly, special thanks are expressed to Omeco S.r.l. for providing the micrographs in this text, thereby contributing a key building block for preparation.

When writing a text, it often happens that one feels the burden of responsibility to transfer the knowledge to others. So many questions arise: will I be able to pass on all the knowledge? Will I be able to find an effective way to do it? Will I be able to make myself understood by the reader? Will I be able to thrill the reader? Will I be able to invent graphic ways to synthesize information effectively? Often these questions are not prompted by an awareness of not knowing how to do things, but by a concern for wanting to do them well and as well as possible. This is because there is a lot of emphasis on the role that one's text can play for the person consulting it. In these moments when so many questions crowd the writer's mind, there are people who can provide instant clarity by reminding the author of the motivations that drove him or her to write. There are people who grasp from the outside the author's passion and expertise and remind him or her how the goal pursued is indeed attainable. People who have sincere friendship and admiration offer their support despite the distance and age difference. In this context, special thanks go to Fausto Capelli. He is a person I admire, whose enthusiasm for technical topics seems inexhaustible and from whom

I have drawn much inspiration over the years. Since we met, he has always had words of praise and admiration for me. Since I told him of my desire to write this text he encouraged me to complete it. Certainly, this text has also taken on a comprehensive structure thanks to his suggestions.

# Contents

# About the Author

**Alessio Malandruccolo** has experience as industrial consultant in Italy and Europe, and is also currently affiliated with the Department of Industrial Engineering of the University of Trento (IT). He gained experience in the steel industry working for a prestigious Italian company, specialized in the production of stainless steels and nickel alloys. At that time, he was involved in R&D, quality control and improvement, failure analysis, and revamping and design of new installations for production of new alloys. He has coordinated product development and industrialization projects for the Italian site of a German company operating in the sheet metal working field, with main focus on the home appliance industry. Subsequently, he started one of the Italian sites of a company operating in the field of mechanical design and engineering consulting. He received his Master degree in Materials Science and Engineering at the University of Trento. He had the opportunity to develop part of his research experience and skills at the European Synchrotron Radiation Facility—ESRF (Grenoble, France)—and at the Diamond Light Source (Didcot, United Kingdom), focusing mainly on X-ray diffraction applied to inorganic materials characterization. Since shortly after graduation, he has been involved in academic teaching as a teaching assistant in Materials Technology and Applied Chemistry at the University of Trento and as a lecturer and teaching assistant in Manufacturing Technology at the Free University of Bolzano. Since 2020, he has also been teaching at high schools on Manufacturing Technology and Applied Science and Technology. He has co-authored publications related to the processing of titanium alloys produced by additive manufacturing. Together with Stefano Gialanella, he is the author of the book *Aerospace Alloys*.

# Abbreviations

| | |
|---|---|
| AES | Auger electron spectroscopy |
| AI | Artificial intelligence |
| AM | Additive manufacturing |
| AOD | Argon oxygen decarburization |
| APF | Atomic packing factor |
| ASME | American society of mechanical engineers |
| ASTM | American society for testing and materials |
| bcc | Body centered cubic |
| bct | Body centered tetragonal |
| BJ | Binder jet |
| CC | Continuous casting |
| CCT | Continuous cooling transformation |
| CCT | Critical crevice temperature |
| CN | Coordination number |
| CNC | Computerized numerical control |
| CPCC | Critical pitting chloride concentration |
| CPT | Critical pitting temperature |
| CRPs | Copper-rich precipitates |
| CSLM | Confocal scanning laser microscopy |
| DBTT | Ductile-brittle transition temperature |
| DED | Direct energy deposition |
| DET | Drop evaporation test |
| DS | Directional solidification |
| DSC | Differential scanning calorimetry |
| DSS | Duplex stainless steel |
| DTA | Differential thermal analysis |
| EAF | Electric arc furnace |
| EB | Electron beam |
| EBM | Electron beam melting |
| EBSD | Electron backscattering diffraction |
| EDS | Energy dispersive spectroscopy |

| | |
|---|---|
| ELC | Extra-low carbon |
| EN | Euronorm |
| EPMA | Electron probe microanalysis |
| ESR | Electroslag remelting |
| FCAW | Flux-cored arc welding |
| fcc | Face centered cubic |
| FGD | Flue-gas desulfurization |
| GB | Grain boundary |
| GMAW | Gas metal arc welding |
| GTAW | Gas tungsten arc welding |
| GTE | Gas turbine engine |
| HAZ | Heat-affected zone |
| hcp | Hexagonal close-packed |
| HDSS | Hyperduplex stainless steel |
| HE | Hydrogen embrittlement |
| IC | Ingot casting |
| IDR | Interdendritic region |
| IOB | Iron-oxidizing bacteria |
| ISS | Ion scattering spectroscopy |
| IUPAC | International union of pure and applied chemistry |
| LF | Ladle furnace |
| MED | Multiple-effect distillation |
| MIC | Microbiologically-induced corrosion |
| MIG | Metal inert gas |
| MOB | Manganese oxidizing bacteria |
| MSB | Microstructural banding |
| MSF | Multi-stage flash |
| NNSC | Near-net shape casting |
| PBF | Powder bed fusion |
| PBR | Pilling-Bedworth ratio |
| PD | Primary dendrite |
| PH | Precipitation hardening |
| PHSS | Precipitation hardening stainless steel |
| PRE | Pitting resistance equivalent |
| PREN | Pitting resistance equivalent number |
| QDS | Quenching during directional solidification |
| R&D | Research and development |
| RE | Rare earth element |
| RO | Reverse osmosis |
| SASS | Superaustenitic stainless steel |
| SCC | Stress corrosion cracking |
| SD | Secondary dendrite |
| SDAS | Secondary dendrite arm spacing |
| SDSS | Superduplex stainless steel |
| SEM | Scanning electron microscopy |

| | |
|---|---|
| SHE | Standard hydrogen electrode |
| SIM | Strain-induced martensite |
| SIMS | Secondary ion mass spectroscopy |
| SLM | Selective laser melting |
| SMSS | Supermartensitic stainless steel |
| SRB | Sulfate-reducing bacteria |
| TCP | Topologically close-packed |
| TEM | Transmission electron microscopy |
| TIG | Tungsten inert gas |
| TRSC | Twin-roll strip casting |
| TTT | Time temperature transformation |
| UNS | Unified Numbering System |
| VAR | Vacuum arc remelting |
| VIM | Vacuum induction melting |
| VOCD | Vacuum oxygen converter decarburization |
| VOD | Vacuum oxygen decarburization |
| WAAM | Wire arc additive manufacturing |
| WIG | Wolfram inert gas |
| WN | Werkstoffnummer |
| XANES | X-ray absorption near edge structure |
| XPS | X-ray photoelectron spectroscopy |

# Chapter 1
# An Introduction to Stainless Steels

**Abstract** It is undeniable that despite the technological progress of the past 250 years, steel still plays a key role in modern technology, and in the public imagination it continues to be the benchmark material to be compared with the latest that have been developed. For a long time, steel has been considered the universal material, and research in the past has led to the development of countless types of this material. As happens quite often in research and development, almost by accident it was noticed that the addition of chromium to steel improved its corrosion resistance properties. This was the incipit for the birth of stainless steels, a family of alloys characterized by an interesting combination of corrosion resistance and the possibility of tailoring mechanical and technological properties. Since that time, major families of stainless steels have been developed, and the continuing need to meet new challenges in applications has led to the formulation of increasingly complex compositions to give rise to new, higher-performance materials. With regard to stainless steels, the main driving force for development is often the corrosion resistance improvement with the goal of approaching, if not exceeding, the performance of a family of alloys that are in some ways close relatives to some stainless steels, but more expansive and based on another metal: nickel alloys. This race for improvement led to the development of high-performance stainless steels, among which the most similar group to nickel alloys stands out: superaustenitic stainless steels. In order to fully understand the characteristics of this group of alloys, a necessary step lays in the analysis of the other types. This turns out to be a key aspect for understanding the differences, superiority and critical issues of these alloys, as well as the possibilities for their use.

**Keywords** Stainless steels · Archaeometallurgy · Steel development · Nomenclature

 
A. Malandruccolo, *Superaustenitic Stainless Steels*, Engineering Materials,
https://doi.org/10.1007/978-3-031-68744-0_1

## 1.1 Historical Background: Steel and Metallurgy

One of the most diffused simple definition for steel is Fe–C-based alloy. However, in common language, steel and the chemical element Fe are often thought to be the same material because people usually associate every metallic object with the most notorious and better-known metal: iron. This is likely to happen due to the fact that steel is a relatively young material, at least that produced by deliberately adjusting the chemical composition and by means of industrial and mass-scale processes; while iron has a millennial legacy in human culture. In fact, Fe and its metallurgy were well known since the ancient times and examples of Fe production from ores date back to 2000 BC. Furthermore, the start of the Iron Age has been recognized to be around 1200 BC (see Fig. 1.1), a moment when iron production from ores started to spread in China, in the Near East, as well as in the Mediterranean area. Before that time, the only possibilities to find iron in a relatively pure form in nature were telluric iron (also known as native iron) and meteoric iron. The first can be found in basaltic rocks with a Ni impurity level of maximum 3 wt%, while the second is the result of the impact of meteorites on Earth and is characterized by a Ni content ranging between 4 and 26 wt%. The latter source is the reason for the poetic name given to Fe-related sciences and activities in Latin languages. In fact, siderurgy derives from the Latin *sidus-sideris*, i.e., star; a concept related to the "magic metal" rained down from the sky [1].

Any metallic material produced at a temperature below its melting point features a characteristic spongy appearance. Iron, chemical element with atomic number equal

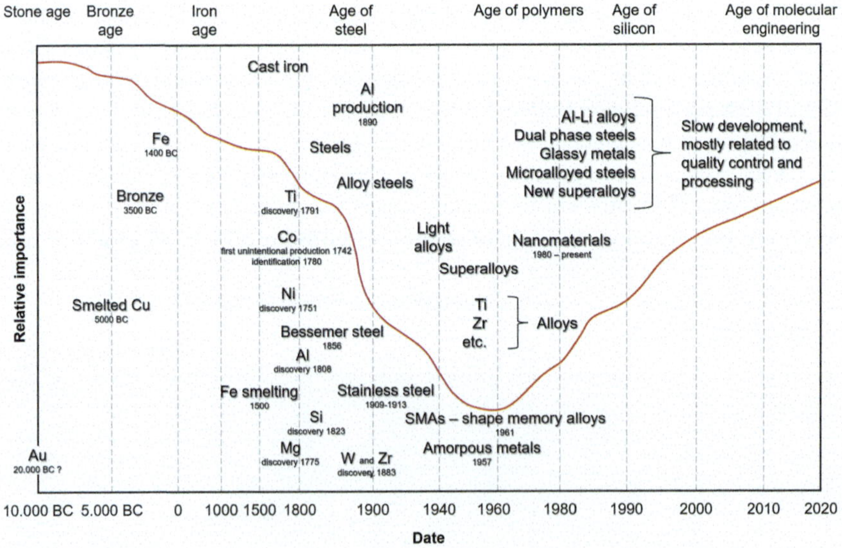


**Fig. 1.1** Relative importance of metals and metal alloys through the eras of human history (redrawn from [1, 2])

to 26 and melting temperature of 1538 °C, also presented itself to our ancestor's early ironworkers in this form. In the past, in order to be able to produce Fe-based artifacts, it was not enough to shape pieces of telluric or native iron. In fact, it became necessary to develop a dual process: transformation of the raw material into Fe sponge and subsequent plastic deformation [3]. As time went by, the technological evolution led to the production of Fe-based materials with higher and higher performance thanks to the tradition and experience handed down by blacksmiths. This process inevitably led to steel, first unconsciously due to the C contamination resulting from the coal-based heating process, and then intentionally with the modern industrial processes made possible by the introduction of coke. To come up with the concept of steel as we know it now, the adoption of coke was a fundamental milestone that eventually led to blast furnaces, the output of which is a particular type of cast iron named *pig iron.* A necessary step after this phase is the reduction of the excessive carbon content (typically higher than about 2 wt%) and this was achieved thanks to the development of converters, in particular the Bessemer converter named after its designer: Sir Henry Bessemer [4].

Prior to the adoption of the Bessemer process, the most common form of malleable Fe–C alloy was wrought iron. In fact, wrought iron was used for structural applications with high produced volumes such as railroad rails and more or less all the today's uses of mild steel (low-C steel) in view of its very low C content. Among the many examples of artifacts of the past built in wrought iron, two of certain interest are the Delhi Pillar and the Eiffel Tower (see Fig. 1.2). In the first case, the wrought iron pillar of Delhi (India) dates back to the fourth century and is more than 7 m in height. The pillar has resisted corrosion over many centuries, raising the interest of corrosion scientists. Its good corrosion behavior results from a favorable environment (no chlorides and no sulfur compounds) together with a good chemistry of the material showing the absence of P and S. This combination led to the formation of a protective and reasonably thick oxide layer (probably magnetite-based). The other more recent example is the *Tour Eiffel*, the wrought lattice tower on the *Champs de Mars* in Paris (France). It was finished in 1889 for the inauguration of the *Exposition Universelle* of 1889. Today the Eiffel Tower is one of the symbols of the French nation, although it can be considered as one of the most famous cases of monument that at the time of construction was subjected to much controversy and outrage, while now is universally appreciated. Wrought iron features high ductility, toughness, good weldability and corrosion resistance in some environments. It is characterized by a high purity degree and consists essentially in pure Fe with less than 0.08 wt% C containing few percent of glassy silicate inclusions coming from slag, which typical elongated shape results from working operations the material was subjected to. Failures of wrought iron show a typical fracture morphology characterized by a wood-like appearance and by the fact that fracture usually occurs along the silicate strings [4].

The development of the Bessemer converter in 1856 is unanimously recognized as the beginning of the modern steelmaking, together with the immediately next open-heart process. Reconnecting now with the incipit of this section, steel is a name, a concept with which everyone is familiar because it is possible to experience its use in an enormous number of products [1, 4]. The origin of the word steel dates

**Fig. 1.2** The Eiffel tower on the Champs de Mars in Paris: an historical example of use of wrought iron (photo property of the Author)

back to a prehistoric West Germanic word, *stakhlam*, that is supposed to be derived from the Germanic base *stakh-* (to be firm). This latter word resembles the Sanskrit *stakati*. The modern German for steel is *Stahl*, while in Russian the word translation is *stalin*. In Latin-derived languages such as Italian and Spanish, steel is translated with the words *acciaio* and *acero*, respectively. The common root seems to be the Latin *aciarium*, name associated in the ancient times to the hardened iron used for the fabrication of tips (Latin *acies*) of swords and knives [1, 5, 6]. Steel can be found in almost every technological field and in every application, from cars in the automotive industry, to aircrafts and satellites in the aerospace, arriving to the deep sea and offshore oil and gas applications. Its extraordinary ability of being tailored and modified has made it a universal material with an almost infinite possibility of grades and chemistries.

## 1.2 Stainless Steels

In Italian, stainless steel is translated as *acciaio inossidabile* (often simply *inox*), i.e., not subjected to oxidation. This nomenclature is obviously used improperly because the superior resistance to corrosion of this class of materials is given by a thin layer of Cr-based oxide that forms on the surface. The German word for stainless steel is *Edelstahl*, a name that combines the adjective *edel* (noble) and the name *Stahl* (steel). This recalls the idea of nobility of a metal or alloy exploited as its better resistance to corrosion, thus resembling the concept of nobility expressed when comparing materials with different standard reduction potentials, where the nobler one resists better to corrosion. The existence of stainless steels is therefore inextricably linked to Cr, because it is the element on which stainless steels rely their superior corrosion resistance. In the following sections the race that historically led to the large-scale production of these materials will be summarized. Emphasis will be deliberately given to the reason for the superior corrosion resistance of stainless steels, carefully explaining the concept of passivity using also the results of recent studies and models. The different types of stainless steels will be presented with the aim of later describing superaustenitics with enough knowledge to allow the reader to understand the most interesting aspects related to this family of alloys. Finally, to conclude the general overview, applications and markets for stainless steels will be analyzed.

### *1.2.1 Fundamentals, Corrosion Resistance and Passivity*

As depicted by Fig. 1.1, stainless steels can be regarded as relatively young materials as their development is linked to the isolation of metallic Cr. In fact, Cr was first identified more than 200 years ago by the French chemist Louis Nicholas Vauquelin, who presented the world's first piece of Cr in 1798 at the French Academy [7]. As a consequence of this, the history of stainless steels dates back to the beginning of the nineteenth century, when great effort was put in developing new alloy steels with improved oxidation resistance. For example, in 1821 the French Pierre Berthier realized that the addition of Cr to Fe made the resulting alloy more resistant against corrosion induced by acid substances [1]. It is interesting to note that the fervent scientific interest in improving steel also involved scientists who would gain prestige for extraordinary merits in other disciplines. This is for example the case of Michael Faraday, known to us for his outstanding groundbreaking work in the field of electricity and magnetism. In their paper "Experiments on the Alloys of Steel, made with a View to its Improvement", James Stodart and Michael Faraday proposed the results of a series of alloying experiments performed on pure Fe and on steel. Several alloying elements were investigated: Ag, Au, Cu, Ni, Pt and Rh. These experiments had a twofold main objective, and a collateral one. The main objective was to verify the possibility of using the abovementioned elements as effective alloying elements for steel with the aim of producing cutting instruments more performing than those made with existing steel. In addition to this, in case there were alloys proving effective improvement in cutting performances, it would have been necessary to evaluate whether they were less prone to oxidation. Finally, the collateral object was to evaluate these combinations for reflecting mirrors. A further investigation of the use of specific alloying elements was presented two years later by the same authors in the publication "On the Alloys of Steel". In addition to the enthusiasm for the success of earlier experiments of the Royal Institution, for the fact that alloys similar to those tested started to be produced in other countries, Stodart and Faraday reported the results of using certain alloying elements in small scale experiments, including Cr. Two combinations of steel with different amounts of Cr were tried. Due to the low content of Cr used (maximum value of about 3 wt%), mechanical properties were improved, but except for some interesting surface features, no better corrosion resistance was obtained as happened instead with the additions of Ni, Pt and Rh [8, 9].

In addition to laboratory experiments, which were carried out using very limited amounts of material, the production of large volumes also began to come up against the limitations of the knowledge of the time. In fact, early attempts to produce ferroalloys to be used in the alloying process for large-scale productions unfortunately resulted in high C contents, which precluded the possibility to effectively produce stainless steels (early trials by Pierre Berthier). Subsequently, an aluminothermic reduction process used for the low-C ferrochromium production route was patented in 1891 by the German chemist Hans Goldschmidt, paving the way for the production of low-C ferrochromium needed in the modern production process of stainless steels [1, 10].

In the 1890s, Sir Robert Abbott Hadfield (the father of Hadfield's manganese steels) dedicated himself to the study of high-Cr steels with C-content in the 1–2 wt% range. His objective was to test their corrosion resistance, although the method adopted implied the use of quite corrosive solution consisting of 50% $H_2SO_4$ for testing corrosion resistance of materials. In fact, at that time it was erroneously believed that this test could be representative of the corrosion resistance of materials and could be used as an accelerated test. Obviously, such a method could not highlight the superior corrosion resistance of stainless steels. As a result of the testing methodology used, Hadfield came to the incorrect and potentially damaging conclusion that Cr actually jeopardizes the corrosion resistance of steels [7].

It is important to state that from here on, the percentages will always refer to the weight percentage (wt%), although only the % symbol will be indicated. In case reference should be made to atomic percentage, it will be adequately expressed by indicating at%. In 1898, thanks to the work of two French scientists, it was shown that high C contents in Fe–Cr alloys are detrimental to corrosion resistance. For what concerns the study of this property, the German Philip Monnartz seems to have been the first to underline the importance of a minimum amount of Cr in steel to establish superior corrosion resistance and to use the term *passivity* for the phenomenon. In fact, in 1911 he published a pioneering work on this subject in the journal *Metallurgie* entitled "The Study of Iron-Chromium Alloys with Special Considerations of their Resistance to Acids" where he listed some interesting conclusions for that time:

- It was highlighted how increasing values of Cr, when approaching 12% from below, led to a drop in the corrosion rate (see also Fig. 1.3). This was particularly evident for tests conducted in $HNO_3$, but also in other acids as well as in waters and atmosphere.

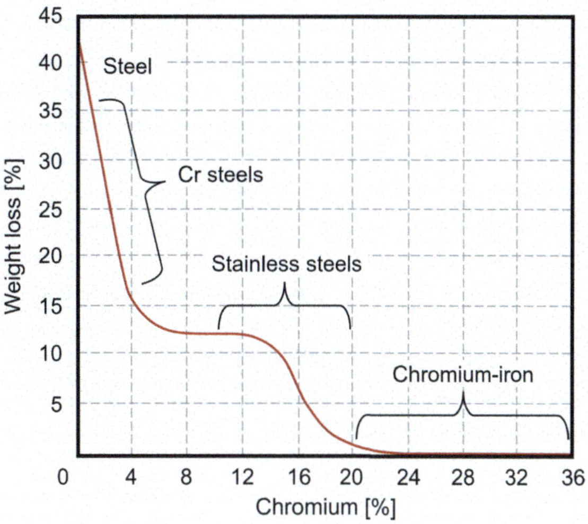


**Fig. 1.3** Effect of the Cr content on the oxidation resistance of steel alloys at 1000 °C (redrawn from [1])

- A low-C content is beneficial to the corrosion resistance of the alloy because free C can form deleterious carbides. However, C can be stabilized by other elements such as Ti, V or W.
- The great increase in corrosion resistance is due to a phenomenon that is called passivity, dependent upon oxidizing conditions.

Monnartz was the first to understand the reasons for what he named the "stainlessness" of high-Cr steels and to recognize that nitric acid can promote passivity [7]. This latter characteristic results from the preferential dissolution of Fe in an external layer of the alloy by the action of $HNO_3$, which results in a consequent increase in the Cr percentage in the surface layer of alloys containing it. This same phenomenon would later take on extreme importance in the pickling processes of steel in modern industry (see also Sect. 4.4). Finally, a little historical fact about the concept of passivity: although the first use of the idea of passivity can be ascribed to Philip Monnartz, the original theory of film formation is again due to Michael Faraday [11].

Research and development on a class of Fe-based alloys with different Cr contents continued for nearly a century, until in 1913 the first heat of an alloy featuring 0.24% C, 0.44% Mn, 0.22% Si and 12.86% Cr was produced under the coordination of the English metallurgist Harry Brearley, head of the Brown-Firth Research Laboratories in Sheffield. There were very active times for research, so much so that practically at the time as Harry Brearley, Frederick Becket and Christian Dantsizen had also started producing alloys with C content in the 0.007–0.015% range and Cr between 14 and 16%. The race for industrial production of stainless steel finally had one of its most important milestones in 1920, when the famous 18–8 austenitic stainless steel appeared on the market [1]. From then on, and for decades afterwards, the development of stainless steels experienced a sort of golden age as new grades emerged frequently. In addition to this, these materials have begun to be used increasingly in an even wider range of engineering fields.

Beyond the metallurgical history of stainless steels, in order to fully understand their properties, it is necessary to know the corrosive processes, and corrosion engineering and science reasons, that can explain the behavior of these materials in various environments and operating conditions. The corrosion resistance of metallic materials strongly depends on the surface condition through the presence of oxides and corrosion products defining the corrosion behavior in practice. In this regard, it is important to make some distinction and to use proper terminology. In fact, the formation of layers on the metallic surface as a consequence of the reaction with the external environment is called *passivation* and it is usually associated to the formation of thick layers, in the order of tens of μm. In the moment when the covering features of the surface layer lead practically to a stop of corrosion, the name to be associated with the phenomenon is *passivity*. In this latter case the thickness of the protective layer is in the order of the nm and the associated name is *passive film* [12]. In any case, often in the literature the terms passivation and passivity are used interchangeably with the care of specifying each time if the concept refers or not to layers of corrosion products that have protective features. The textbooks of McCafferty [13, 14] and of Pedeferri [12], used as references in this paragraph, are particularly recommended for a comprehensive covering of the mechanism according to which passivation and passivity take place, especially the part regarding the topics related to electrochemistry. Nevertheless, the textbooks cited above also provide an excellent knowledge base for the study of corrosion processes to which stainless steels are also subjected. In addition, the most relevant aspects of corrosion resistance, corrosive phenomena and environments will also be presented in Chap. 3 of this text.

As already declared, the outstanding corrosion resistance of stainless steels is the consequence of the presence of a thin oxide film on the surface (i.e., a passive film), with a thickness in the nanometric range. The study of passive films on stainless steels is often carried out using anodic polarization techniques (electrochemical polarization). Results are given in form of anodic polarization curves that give information about the electrochemistry behind the passivation/passivity phenomenon. As indicated in the reference textbooks of Pedeferri [12] and McCafferty [13, 14], aqueous corrosion phenomena are known to be electrochemical processes with electron and ion transfer. They are therefore characterized not only by an electrical potential associated to the chemical reactions (also interpreted as thermodynamic indication of spontaneity), but also by current and current densities values. By greatly simplifying the discussion on this specific topic, it can be stated that since the formation of passivation/passivity layers is also a corrosive phenomenon, a current density can be associated to it (further details in Chap. 3). Current density associated with a corrosive phenomenon is usually taken as indicator of the kinetics of the phenomenon itself: higher current densities imply more intense phenomena with faster kinetics. In fact, the current density is directly proportional to the mass of metal/alloy corroded [12]. For the interpretation of the anodic polarization curves there are two current density values to refer to: the critical current density of passivation ($i_c$) and the passive current density ($i_p$). The $i_c$ value is the current density value beyond which an abrupt drop in

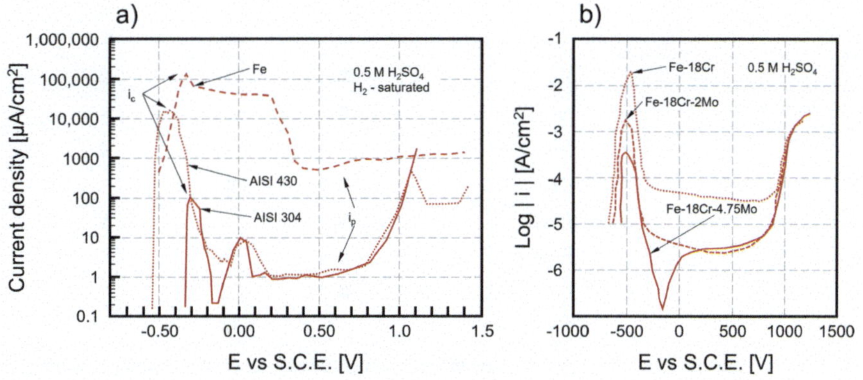


**Fig. 1.4** Anodic polarization curves for various metallic materials: **a** anodic polarization curves for pure Fe, AISI 304 stainless steel (Fe–18Cr–8Ni) and AISI 430 stainless steel (Fe–18Cr) in 0.5 M $H_2SO_4$ acidic solution; **b** effect of Mo additions on the anodic polarization of Fe–18Cr alloys in 0.5 M $H_2SO_4$ (redrawn from [13])

the current density happens and thus a sharp reduction in the kinetics of the corrosive process. Anodic polarization curves of passive metals and alloys always show an initial increase in current density up to a critical value because the formation of the passive film, which is a corrosion process, is taking place. Once the passive film is formed, as mentioned in the definition of passivity, a practical stop of the corrosive process takes place. For this reason the passive current density ($i_p$) is lower than $i_c$ or than the current density of a non-passive metal or alloy in the same potential region (see Fig. 1.4).

An indication of how the concept just explained can be applied to obtain information about passive film formation and associated corrosion resistance can be inferred from Fig. 1.4a. Figure 1.4a shows the comparison of anodic polarization curves of various materials placed in an acidic solution (0.5 M $H_2SO_4$). The materials are pure Fe, a ferritic stainless steel (AISI 430, see Sect. 1.2.2) and an austenitic stainless steel (AISI 304, see Sect. 1.2.4). By looking at Fig. 1.4a it is easy to recognize that the $i_c$ value for AISI 304 is the lowest among the three, indicating that it is easier to passivate AISI 304 rather than AISI 430. Furthermore, among the three metallic materials pure Fe has the highest $i_c$ value, i.e., it is more difficult to passivate pure Fe rather than the two stainless steels. Concerning the values of the passive current density ($i_p$), AISI 304 and AISI 430 show similar values that are considerably lower than that of pure Fe. Hence, it can be simplistically concluded that stainless steels are more corrosion resistant than pure Fe due to the easy formation of a passive film and that between AISI 304 and AISI 430, the one that performs better in the test environment is AISI 304. This result is not so much caused by the associated $i_p$ value, but rather to the greater ease of formation of the passive film, i.e., a lower value of $i_c$. The effect of the addition of specific alloying elements can also be studied by electrochemical passivation experiments and by observing the anodic polarization curves. Considering binary alloys with passive behavior such as high-Cr Fe–Cr

**Table 1.1** Effect of the addition of Cr, Mo and Ni on different metals and alloy systems in terms of corrosion resistance

| Metal or alloy | Cr | Mo | Ni |
|---|---|---|---|
| Fe | ● | ○ | ● |
| Ni | ● | ○ | X |
| Mo | ○ | X | ● |
| Cr | X | ● | ○ |
| Fe–Ni | ● | X | X |
| Fe–Cr | X | ● | X |
| Fe–Ni–Mo | ● | X | X |
| Fe–Cr–Ni | X | ● | X |

The symbol ● indicates positive effects, the symbol ○ indicates no positive effects; the symbol X indicates that the combination has not been studied, or the element is combined with itself [13]

alloys, the study can be performed by systematically adding different amounts of single alloying elements. The criterion according to which a specific element can be classified beneficial is its ability to reduce $i_c$ and/or $i_p$. To mention the well-known case of Mo, its addition is unanimously considered beneficial to corrosion resistance and the effects on the passive behavior can be derived from Fig. 1.4b. In fact, Fig. 1.4b shows that progressive addition of Mo to a Fe–Cr alloy leads to a reduction of both $i_c$ and $i_p$. Using this same method applied to other alloys and with different alloying elements, it is then possible to understand whether or not a given element has positive effects. A simple summary of the abovementioned concept is given in Table 1.1, where it can be seen that the addition of Mo has positive effect on the alloy system only in presence of Cr. Furthermore, the effect of the addition of both Cr and Ni are also reported [13]. It has been shown that anodic polarization curves give interesting information about the passive behavior of a metallic material and about the addition of specific alloying elements. However, no information can be derived about the nature, i.e., chemical composition, of the passive film. Other techniques are required to obtain such information.

One of the possible methods that may be employed is X-ray photoelectron spectroscopy (XPS). With reference to stainless steels, when the material contains Mo, tetravalent ions ($Mo^{4+}$) are usually incorporated in the part of the passive film that is closer to the bulk alloy, i.e., the inner part. Differently, hexavalent cations ($Mo^{6+}$) are there in the outer layer of the passive film. Recent investigations suggested that molybdate ions produce a bipolar passive film by acting as cation-selective spices, while the inner layer of the passive film contains $Cr_2O_3$, which is anion-selective. This distribution of chemical compounds leads to a specific behavior of the passive film: the outward flux of cations through the passive film produced by metallic dissolution at the metal/film interface is not favored and similar situation occurs for the inward flux of anions (like the dangerous $Cl^-$), which is not promoted too. According to this latter explanation, since mass transport is severely hindered, it is difficult for the corrosive process to continue, and this mechanism helps explaining why the formation of a thin surface oxide layer on stainless steels results in a higher corrosion resistance in several environments and in various service conditions [13].

The example cited of application of XPS for the investigation of the nature of passive films is just one of the possibilities that can be considered for this purpose. In fact, the results of various examinations, both in-situ and ex-situ, can be combined to obtain a more complete picture of the situation. In their paper, Olsson and Landolt [11] pointed out that data collected from both in-situ and ex-situ characterization experiments agree in terms of overall results showing that, for example, at short times, charge transfer of the metal/film or the film/solution is a limiting factor for film growth on stainless steel. The employed techniques were: XPS (X-ray photoelectron spectroscopy), AES (Auger electron spectroscopy), SIMS (secondary ion mass spectroscopy) and ISS (ion scattering spectroscopy). Early and intensive research activities have pointed out that the passive film growth occurs in a relatively short time (seconds or minutes) and this explains the "instantaneous" formation character usually associated to the protective layer present on the surface of stainless steel. On the other hand, long range ordering is considerably slower than film growth and usually takes several hours. It is now known that the passive film formed on stainless steel surface has a dynamic behavior because it is able to adapt itself to changes in potential or in anion concentrations in the electrolyte. Furthermore, once the film is formed the reaction rate between the alloy and the environment lowers itself by several orders of magnitude (see Fig. 1.3 as reference for this behavior for what concerns high-temperature oxidation and Fig. 1.4 for the reduction in the corrosion kinetics expressed as current density). This is for sure one of the features that allow stainless steels to resist against corrosion attacks in various environments [11].

Unfortunately, according to what has been explained previously about the dynamic nature of the passive film and its dependence on the alloy chemistry, it is not possible to create a comprehensive model describing the characteristics of each individual passive film for each of the many existing grades of stainless steel. Surface analyses allowed to propose, for example, a model for passive film formed on the surface of austenitic stainless steels in contact with acidic solutions after anodic polarization. The model is based on a three-layer structure when the outer part consists of a hydroxide film formed on an oxide layer. The bulk material is separated from the oxide layer by means of a Ni-enriched layer, also called the "apparent" metal (note that this model is essentially based on XPS measurements). The Ni enrichment is the consequence of a selective oxidation process of both Cr and Fe during anodic polarization. A comparison between the oxide layer and the "apparent" metal layer shows a strong enrichment in Ni in the "apparent" metal layer, whereas the oxide layer is enriched in Cr with practically total absence of Ni (see Fig. 1.5). Other techniques can be employed for the study of passive films, a further example is ion scattering spectroscopy (ISS) that has been applied for the analysis of the passive film on a Fe–Cr alloy by Calinski and Strehblow [15]. Results of ISS pointed out that in the center of the passive film a strong Cr enrichment is found. On the other hand, the metal/film interface is characterized by a chemical composition that essentially follows the composition of the bulk material [15]. In addition to the ex-situ techniques, X-ray adsorption near edge structure (XANES) has been recently employed for the investigation of chemical composition of passive films on Fe–Cr alloys. Obtained results are in good agreement with the already cited ones, confirming Cr enrichment in the passive film as a consequence of selective dissolution of Fe. In addition to the study of the chemical composition, however, it is worth remembering that the presence and features of passive films are influenced by several parameters related to the environment: temperature, pH and presence of halides in the electrolyte (for stainless steel, reference is made in particular to Cl compounds). For detailed description of each one of these parameters the reader may refer to the references of this chapter and to Chap. 3. Of course, alloy microstructure and composition strongly influence the passive film composition and properties, as mentioned above with regard to Mo (see also Appendix A for other details about Mo and the other alloying elements) [11]. Furthermore, the passive film can grow or dissolve and has also the ability to adsorb or incorporate anions. Furthermore, when talking about the behavior in aqueous environments it is important to state that a passive film is subjected to a constant exchange of spices with the electrolyte and the environment. This affects both its thickness and composition, altering them. It has been demonstrated that a critical parameter associated to the stability of the passive film is the time the film needs for responding to an environmental change and, in addition to ex-situ techniques, in-situ analyses gave the possibility to determine several time-dependent parameters [11].

Although many results have already been achieved, it is a matter fact that surface analysis of passive films on stainless steels continues to be an interesting field for intense experimental activity. This is for instance the case of the graph theory model proposed by Edward McCafferty. As shown in Table 1.1, addition of Cr may have

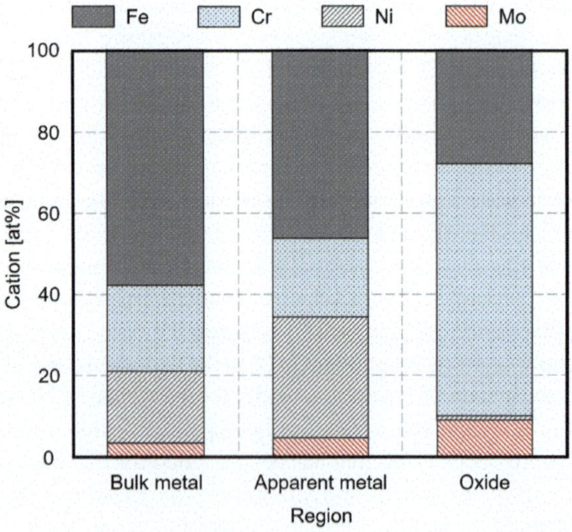


**Fig. 1.5** Three-layer model of the results of XPS analyses on a passive film formed in the high passive region in 0.1 M HCl + 0.4 M NaCl solution. See main text for the detailed description (redrawn from [11])

positive effects also in other alloy systems in addition to stainless steels. The protective action of Cr is also associated to a minimum concentration of 13.0 at% in binary Fe–Cr systems. It is important to note that for other alloy systems there exist also a critical threshold concentration of Cr for a passive layer to develop. The graph theory model aims to show and demonstrate that there is a minimum concentration of $Cr^{3+}$ ions in the oxide film so as to enable the formation of a continuous network of Cr–O–Cr bridges in the passive film. Such condition in Fe–Cr alloys is provided by the 13.0 at% Cr. Detailed information about the graph theory is given in the textbook of McCafferty [14] in the reference list and also in the papers of the same author in the further reading list. The method has already been successfully applied to 18 different binary alloy systems, some of the results of which are given in Table 1.2.

The method just mentioned was later extended to the Fe–Cr–Ni system for the study of ternary alloys, with particular interest in Fe–18Cr–8Ni (AISI 304) stainless steel. Results showed that $Ni^{2+}$ cations participate to the formation of the passive film, the same holds for a certain amount of $Fe^{3+}$ ions. Furthermore, the Cr–O–Cr network becomes more stable and the penetration of dangerous $Cl^{-}$ anions through the tangled network is limited [14]. The corrosion resistance of stainless steels is not the only feature that makes them some of the most versatile Fe-based alloys. As it will be discussed in Sect. 1.3, the field of application of stainless steels is very broad. One of the reasons behind the versatility of these materials is that, like other Fe-based alloys and with steels in general, a truly remarkable number of grades can be produced from as many combinations of alloying elements. For this reason, stainless steels are classified into five main groups according to their composition and resulting microstructure: ferritic, martensitic, austenitic, PH (precipitation hardened or precipitation hardening stainless steels) and duplex (also known as dual phase or austenitic-ferritic stainless steels) [1]. In addition to these groups, there are also subgroups of almost every family mentioned: superduplex and hyperduplex, superferritic, supermartensitic and superaustenitic stainless steels. These subgroups cannot be classified as standalone macro-groups in their own right as they are special cases (in terms of chemical composition and properties) of one of the five abovementioned

**Table 1.2** Selected results of the application of the graph theory on various alloy systems

| Alloy system | Calculated cationic mole fraction in the oxide | Corresponding alloy composition (at.%) | Observed critical alloy composition (at.%) |
|---|---|---|---|
| Fe–Cr | 0.30 $Cr^{3+}$ | 13.0 | 13.0 |
| Ni–Cr | 0.29 $Cr^{3+}$ | 7.0 | 8.0–11.0 |
| Al–Cr | 0.31 $Cr^{3+}$ | 48.0 | 40.0–50.0 |
| Co–Cr | 0.23 $Cr^{3+}$ | 6.0 | 9.0–11.0 |
| Cr–Mo | 0.16 $Mo^{4+}$ | 21.0 | 20.0 |
| W–Zr | 0.21 $Zr^{4+}$ | 11.0 | 15.0 |

Note that the corresponding alloy composition data are obtained from surface analysis and that the observed alloy composition refers to corrosion data [14]

families. Moreover, the volumes produced per year would not be such as to justify further subdivision of the main families anyway (see Sect. 4.8).

Cr is definitely the element around which the understanding of stainless steels must be developed. Further evidence of this lays in the study and prediction of phases that can form in relation to stainless steels that, in addition to the basic knowledge of the Fe–C phase diagram, must necessarily pass through the evaluation of the influence of this element. A good starting point in this context is the analysis of the Fe–Cr equilibrium phase diagram given in Fig. 1.6. From the simple visual analysis of the Fe–Cr binary phase diagram, it is possible to note the limited extension of the so-called $\gamma$-loop of the austenitic phase (*fcc*), confirming the $\alpha$-stabilizing nature of Cr (see also Appendix A for what concerns phase stabilization and carbide-forming attitude of alloying elements). Furthermore, an increase in Cr content leads to a lowering of the Curie temperature, i.e., the temperature at which a magnetic transition occurs in pure Fe and above which a paramagnetic behavior is observed, in spite the retention of a *bcc* structure. In addition to this, the possibility for sigma ($\sigma$) phase to form is observed for certain Cr content and temperature intervals. The presence of $\sigma$ is associated to deleterious effects in the material, as will be extensively explained in Sect. 2.4.6.2. Last but not least, the Fe–Cr system is also characterized by the formation of a deleterious phase at a specific temperature that leads to a phenomenon known as 475 °C embrittlement. This embrittlement is associated to the formation of a *bcc* Cr-rich phase named $\alpha'$, whose presence is typical of ferritic and duplex grades and is associated with a decrease in ductility and toughness, as well as a reduced corrosion resistance. Further details regarding the $\alpha'$ phase will be discussed in Sect. 1.2.2 and in Appendix B. However, in order to avoid confusion, it is also important to state at this point that $\alpha'$ is usually the name also associated with another microconstituent, namely one of the two strain-induced martensite (SIM) types that may form in austenitic stainless steels (see Sect. 1.2.4) [1].

The analysis proposed for Cr could also be done for the other alloying elements. However, to further explore the topic of binary phase diagrams, the reader may refer to the two textbooks of Massalski found within the reference list and further reading at the end of this chapter. Further information about the effects of each of the alloying elements that can be found in steels can be derived from Appendix A. Moreover, additional information about Mo, Ni and N will be discussed more in details in upcoming sections and in Chap. 2. Although the study of binary phase diagrams is important when dealing with metallic alloys, it is well known that their use is not sufficient to fully understand the nature of transformations and phases typical of stainless steels. In fact, stainless steels are complex multicomponent systems that need equally complex thermodynamic calculations to be described at the equilibrium. Furthermore, the real transformation kinetics usually deviate greatly from equilibrium conditions and this cannot be described by phase diagrams. These limitations lead to the impossibility of having phase diagrams capable to describe systems with more than four relevant components. Further limitation derives from the difficulties of considering the effects of cooling rate variations typical of heat treatments or solid-state diffusion of some elements (also related to the precipitation of intermetallic phases) [10, 17]. As it

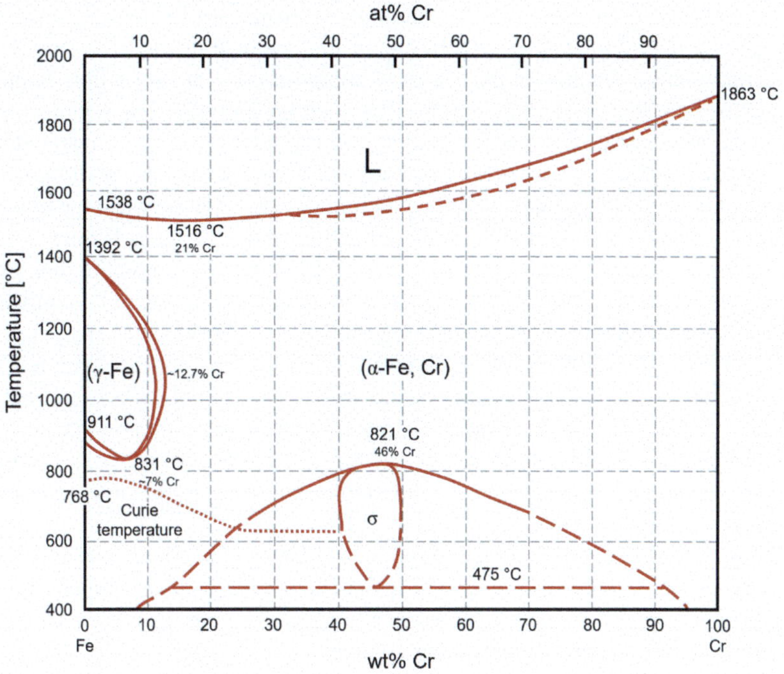


**Fig. 1.6** Binary Fe–Cr equilibrium diagram (redrawn from [16]) [1, 10]

is the case of the study of other complex multicomponent systems (e.g., superalloys) fundamental help has come from the application of computational thermodynamics and analysis of transformation kinetics through the development of effective tools and models based on PHACOMP, ThermoCalc©, DICTRA and other modeling approaches. Again, with reference to superalloys, thanks to the successful application of such methodologies new criteria were also developed for the definition of the stability range of deleterious TCP (topologically close-packed) phases, which will be presented with reference to stainless steels in the next sections and in Chap. 2 [1]. In order to better understand the following sections and what will be discussed in Chap. 2, the reader is encouraged to refer to texts related to computational thermodynamics of materials, or to user guides or manuals made available by companies that develop computational thermodynamics software.

Although very useful and of great support and precision, computational thermodynamics methods are not yet so widespread in practical industrial field due to their complexity and to the high level of knowledge needed for their use, thus making their usage not always accessible to professionals. In fact, ThermoCalc© is used worldwide for educational and academic purposes and in research laboratories, with less extent in industry. In practical context and in the absence of computational thermodynamics software, phase diagrams provide partially complete information due to the already discussed limitations associated with them. Therefore, some of the most popular tools still used, especially for thermal treatments, are TTT and CCT curves. With specific regard to stainless steels, one of the practical methods for predicting microstructure as a function of the chemical composition is the Schaeffler diagram. As it can be seen from Fig. 1.7, the Schaeffler diagram is divided into regions representing the microstructures that can be detected for what concerns stainless steels. The method proposed to obtain subdivision, and consequently the position of the specific alloy in the diagram, is based on an initial distinction between the alloying elements. As written in Appendix A, classification of the elements is made on the basis of the α-stabilizing or γ-stabilizing effect. Elements with α-stabilizing effect are used calculate the position of the alloy on the x-axis according to Cr equivalent value, while γ-stabilizing elements are used to find the coordinate on the y-axis Ni equivalent value. Schaeffler method allows to obtain a quite rough estimation of the microstructure and is widely used in practice where, for example, it is useful for the prediction of the phases obtained after welding processes. Another possible application is the prediction of the phase balance in duplex stainless steels during chemical composition tuning while refining the liquid alloy prior to casting, thus foreseeing the result after solidification.

In order to increase the accuracy of the Schaeffler diagram, an extended form of the equations for both Cr and Ni equivalent calculation has been developed over the years. This was made in order to consider the largest amount possible of alloying elements that result from the increasing complexity of the chemistry of the alloys. The two extended equations are given in Eqs. 1.1 and 1.2 [19]:

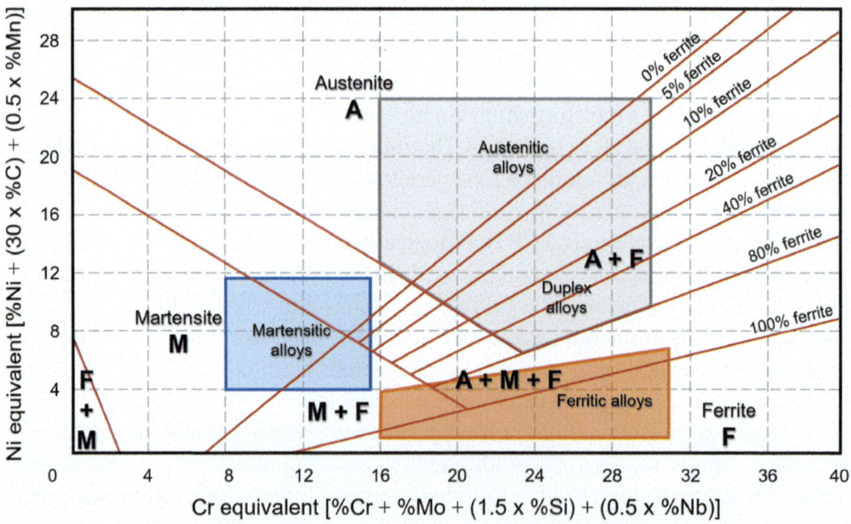


**Fig. 1.7** Schaeffler diagram for stainless steels with evidence of compositional ranges of austenitic, martensitic, ferritic and duplex alloys (redrawn from [10, 18])

$$\mathrm{Ni}_{eq} = \%\mathrm{Ni} + \%\mathrm{Co} + (30 \times \%\mathrm{C}) + (25 \times \%\mathrm{N}) + (0.5 \times \%\mathrm{Mn}) + (0.3 \times \%\mathrm{Cu}) \tag{1.1}$$

$$\mathrm{Cr}_{eq} = \%\mathrm{Cr} + (2 \times \%\mathrm{Si}) + (1.5 \times \%\mathrm{Mo}) + (5 \times \%\mathrm{V}) + (5.5 \times \%\mathrm{Al}) + (1.75 \times \%\mathrm{Nb}) + (1.5 \times \%\mathrm{Ti}) + (0.75 \times \%\mathrm{W}) \tag{1.2}$$

Obviously, due to the intended pragmatic use of the tool, the Schaeffler diagram cannot be fully exhaustive and precise in every situation. In fact, it shows limitations in predicting the amount of delta (δ) ferrite (a parameter that can be useful, for example, in welding practices). Because of these limitations, other methods have been developed, such as the DeLong method, which is used as a standard procedure by the International Welding Institute [10]. Having completed this first introductory part in preparation for understanding the peculiarities of stainless steels, the following paragraphs will be dedicated to a quick overview of the five types. A presentation organized in such a way as to highlight the origins, chemical composition, and the most important and characterizing properties will be given. This will be done with the intention of guiding the reader to the heart of this work: superaustenitic stainless steels.

### *1.2.2 Ferritic Stainless Steels*

Ferritic stainless steels are generally perceived as one of the least complex and cheapest families of stainless steels. This can be easily understood by simply looking at their chemical composition, as well as the heat treatment and processing possibilities. In the simplest way, ferritic stainless steels are Fe–Cr–C alloys where Cr content must be such that to exceed the threshold necessary for the alloy to become stainless. At the same time, Cr must not combine with C, thus forming precipitate carbides (see Appendix B) and leading to the deleterious consequence of localized decrease in the Cr content in the ferritic matrix [17]. The beginning of the history of ferritic stainless steels dates back to 1909 when the Frenchman Albert Marcel Germain René Portevin published a study on martensitic stainless steels and on ferritic grades containing 17.4% Cr and 0.12% C (a kind of ancestor of AISI 430). However, the real birth of ferritics occurred few years later thanks to the work of Christian Dantsizen and Frederick Becket. Ferritic stainless steels were first used and marketed around 1920, the year in which Brown Bayley's Steel Work Ltd of Sheffield produced a small heat of a steel containing 11.7% Cr and 0.07% C [10]. At the beginning of their era, ferritic stainless steels were characterized by high Cr values because it was hard to obtain very low C contents both due to the absence of low-C ferrochromium alloying compound and AOD process. In fact, ferrite is essentially not able to solubilize C at room temperature, and this can be easily inferred by looking at the Fe–C phase diagram (see also Sect. 2.2). As a consequence of this, early ferritic stainless steels might be subjected to formation of Cr carbides as long as C is available. The excess

Cr had the function of combining with C during those processes where this reaction would be unavoidable, i.e., cooling from high temperature or thermal treatments, and preventing locally the Cr concentration from falling below the critical threshold value for stainlessness. Moreover, the excess of Cr also served to counteract the austenitizing action of other elements that, at that time, could not be carefully controlled during melting and refining processes, i.e., C and N. The development (late 1950s) and subsequent advent of the AOD process in 1967 (first commercial use) changed the world of stainless steels and, therefore, that of ferritics allowing compositional control at a level that was not previously possible [17]. Thanks to the introduction of AOD and VOD processes, the more efficient C reduction, as well as a better control of the overall chemistry, new perspectives for ferritic stainless steels were allowed. In particular, it was possible to reduce the level of C + N to a value low enough to allow their potentially negative effects to be neglected by a small and controlled addition of Nb and Ti. This allows Nb and Ti to react with C, thus hindering Cr to form carbides. The name associated to this practice is stabilization. Steels that take advantage of this procedure are called stabilized grades [1]. A possible drawback of Ti stabilization is that unless the N content can be very low, Ti may react with N in the molten alloy causing the formation of primary TiN. The nitrides tend to agglomerate, thus leading to the formation of clusters and eventually lead to problems during casting, as well as quality issues related the surface defects in billets and ingots after solidification. A possibility for reducing this deleterious phenomenon is a dual stabilization with Ti and Nb, coupled with very low N content (below 0.02%).

As a consequence of the technological breakdown caused by AOD, it is possible to identify at least two groups of ferritic stainless steels: early high-C grades such as AISI 430 and AISI 446 (see Table 1.3) and modern stabilized ferritic stainless steels such as AISI 409 and AISI 439.

Among the various possibilities for classifying ferritic stainless steels, a further one in addition to that related advent of the AOD process is to group them into three generations:

- First generation such as AISI 430 and AISI 446. These steels were not fully ferritic in their composition and over any temperature range. They were developed during the first part of the twentieth century and were characterized by relatively high C content with consequent need for high Cr values in their chemical analysis (see Table 1.3). Solution annealing treatments are performed typically in the 705–830 °C range.
- Second generation such as AISI 409. In this case the alloys are characterized by lower C and N content and by the addition of Cr stabilizers such as Nb and Ti. The structure is fully ferritic at any temperature (Ti also gives a contribution in this regard because it is a ferrite-stabilizing element). Solution annealing treatments are performed typically in the 870–925 °C range.
- Third generation ferritic stainless steels that were first introduced in the market around 1970. Their existence derives from the possibility of reaching a more efficient decarburization and low N content. In fact, the typical C and N content is in the order of 0.02% or lower. Furthermore, these steels are Nb- or Ti-stabilized and

**Table 1.3** Chemical composition of most common ferritic stainless steels according to the AISI nomenclature [10, 17, 20]

| Type | C | Cr | Mo | Mn | Ni | Others |
|---|---|---|---|---|---|---|
| 405 | 0.08 | 11.5–14.5 | – | – | – | Al = 0.1–0.3<br>P = 0.04<br>S = 0.03 |
| 409 | ≤ 0.08 | 10.50–11.75 | – | ≤ 1.0 | ≤ 0.5 | Ti = 6(C + N) to 0.75 max |
| 409Cb | 0.02 | 12.5 | – | – | 0.2 | Nb = 0.4 |
| 430 | ≤ 0.12 | 16.0–18.0 | – | ≤ 1.0 | ≤ 0.75 | Si ≤ 1.0 |
| 430F | 0.12 | 16.0–18.0 | 0.6 | – | – | P = 0.06<br>S = 0.15 min |
| 430FSe | 0.12 | 16.0–18.0 | – | – | – | Se = 0.15 min |
| 434 | 0.12 | 16.0–18.0 | 0.75–1.25 | – | – | P = 0.04<br>S = 0.03 |
| 436 | 0.12 | 16.0–18.0 | 0.75–1.25 | – | – | Nb + Ta = 5 x %C |
| 439 | 0.07 | 17.0–19.0 | – | – | 0.5 | Ti = 0.20 + 4(C + N) min to 1.0 max |
| 441 | 0.02 | 18.0 | – | – | 0.3 | Nb = 0.7<br>Ti = 0.1–0.6 |
| 444 | ≤ 0.025 | 17.5–19.5 | 1.75–2.50 | ≤ 1.0 | ≤ 1.0 | Nb + Ti ≥ 0.20 + 4(C + N) |
| 446 | ≤ 0.20 | 23–27 | – | ≤ 1.50 | – | N ≤ 0.25<br>Si ≤ 1.0 |

contain also another important ferrite-forming element that, as already explained in previous sections, brings positive effects for what concerns corrosion resistance: Mo. Solution annealing treatments are performed typically in the 955–1100 °C range [10].

Due to their unstabilized nature, early generations may not be fully ferritic due to the C level that can lead to the existence of high-temperature austenite under certain conditions. In the absence of Ni, this austenite may transform into martensite during rapid cooling, i.e., quenching after annealing treatments. For this reason, these grades are always used in the annealed condition thanks to a long sub-critical annealing treatment that promotes Cr redistribution and avoids martensite formation upon quenching (rapid cooling is necessary to avoid formation of deleterious intermetallics such as the σ phase, see Appendix B).

As a consequence of their chemical composition (see Table 1.3), ferritics can be subjected to the precipitation of harmful phases, in particular the most dangerous are σ and α′ (but also χ phase and Laves phases can form, see Appendix B). High-Cr σ phase can form when the material is exposed to the 500–800 °C temperature range, it is a brittle *bct* phase belonging to the family of TCP (topologically-close packed) phases that causes a drop in the toughness of the alloy (see Appendix B and Sect. 2.4.6). Furthermore, σ phase formation causes a decrease in the corrosion resistance due to the Cr depletion associated with its formation. Cold working operations might promote σ phase formation. This happens because σ formation requires substitutional diffusion of Cr and cold working practices are able to enhance substitutional diffusion. σ phase may be redissolved thanks to solution annealing, however its complete removal is difficult and not immediate. Therefore, complete homogeneity recovery is quite complicate [17].

Grades with more than 18% Cr can be subjected to the formation of the deleterious *bcc* coherent α′ phase, which presence is associated with the so-called 475 °C embrittlement (this was the early name because at the beginning the true nature behind the embrittlement was not totally understood) [17]. Lattice of α′ is characterized by Cr and Fe atoms in an ordered matrix in which the two metals occupy sites that are equivalent to two interlocking simple cubic matrices. Unfortunately, as far as the detectability of the α′ phase is concerned, the similarities between Fe and Cr atoms (atomic size) and the small precipitate size make direct observation and discrimination between α ferrite and Cr-rich α′ very difficult, also when TEM (transmission electron microscopy) is employed. For this reason, chemical microanalysis techniques, as well as structural analysis techniques via X-ray diffraction are rather ineffective. As a consequence of this, special techniques must be employed [21]. α′ turns out to be coherent because its lattice closely matches that of ferrite. Consequently, presence of α′ causes hardening and embrittlement as demonstrated by an increase in tensile strength, yield strength and hardness; while impact resistance and elongation values decrease. This properties variation is the result, among other phenomena, of a change in the behavior of the alloy with respect to dislocation movement. The presence of α′ seems to reduce the stacking fault energy, thus making dislocation cross-slip more difficult. Another evidence of the embrittlement of α′-containing ferrite matrix is the typical cleavage-type brittle fracture at room temperature that characterizes the material. The embrittling effects of α′ strongly depend on the chemical composition of the alloy. In particular, the higher the Cr content, the stronger the detrimental effect on the mechanical properties would be. Corrosion resistance is also reduced due to the associated Cr depletion in the ferrite matrix. Although there is generally agreement regarding the deleterious effects on both mechanical and corrosion resistance properties and their associated triggers, there still seems to be no complete certainty about the mechanism of α′ formation. In particular, no full agreement on the fact that the formation occurs according to spinodal decomposition or nucleation and growth. Some author suggested that both mechanisms are possible and that also the formation mechanism is chemistry-dependent and that, for example, addition of specific elements such as Ni leads to a change in the precipitation mode from nucleation and growth toward spinodal decomposition [10, 17].

It is well known that the toughness of ferritic stainless steels is a critical issue. This results, for example, in a certain susceptibility to crack formation or brittle fracture for continuous casting billets that have not undergone heat treatment. Like every other *bcc* structures, ferritic stainless steels are characterized by limited slip systems, high stacking fault energy and lattice anisotropy. As a consequence of this latter feature, they have pronounced mechanical anisotropy after deformation. Mechanical properties of ferritics are more or less comparable to those of austenitic stainless steels (see Tables 1.4 and 1.8), however the *bcc* structure is characterized by a ductile-to-brittle transition associated with a specific limiting temperature (DBTT, ductile–brittle transition temperature) that limits the use at relatively low temperatures. DBTT in ferritics strongly depends on the content of interstitial elements and therefore high C and N content have detrimental effects and tend to raise its value [10, 17].

As already stated, the overall mechanical behavior depends strongly on C and N presence, as well as substitutional elements (Cr has the main consequence of producing some solid solution hardening). Ferritic stainless steels cannot be hardened by means of heat treatments and are also characterized by a lower strain hardening coefficient than austenitic grades.

Corrosion resistance is of course chemistry-dependent. Resistance against generalized corrosion, localized corrosion and crevice corrosion is promoted by Cr, Mo and N (see Appendix A), however ferritic stainless steels are not able to contain high amount of N due to its limited solubility in ferrite (the opposite occurs with austenitics). Therefore, the contribution of this element to corrosion resistance is reduced. Consequently, primary importance is given to Cr and Mo. As a consequence of this, any element that can combine with Cr or Mo might have deleterious consequences on corrosion resistance due to depletion of these two elements from the matrix. Ferritics are characterized also by a low thermal expansion coefficient; to give an idea, that of ferritic grades is around 10–10.5 $\times$ $10^{-6}$ $K^{-1}$, while the rest of stainless steel grades have values between about 13.5 $\times$ $10^{-6}$ $K^{-1}$ and 17.0 $\times$ $10^{-6}$ $K^{-1}$ (see Sect. 3.1). This feature leads to the possibility for the high-temperature scale developed on the surface to be less subjected to spalling. This latter characteristic makes them good candidates for environments characterized by thermal cycling around temperatures that are not excessively high.

Ferritic stainless steels show excellent corrosion resistance against generalized corrosion, crevice, pitting and are almost immune to SCC in media and environments containing chlorides. An example of further development of these materials was the introduction on the market of a sub-family called *superferritic stainless steels*. Thanks to the high content of Cr and Mo, both ferrite-forming elements, Ni addition became possible without the risk of austenite formation. This has also the great advantage of increasing toughness and corrosion resistance under reducing conditions, as well as crevice and pitting corrosion resistance (see Sect. 1.2.7 for further details). For what concerns the uses, further details will be discussed in Sect. 1.3.

### *1.2.3 Martensitic Stainless Steels*

The development of the very first commercial martensitic stainless steel can be probably ascribed to Harry Brearley in 1913, while attempts and laboratory experiments on alloys whose chemistry was martensitic date back to some years before. Martensitic stainless steels can be defined as Fe–Cr–C alloys characterized by a Cr range between 11.5 and 18%, C in the 0.1–1.2% range, while the intentional addition of

**Table 1.4** Mechanical properties of selected ferritic stainless steels with different designations

| Type | EN | UNS | Density (kg/dm$^3$) | UTS (MPa) | Yield strength (MPa) | Elongation (%) | Reduction in area (%) | Hardness | Condition |
|---|---|---|---|---|---|---|---|---|---|
| 405 | 1.4002 | S40500 | 7.60 | 480 | 275 | 30.0* | 60.0 | 150 HB | Annealed |
| 405 | 1.4002 | S40500 | 7.60 | 585 | 480 | 20.0* | 60.0 | 185 BH | Cold drawn |
| 409 | – | S40900 | 7.70 | 450 | 240 | 25.0* | – | 75 HRB | Annealed |
| 409Cb | – | S40940 | 7.70 | 450 | 240 | 50.0** | 70.0 | 130 HRB | Annealed |
| 430 | 1.4016 | S43000 | 7.75 | 415–480 | 205–275 | 20.0–22.0 | 45.0 | 88 HRB max | Annealed |
| 430F | – | S43020 | 7.70 | 485–585 | 275 | 20.0 | 45.0 | – | Annealed |
| 434 | 1.4113 | S43400 | – | 530–545 | 365–415 | 23.0–33.0 | 78.0 | 90 HRB max | Annealed |
| 436 | 1.4526<br>1.4513 | S43600 | – | 530 | 365 | 23.0 | – | 83 HRB max | Annealed |
| 439 | 1.4510 | S43035 | 7.70 | 415 min | 205 min | 22.0 min | – | 89 HRB max | Annealed |
| 441 | – | S44100 | 7.70 | 470 | 250–300 | 18.0 min | – | – | Annealed |
| 444 | 1.4521 | S44400 | 7.75 | 415 | 275 | 20.0 | – | 95 HRB max | Annealed |
| 446 | 1.4749 | S44600 | 7.50 | 550 | 345 | 25.0* | 45.0 | 86 HRB | Annealed |
| 446 | 1.4749 | S44600 | 7.50 | 605 | 462 | 26.0* | 64.0 | 96 HRB | Annealed at 815 °C + cold drawn |

*UTS* ultimate tensile strength. * refers to elongation in 50 mm, ** refers to 4d elongation evaluation [10, 17, 20, 22]

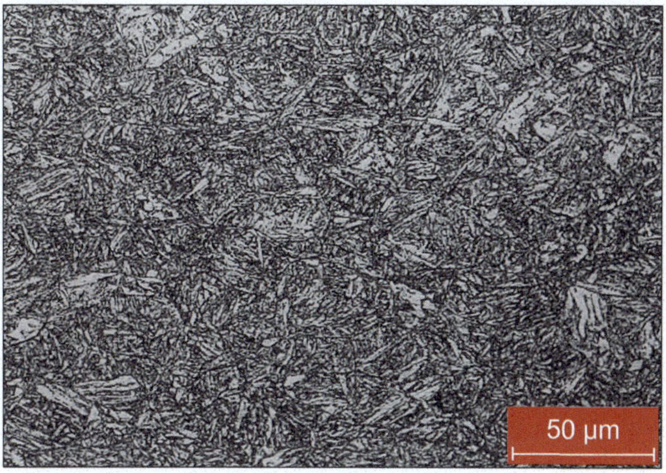


**Fig. 1.8** Optical micrograph of martensitic stainless steel [23]

other elements is limited, or even absent. Furthermore, they are ferromagnetic and characterized by good corrosion resistance in mild environments, definitely lower than that of austenitic and ferritic grades. Obviously, as the name suggests, their main feature is the martensitic microstructure (*bct*) obtained as a consequence of martensitic transformation resulting from the cooling process they have been subjected to (see Fig. 1.8) [1].

Among the various possibilities in this regard, martensitic stainless steels can be classified in three groups and the classification is made according to the C content and to the relevant application:

- Low-C martensitic stainless steels for gas turbines. They are usually characterized by a fine martensitic structure.
- Medium-C martensitic stainless steels for cutlery.
- High-C martensitic stainless steels for wear-resistant applications usually characterized by ultrafine martensitic structure ad primary carbides.

As it will be discussed in the following, chemical composition of martensitic stainless steels strongly depends on the equilibrium between γ- and α-stabilizing action of the alloying elements, as it can be inferred by considering the various combinations given in Table 1.5 [1]. In order to obtain a martensitic structure, the austenitic field must be wide enough, in terms of temperature and compositional range, to allow formation of martensite after quenching. As a consequence of this, since Cr content is kept as high as possible to ensure corrosion resistance, its α-stabilizing effect must be counterbalanced with proper C content (γ-stabilization effect, see Appendix A). Furthermore, the addition of other elements must consider the abovementioned equilibrium. For instance, thanks to its γ-stabilizing attitude, Ni has a great importance because it allows a wider range of Cr concentrations to be used. This is the reason why Ni-bearing grades such as AISI 414 and AISI 431 are characterized by a lower C content. In fact, C reduction is the direct consequence of the austenite-stabilizing effect of Ni and of the needed equilibrium between Ni and C itself (see also Fig. 1.10). Ni addition leads to an increase in both toughness and corrosion resistance, while Mo is added to improve corrosion resistance and mechanical properties, always considering its strong α-stabilizing action. Finally, Nb, Si, V and W are used to modify the response to the tempering treatment [1].

In order to provide the reader with basic knowledge about martensitic transformation to understand this section, to be able to independently study the topic and to better understand any comparison in the next sections, it is worth to remember that formation of martensite occurs by deformation of the austenite lattice with no associated diffusion of atoms. The deformation process leads to a change in the shape for the transformed region. This change consists of a large shear and a volume expansion. For the abovementioned reasons, the martensitic transformation is said to be a diffusionless (even C- or N-supersaturated, or not) and shear transformation associated with a highly crystallographic character resulting from the specific deformation of austenite that generates it [24]. It is worth saying that C- or N-supersaturation condition is not necessary for the transformation to happen. In fact, martensitic transformation may occur also in other alloy systems in addition steels. Some examples are: Ti alloys, binary systems (e.g., In-Tl, Gd-Ce) and for specific compounds such as $ZrO_2$ [17, 24]. By simplifying the discussion with reference to steels, the driving force for the martensitic transformation to happen is the great difference between the free energy

**Table 1.5** Chemical composition of martensitic stainless steels [1]

| Type | C | Cr | Mn | Si | Ni | P | S | Others |
|---|---|---|---|---|---|---|---|---|
| 403 | 0.15 | 11.5–13.0 | 1.0 | 0.5 | – | 0.04 | 0.03 | – |
| 410 | 0.15 | 11.5–13.5 | 1.0 | 1.0 | – | 0.04 | 0.03 | |
| 410S | 0.08 | 11.5–13.5 | 1.0 | 1.0 | 0.60 | 0.04 | 0.03 | – |
| 414 | 0.15 | 11.5–13.5 | 1.0 | 1.0 | 1.25–2.50 | 0.04 | 0.03 | – |
| 416 | 0.15 | 12.0–14.0 | 1.25 | 1.0 | – | 0.06 | 0.15 min | Mo = 0.60 |
| 416Se | 0.15 | 12.0–14.0 | 1.25 | 1.0 | – | 0.06 | 0.06 | Se = 0.15 min |
| 420 | 0.15 min | 12.0–14.0 | 1.0 | 1.0 | – | 0.04 | 0.03 | – |
| 420F | 0.15 min | 12.0–14.0 | 1.25 | 1.0 | – | 0.06 | 0.15 min | Mo = 0.60 |
| 422 | 0.20–0.25 | 11.5–13.5 | 1.0 | 0.75 | 0.5–1.0 | 0.04 | 0.03 | Mo = 0.75–1.25 W = 0.75–1.25 V = 0.15–0.30 |
| 431 | 0.20 | 15.0–17.0 | 1.0 | 1.0 | 1.25–2.50 | 0.04 | 0.03 | – |
| 440A | 0.60–0.75 | 16.0–18.0 | 1.0 | 1.0 | – | 0.04 | 0.03 | Mo = 0.75 |
| 440B | 0.75–0.95 | 16.0–18.0 | 1.0 | 1.0 | – | 0.04 | 0.03 | Mo = 0.75 |
| 440C | 0.95–1.20 | 16.0–18.0 | 1.0 | 1.0 | – | 0.04 | 0.03 | Mo = 0.75 |

of ferrite (*bcc*) with respect to that of austenite (*fcc*), where the lower value is associated to ferrite. This difference in free energy increases with decreasing temperature until spontaneous transformation occurs in the material through a certain sequence at a remarkable speed. The result is the formation of martensite, which composition is identical to that of the parent phase (austenite), but whose crystal lattice, which should have been that of ferrite (*bcc*, more energetically convenient), is the result of a tetragonal distortion of the *bcc* one [17]. $M_s$ expresses the temperature at which the martensitic transformation starts and the *bcc* lattice becomes more energetically convenient. However, this temperature is too low to allow diffusion transformations to happen. Consequently, a rearrangement of the atoms in the *fcc* lattice occurs leading to the formation of a tetragonally distorted lattice (*bct*). In this regard, C plays an important role because it is well known that interstices in the *bcc* structure are smaller in *fcc*, even though the *bcc* is expanded with respect to the *fcc* lattice. The extent of distortion is proportional to the C content. In fact, at low C levels accommodation of the distortion occurs from site to site, but above 0.022% C this process cannot further occur and therefore the tetragonal distortion takes place. This latter condition leads to C to be supersaturated in the as-formed martensite [17].

Moreover, the speed of the transformation interface is such that it exceeds the diffusion rate of C in austenite, as it can be inferred from Eq. 1.3:

$$v_{\alpha'} > \left(\frac{D_C^{\gamma}}{\lambda}\right) \tag{1.3}$$

where

- $v_{\alpha'}$ is the advancement speed of the martensitic transformation.
- $D_C^{\gamma}$ expresses the coefficient of C diffusion.
- $\lambda$ indicates the distance between interstices in which C resides.

The term to the right of the inequality sign expresses the speed that C can reach while diffusing in the lattice. As a consequence of what is stated in Eq. 1.3, C results trapped within the crystal lattice. The growth of martensite in steel can reach very high speed, that may approach a substantial fraction of the speed of sound (more than 300 m/s). Very often, in practice, martensitic transformation is said to be so fast that it takes on an almost instantaneous character. Transformations that involve diffusion are much slower, for instance it has been reported that the fastest solidification process recorded was that occurred with pure Ni at a rate of 80 m/s. As a consequence of the high rate associated with the martensitic transformation, it becomes even more evident that this phenomenon is inconsistent with diffusion [24]. On the other hand, as depicted by the Koistinen and Marburger equation, the extent of the transformation turns out to be not time-dependent:

$$1 - V_{\alpha'} = e^{[\beta(M_s - T_q)]} \tag{1.4}$$

where

- The term $1 - V_{\alpha'}$ indicates the volume fraction of untransformed austenite, hence $V_{\alpha'}$ indicates the volume fraction of martensite.
- $\beta$ is a constant equal to $-0.011$.
- $M_s$ is the martensite start temperature.
- $T_q$ is the quenching temperature, the temperature below $M_s$ to which the material is cooled.

The fact that time is not present in the equation suggests the athermal nature of the process, i.e., the volume fraction of generated martensite only depends on the undercooling below $M_s$. The athermal definition could be essentially considered a pragmatic consequence of the fact that the transformation occurs on time scales that are remarkedly shorter than what can be observed by ordinary perception. For a more comprehensive dissertation regarding the possible time-dependence of martensitic transformation, the reader may refer to the textbook by Bhadeshia and Honeycombe that can be found in the reference list of this chapter [24].

Both Ni presence (e.g., AISI 431) and high C concentrations (e.g., AISI 440C) may lead to the formation of considerable volume fractions of retained austenite after quenching. Retained austenite is defined as the untransformed austenite when the $T_q$ value (see Eq. 1.4) is equal to room temperature. The conversion to martensite can be forced by cooling the material at subzero temperatures because the presence of retained austenite is a consequence of a martensite finish ($M_f$) value below room temperature ($M_f$ is frequently defined as the temperature at which 95% of martensite has been originated from the parent austenite). In fact, in some cases the $M_f$ temperature may decrease down to cryogenic values. On the other side, low-C martensitic grades are characterized by relatively high values of both $M_s$ and $M_f$ temperatures. For instance, AISI 410 (see Table 1.5) has the $M_s$ temperature at 350 °C, while the $M_f$ point is at 250 °C [1]. Alloying elements with the ability of stabilizing austenite tend to lower the $M_s$ temperature, except for Co. For this reason, there are some limitations in the use of alloying elements and in their overall content. For example, elements that promote corrosion resistance (i.e., Cr and Mo) have α-stabilizing action and as already mentioned, an increase in their concentration must be balanced with the addition of γ-stabilizers. This leads inevitably to a lowering of $M_s$ with all the possible associated negative consequences (e.g., high retained austenite content) [17].

The typical morphology of the microstructure of martensite (plates or laths) is associated to constraining action of the volume surrounding it and the shape is essentially the result of the minimization of the strain energy associated to deformation [24]. The most important and interesting feature of martensitic stainless steels is, of course, their attitude to develop a martensitic structure after quenching, thus remarkedly increasing their strength (see Table 1.6 for typical mechanical properties and related treatments). This is possible thanks to proper thermal treatments, whose processing parameters must be carefully designed and optimized as a function of the specific steel grade, desired mechanical properties, size and shape of the workpiece etc. Generally speaking, martensitic stainless steels can be treated in the same way as plain C, low alloy and alloy steels, meaning that at least three stages must be employed: austenitizing, quenching and tempering. A detailed description regarding the heat treatment of martensitic grades will not be provided since it is not consistent with the purpose of this section, however the reader may refer to some of the texts in the reference list (e.g., the textbook of Totten [10] or Campbell [20]).

As it is well known, and already partially discussed, addition of alloying elements affects the hardenability of steels. This is particularly true for Cr, as this element has the ability to cause a significant shift of the TTT curves to the right. As a consequence of this, martensitic stainless steels are highly hardenable, also by means of cooling media that are not so drastic (e.g., oil or air) and transformation might occur directly after hot working operations during cooling down to room temperature. Conversely, high-C types tend to form a martensite characterized by high residual stresses, thus more prone to cracking due to internal stresses and associated strain. For this latter reason, after hot working it is common practice to carry out directly the tempering treatment [1]. Another possibility, when direct tempering is not suitable, is to keep the material at a temperature of few hundred degrees above room temperature in order to limit the deleterious effect of residual stresses. Tempering treatments are employed to produce a less-strained, softer (i.e., lower hardness) and tougher martensite as a consequence of C diffusion caused by the temperature increase associated with this specific thermal treatment. An important characteristic that these steels should possess is that the temperature range at which austenite forms on heating should be sufficiently high to allow tempering treatments above the temper embrittlement temperature range [17].

In view of their high C concentration, martensitic stainless steels show lower corrosion resistance when compared to austenitics or ferritics. This feature appears to be easy to understand due to the tendency of martensitic stainless steels to form Cr carbides, especially the $Cr_{23}C_6$ carbide (see Appendix B for further details). As a consequence of this, the corrosion resistance results somehow jeopardized and definitely lower in comparison to alloys with same Cr content, but without carbides [1]. In addition to this, and differently from ferritic grades, martensitic stainless steels are always susceptible to SCC (stress corrosion cracking) when the hardness value exceeds HRC 22 [17].

For what concerns their use, martensitic stainless steels find application in fields where tribological features (i.e., wear resistance), the relatively high mechanical properties and a sufficient corrosion resistance are the main requirements. However, for further details the reader may refer to Sect. 1.3.

### *1.2.4 Austenitic Stainless Steels*

It is believed that the very first commercial austenitic stainless steel was developed between 1909 and 1912 by Edward Maurer and Benno Strauss at the Krupp Research Laboratories. As it is the case with the other stainless steels types, the name of this class of alloys derives from the type of stable phase at room temperature, the $\gamma$ (*fcc*) phase called austenite (further details in Sect. 2.2). These steels usually offer an excellent combination of ductility, toughness, corrosion resistance and weldability

**Table 1.6** Mechanical properties of selected martensitic stainless steels with different designations

| Type | EN | UNS | UTS (MPa) | Yield strength (MPa) | Elongation in 50 mm (%) | Reduction in area (%) | Hardness | Condition |
|---|---|---|---|---|---|---|---|---|
| 403 | 1.4006 | S40300 | 517 | 276 | 35.0 | 70.0 | 82 HRB | Annealed |
| 403 | 1.4006 | S40300 | 765 | 586 | 23.0 | 67.0 | 97 HRB | Tempered |
| 410 | 1.4006 | S41000 | 1089 | 1006 | 13.0 | 70.0 | – | Oil quenched from 980 °C and tempered at 540 °C |
| 414 | 1.4008 | S41400 | 793 | 621 | 20.0 | 60.0 | 235 HB | Annealed |
| 414 | 1.4008 | S41400 | 896 | 793 | 15.0 | 58.0 | 270 HB | Cold drawn |
| 414 | 1.4008 | S41400 | 1006 | 800 | 19.0 | 58.0 | – | Oil quenched from 980 °C and tempered at 650 °C |
| 420 | * | S42000 | 655 | 345 | 25.0 | 55.0 | 195 HB | Annealed |
| 420 | * | S42000 | 758 | 690 | 14.0 | 40.0 | 228 HB | Annealed and cold drawn |
| 431 | * | S43100 | 862 | 655 | 20.0 | 55.0 | 260 HB | Annealed |
| 431 | 1.4057 | S43100 | 896 | 758 | 15.0 | 35.0 | 270 HB | Annealed and cold drawn |
| 431 | 1.4057 | S43100 | 834 | 738 | 20.0 | 64.0 | – | Oil quenched from 980 °C and tempered at 650 °C |
| 440C | 1.4125 | S44004 | 758 | 448 | 14.0 | 25.0 | 97 HRB | Annealed |
| 440C | 1.4125 | S44004 | 862 | 690 | 7.0 | 20.0 | 260 HB | Annealed and cold drawn |
| 440C | 1.4125 | S44004 | 1965 | 1896 | 2.0 | 10.0 | 580 HB | Hardened and tempered at 315 °C |

UTS = ultimate tensile strength. * indicates more than one possible alternative designation [1]

[1]. Austenitics are the most common stainless steels type. This is due to their peculiar features. First of all, the possibility of covering a wide range of mechanical properties that can be modified, in the first place, by cold working. In addition to this, they are characterized by a well predictable corrosion behavior in any environment, as well as a corrosion resistance level that is generally high [17]. As it can be seen from Appendix A, but also indirectly from Fig. 1.7, specific alloying elements with a $\gamma$-stabilizing effect are the reason why for the existence of the austenitic phase at room temperature.

Early examples of austenitic stainless steels were not characterized by a very refined chemical composition, partly due to the absence of the AOD process until the 1960s. As a consequence of this, at the beginning of their history these alloys were prone to intergranular corrosion caused by the precipitation of Cr carbides, especially the Cr-rich $M_{23}C_6$ (see Appendix B). As it is well known, this degradation process is named sensitization and causes a decrease in the corrosion resistance as a consequence of the Cr depletion caused by the carbides' formation. Several strategies have been developed over the years to address this problem. First and foremost is the improvement of production technology, i.e., melting and refining, to limit the C content (see Fig. 1.9 to understand the effect of decreasing C content on the precipitation of $M_{23}C_6$ carbides). Secondly, the development of low-C grades, those that are known as L-types such as AISI 304L and AISI 316L. Finally, as already discussed for ferritic grades, a further step forward was the introduction on the market of Ti- and Nb-stabilized grades such as AISI 321 and AISI 347, respectively [1].

It is worth noting that the chemical composition (see Table 1.7) is designed to suppress the austenite-to-ferrite transition on cooling, thus keeping the alloy fully

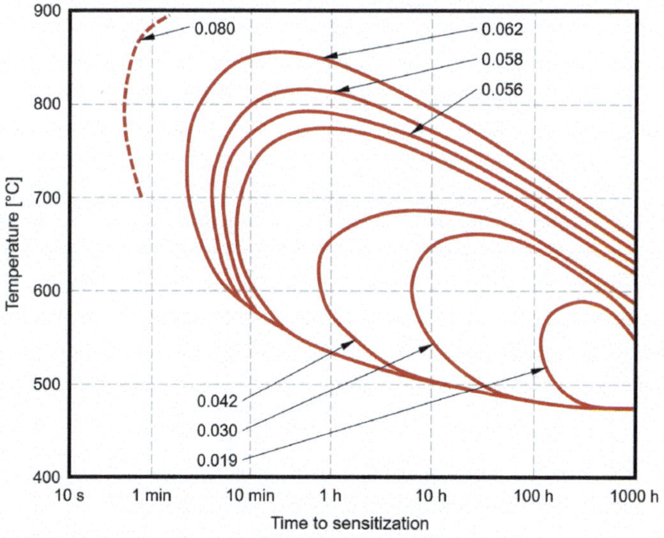


**Fig. 1.9** Precipitation curves for the $M_{23}C_6$ carbide as a function of C content (redrawn from [17])

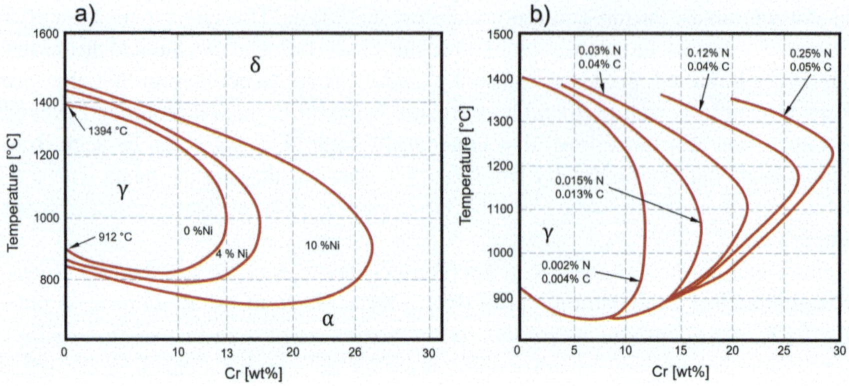


**Fig. 1.10** Variation of the amplitude of the so-called γ loop as a function of the chemical composition. **a** Influence of the Ni content; **b** influence of both N and C content (redrawn from [1])

austenitic from solidification to room temperature. In order to exploit this need, a common feature of most austenitics is the presence of a significant percentage of Ni, with the specific aim of stabilizing the γ phase (see Fig. 1.10a). Furthermore, Ni addition leads to an increase in both toughness and corrosion resistance. For some grades Mn can partially replace Ni, thus limiting its amount to about 6%. In fact, Mn is also an austenite stabilizer and the types of stainless steels (AISI 2xx series) that employ it in place of Ni have a Mn percentage in the 5.5–15.5% range [1].

C plays a very important role in austenitics, in fact if on one hand it helps the stabilization of the γ phase, on the other hand its presence leads to the deleterious formation of Cr carbides due to the high affinity between the two elements (see also Fig. 1.9). Carbides formation occurs as a result of this affinity and whenever, even locally, conditions of supersaturation are reached while the diffusion rates are such that for C and Cr so segregate into precipitates. For what concerns supersaturation related to C in austenite, it is worth remembering that at solidification the solubility limit of C is over 0.4%, however this value markedly decreases with decreasing temperature. In fact, at room temperature a very tiny quantity of C is soluble in the austenite that characterizes stainless steels. All these features, combined with the fact that bulk diffusion is much slower than GB diffusion, lead to preferential precipitation of carbides in the grain boundary region. Furthermore, as already mentioned for what concerns ferritics, since C diffusion is several orders of magnitude more rapid than that of Cr, diffusion of C and combination with Cr occurs according to an in-situ process, leading to Cr depletion near the GBs. This affects obviously the corrosion resistance of the material. In order to get rid of this situation, it is necessary to decompose the carbides and allow Cr to diffuse by means of solution annealing treatments. However, it is interesting to note that the time required for dissolution is several orders of magnitude longer than the time window useful for precipitation.

Two other very important alloying elements are Mo and N. They have two different effects on the γ phase: Mo inhibits its stabilization, being an α-stabilizer, while

**Table 1.7** Chemical composition of selected austenitic stainless steels [1]

| Type | C | Cr | Ni | Mn | Si | P | S | Others |
|---|---|---|---|---|---|---|---|---|
| 201 | 0.15 | 16.0–18.0 | 3.5–5.5 | 5.5–7.5 | 1.0 | 0.06 | 0.03 | N = 0.25 |
| 202 | 0.15 | 17.0–19.0 | 4.0–6.0 | 7.5–10 | 1.0 | 0.06 | 0.03 | N = 0.25 |
| 301 | 0.15 | 16.0–18.0 | 6.0–8.0 | 2.0 | 1.0 | 0.045 | 0.03 | – |
| 303 | 0.15 | 17.0–19.0 | 8.0–10.0 | 2.0 | 2.0–3.0 | 0.20 | 0.15 min | Mo = 0.6 |
| 303Se | 0.15 | 17.0–19.0 | 8.0–10.0 | 2.0 | 2.0–3.0 | 0.20 | 0.06 | Se = 0.15 min |
| 304 | 0.08 | 18.0–20.0 | 8.0–10.5 | 2.0 | 1.0 | 0.045 | 0.03 | – |
| 304H | 0.04–0.10 | 18.0–20.0 | 8.0–10.5 | 2.0 | 1.0 | 0.045 | 0.03 | – |
| 304L | 0.03 | 18.0–20.0 | 8.0–10.5 | 2.0 | 1.0 | 0.045 | 0.03 | – |
| 304LN | 0.03 | 18.0–20.0 | 8.0–10.5 | 2.0 | 1.0 | 0.045 | 0.03 | N = 0.10–0.16 |
| 304N | 0.08 | 18.0–20.0 | 8.0–10.5 | 2.0 | 1.0 | 0.045 | 0.03 | N = 0.10–0.16 |
| 309 | 0.20 | 22.0–24.0 | 12.0–15.0 | 2.0 | 1.0 | 0.045 | 0.03 | – |
| 309S | 0.08 | 22.0–24.0 | 12.0–15.0 | 2.0 | 1.0 | 0.045 | 0.03 | – |
| 310 | 0.25 | 24.0–26.0 | 19.0–22.0 | 2.0 | 1.5 | 0.045 | 0.03 | – |
| 310S | 0.08 | 24.0–26.0 | 19.0–22.0 | 2.0 | 1.5 | 0.045 | 0.03 | – |
| 316 | 0.08 | 16.0–18.0 | 10.0–14.0 | 2.0 | 1.0 | 0.045 | 0.03 | Mo = 2.0–3.0 |
| 316F | 0.08 | 16.0–18.0 | 10.0–14.0 | 2.0 | 1.0 | 0.20 | 0.10 min | Mo = 1.75–2.5 |
| 316L | 0.03 | 16.0–18.0 | 10.0–14.0 | 2.0 | 1.0 | 0.045 | 0.03 | Mo = 2.0–3.0 |
| 316LN | 0.03 | 16.0–18.0 | 10.0–14.0 | 2.0 | 1.0 | 0.045 | 0.03 | Mo = 2.0–3.0 N = 0.10–0.16 |
| 316N | 0.08 | 16.0–18.0 | 10.0–14.0 | 2.0 | 1.0 | 0.045 | 0.03 | Mo = 2.0–3.0 N = 0.10–0.16 |
| 317 | 0.08 | 18.0–20.0 | 11.0–15.0 | 2.0 | 1.0 | 0.045 | 0.03 | Mo = 3.0–4.0 |
| 317L | 0.03 | 18.0–20.0 | 11.0–15.0 | 2.0 | 1.0 | 0.045 | 0.03 | Mo = 3.0–4.0 |
| 321 | 0.08 | 17.0–19.0 | 9.0–12.0 | 2.0 | 1.0 | 0.045 | 0.03 | Ti = 5 × %C min |
| 330 | 0.08 | 17.0–20.0 | 34.0–37.0 | 2.0 | 0.75–1.5 | 0.04 | 0.03 | – |

(continued)

**Table 1.7** (continued)

| Type | C | Cr | Ni | Mn | Si | P | S | Others |
|---|---|---|---|---|---|---|---|---|
| 347 | 0.08 | 17.0–19.0 | 9.0–13.0 | 2.0 | 1.0 | 0.045 | 0.03 | Nb = 10 × %C min |
| 348 | 0.08 | 17.0–19.0 | 9.0–13.0 | 2.0 | 1.0 | 0.045 | 0.03 | Co = 0.2 Nb = 10 × %C min Ta = 0.10 |

N promotes it (see Fig. 1.10b). As already mentioned, Mo has positive effects on corrosion resistance and self-healing properties of the passive film. In addition to this, Mo plays the role of mechanical properties enhancer thanks to its solid-solution hardening effect, however it does not improve oxidation resistance (see Sect. 3.3.10). On the other hand, N has also the function of improving mechanical properties, pitting resistance and its presence delays the formation of the σ phase in those grades that possess a chemical composition prone to its precipitation.

A final aspect of Mo, relevant to the main part of this text, is that in high amounts (> 4%) and in combination with high Cr, N and Ni contents it gave rise to a class of austenitics that was given the name of superaustenitics at the turn of the 1970s and 1980s [22]. This class of materials was introduced mainly as a result of research and development on applications where a high corrosion resistance was the main issue. For standard austenitics, addition of REs (rare earth elements) produce a more stable and oxidation resistance surface oxide layer, while Cu leads to improvements in the cold-workability and corrosion resistance against sulfuric acid [17]. As it can be inferred from Appendix B, austenitic stainless steels are also prone to intermetallic phases formation. Deleterious intermetallic phases can form in specific grades as a consequence of their chemical composition in combination with sufficiently slow cooling processes or operating temperatures that are high enough to promote the formation of such phases. The most relevant is the σ phase, but also χ and Laves phases may form (see Appendix B for further details).

As mentioned, austenitic stainless steels possess this structure at any temperature from solidification downward. However, for particular conditions it is possible that these steels develop a certain percentage of martensite called strain-induced martensite (SIM) inside them. By greatly simplifying, as the name suggests, this martensite is induced by particular deformation conditions. It can form below a specific temperature ($M_d$) when a threshold level of strain has been overcome, thus making martensite more energetically-convenient than austenite [1]. This type of martensite could be generated in materials and components that have undergone cold working (e.g., cold drawing) and can be readily detected, even with NDT (nondestructive testing) techniques thanks to the fact that compared to austenite, SIM is magnetic. In order to eliminate it, particular annealing treatments can be employed, leading to its reversion into austenite [1].

Austenitic stainless steels cannot be hardened by means of thermal treatments, since they do not undergo phase transitions on cooling. However, strengthening can be exploited thanks to cold working operations (e.g., cold drawing; see also Table 1.8). Stainless steels belonging to this family of alloys are usually supplied in the annealed condition. It is important to underline that the annealing treatment must be followed by a cooling process with a cooling rate high enough to avoid/limit carbides formation, i.e., the cooling trajectory must be such that the carbides' transformation curves in the relevant TTT diagrams are not crossed (see Fig. 1.9). Most common heat treatments for austenitic stainless steels are solution annealing and stress relief annealing, however also other treatments are possible. The choice depends on the specific grade, on the requirements and on the previous thermal, thermo-mechanical and deformation history of the material. Typical mechanical properties of selected austenitic stainless steels are given in Table 1.8. It is interesting to mention that, in view of their *fcc* structure, austenitic stainless steels do not have a ductile-to-brittle transition like ferritics. Therefore, their operative range in terms of temperature can be extended down to values below − 200 °C. As it has already been declared, cold working processes may be employed for increasing mechanical properties with specific reference to the yield strength. However, a reduction in ductility must be expected as a consequence of these processes.

As already declared, corrosion resistance is one of the key properties according to which austenitic stainless steels are selected by the designer. Of course, the specific use must take into consideration the chemical composition of the steel and the way it influences the performance in the specific environment. In fact, if on one hand the corrosion resistance of austenitic grades is generally high, it is worth noting that austenitic stainless steels are prone to SCC in environments containing Cl compounds. Furthermore, if compared with ferritic grades, the resistance of austenitic stainless steels to cyclic oxidation is lower due to a thermal expansion coefficient that does not fit properly that of the oxide layer. As a consequence of this, the higher expansion of the substrate tends to promote cracking and subsequent spallation of the oxides formed on the surface [17].

The range of application of austenitic stainless steels is one of the widest among all the materials. Anyway, some examples of use are architecture and construction (e.g., AISI 304L and AISI 316L), exhaust components in the automotive industry, marine, aerospace, oil and gas, chemical and medical sectors. Stabilized grades are usually employed in special engineering fields, for instance the AISI 321 is used in aerospace, while AISI 347 finds applications in refineries.

### *1.2.5 Precipitation Hardening Stainless*

This family of stainless steels takes its name from a particular heat treatment (aging) used to intentionally induce the precipitation of second phases to improve the mechanical properties of the material. Precipitation hardening stainless steels (also known as PH, PHSSs or precipitation-hardenable stainless steels) can be classified

**Table 1.8** Minimum tensile properties at room temperature of selected austenitic stainless steels

| Type | EN | UNS | Tensile strength (MPa) | Yield strength (MPa) | Elongation (%) | Reduction in area (%) | Condition |
|---|---|---|---|---|---|---|---|
| 303 | 1.4305 | S30300 | 515–585 | 205 | 40.0 | 50.0 | Annealed |
| | | | 790 | – | – | – | Cold-finished |
| 303Se | – | S30323 | 515–585 | 205 | 40.0 | 50.0 | Annealed |
| | | | 790 | – | – | – | Cold-finished |
| 304 | 1.4301 | S30400 | 515–620 | 205–310 | 40.0 | 50.0 | Annealed |
| | | | 620 | 310 | 30.0 | 40.0 | Cold-finished |
| 304L | 1.4306 1.4307 | S30403 | 450–480 | 170 | 40.0 | 50.0 | Annealed |
| | | | 620 | 310 | 30.0 | 40.0 | Cold-finished |
| 309 | 1.4828 | S30900 | 515 | 205 | 30.0 | 40.0 | Annealed |
| | | | 620 | 310 | 30.0 | 40.0 | Cold-finished |
| 310 | 1.4845 | S31000 | 515 | 205 | 30.0 | 40.0 | Annealed |
| | | | 620 | 310 | 30.0 | 40.0 | Cold-finished |
| 316 | 1.4401 | S31600 | 515 | 205 | 40.0 | 50.0 | Annealed |
| | | | 620 | 310 | 40.0 | 40.0 | Cold-finished |
| 316L | 1.4404 | S31603 | 450–480 | 170 | 40.0 | 50.0 | Annealed |
| | | | 620 | 310 | 30.0 | 40.0 | Cold-finished |
| 317 | 1.4445 | S31700 | 515 | 205 | 35.0 | 50.0 | Annealed |
| | | | 620 | 310 | 30.0 | 40.0 | Cold-finished |
| 321 | 1.4541 | S32100 | 515 | 205 | 35.0 | 50.0 | Annealed |
| | | | 620 | 310 | 30.0 | 40.0 | Cold-finished |
| 347 | 1.4550 | S34700 | 515 | 205 | 40.0 | 50.0 | Annealed |
| | | | 620 | 310 | 30.0 | 40.0 | Cold-finished |

Cold-finished condition refers to material subjected to a certain amount of cold-work [1, 18]

as a group of alloys characterized by secondary hardening as a consequence of the precipitation of strengthening phases that are not carbides. This particular feature is made possible by a chemical composition featuring the presence of specific alloying elements capable to form strengthening precipitates as a consequence of exposing the material at specific temperatures (i.e., aging treatment) [1]. The metallurgical principle on which precipitation hardening is based is that a solution annealed alloy (also supercooled solid solution) modifies its metallurgical structure through aging. The resulting main advantage is that the material can be manufactured in the annealed condition and reinforced at a later stage. Of course, the basic requirement for precipitation hardening to happen in any alloy system is that the solid solubility limit should decrease with decreasing temperature. Thanks to this interesting possibility the problems associated with high-temperature treatments (e.g., the distortion that can occur on quenching) can be overcome. Among the reasons that led to the development of

PH stainless steels there is certainly the desire to produce materials that could ensure high mechanical properties, overcoming the low toughness of high-C martensitic grades, while maintaining a corrosion resistance comparable to that of austenitics [1]. Although the first PH stainless steel was discovered by William Kroll in 1929, the first PHSS to be introduced commercially was the Stainless W (see Table 1.9 for its chemical composition) produced by the Carnegie-Illinois Steel Company of Pittsburg [7]. As it can be inferred from data given in Table 1.9, PH stainless steels can be clustered in three main groups as a function of the parent phase in which precipitation hardening takes place:

- Martensitic. They are characterized by a fully martensitic structure after cooling from solution annealing. The aging treatment is performed directly on the martensitic structure and leads to the formation of the strengthening intermetallic phase. For these reasons they are also called *single-treatment* alloys.
- Semiaustenitic. They are characterized by an austenitic structure after solution annealing. However, this austenite is thermodynamically unstable and, after a subsequent treatment route composed of austenite conditioning (also known as *trigger anneal*) and transformation (see Fig. 1.11), martensite formation is induced in the material. Finally, aging is performed on the obtained martensite. As a consequence of the specific treatment sequence, these steels are also called *double-treatment* alloys.
- Austenitic. These steels are characterized by an austenitic structure at any temperature from solidification downward. The treatment sequence is composed of solution annealing and aging (see Fig. 1.11) [1].

Compared to the early types, modern PHSSs have a more refined chemical composition aimed at maximum limitation of the presence of deleterious δ ferrite and retained austenite [17]. As seen in previous sections, it is clear that the stabilization of one phase over another (e.g., austenite rather than martensite) is primarily dependent on chemical composition. The same is true for the precipitates that can form, so by analyzing the compositions of PHSS steels it is possible to understand both which group they belong to, and the type of precipitate that will reinforce them (see Table 1.9). To avoid confusion and to give the reader explanation on what could easily be a question in this context, it is worth to note that at least two of the three members of the family of PH stainless steels could, to a first approximation, be included in two of the previously introduced families (i.e., martensitic and austenitic stainless steels). However, the very particular reinforcement mechanism by precipitation of intermetallics that characterizes them, has made it preferable to adopt a classification that would highlight its importance and not only based on the stable phase at room temperature. It is well known that precipitation hardening is a metallurgical phenomenon whose products are dependent on the chemical composition of the alloy (see, for example, Al–Cu alloys, Ni-based superalloys, maraging steels, Ti-alloys etc.), and therefore the three groups will also differ in the type of reinforcing precipitate. Taking three of the best-known representatives of these steels as reference, it is possible to see this latter difference as follows:

**Table 1.9** Chemical composition of selected PHSS

| Type | Trade or common name | AISI grade | C | Cr | Ni | Mo | Mn | Si | % Others |
|---|---|---|---|---|---|---|---|---|---|
| M | 13-8Mo | – | ≤ 0.05 | 12.25–13.25 | 7.5–8.5 | 2.0–2.5 | ≤ 0.2 | ≤ 0.1 | Al = 0.90–1.35 N = 0.01 |
| M | 15-5PH | – | ≤ 0.07 | 12.5–14.0 | 3.5–5.5 | – | ≤ 1.0 | ≤ 1.0 | Cu = 2.5–4.5 Nb = 0.15–0.45 |
| M | 17-4PH | AISI 630 | ≤ 0.07 | 12.5–17.5 | 3.0–5.0 | – | ≤ 1.0 | ≤ 1.0 | Cu = 3.0–5.0 Nb = 0.15–0.45 |
| M | Custom 450 | – | ≤ 0.05 | 14.0–16.0 | 5.0–7.0 | 0.5–1.0 | ≤ 1.0 | ≤ 1.0 | Cu = 1.25–1.75 Nb ≥ 8 × %C |
| M | Stainless W | AISI 635 | ≤ 0.08 | 16.0–17.0 | 6.0–7.5 | – | ≤ 1.0 | 0.5 | Al = 0.4 Ti = 0.4–1.2 |
| SA | 15-7Mo | AISI 632 | ≤ 0.09 | 14.0–16.0 | 6.50–7.75 | 2.0–3.0 | ≤ 1.0 | ≤ 1.0 | Al = 0.75–1.50 |
| SA | 17-7PH | AISI 631 | ≤ 0.09 | 16.0–18.0 | 6.50–7.75 | – | ≤ 1.0 | ≤ 1.0 | Al = 0.75–1.50 |
| SA | AM355 | AISI 634 | 0.13 | 15.5 | 4.3 | 2.75 | 0.95 | - | N = 0.1 |
| A | 17-10P | – | 0.07 | 17.0 | 10.0 | – | 0.75 | 0.5 | P = 0.3 |
| A | 17-14CuMo | AISI 653 | 0.12 | 16.0 | 14.0 | 2.5 | 0.75 | 0.5 | Cu = 3.0 Nb = 0.4 Ti = 0.3 |
| A | A286 | AISI 660 | ≤ 0.08 | 13.5–16.0 | 24.0–27.0 | 1.0–1.5 | ≤ 2.0 | ≤ 1.0 | Al = 0.35 B = 10–100 ppm Ti = 1.90–2.35 V = 0.1–0.5 |
| A | HNM | – | 0.3 | 18.5 | 9.5 | – | 3.5 | – | – |

*A* austenitic, *M* martensitic, *SA* semiaustenitic [1]

- 17-4PH: also known as AISI 630 (1.4542). Strengthening is given by the formation of Cu-rich precipitates (CRPs, see also Appendix B). In addition to this, the material is characterized by martensitic transformation, which further improves mechanical properties.

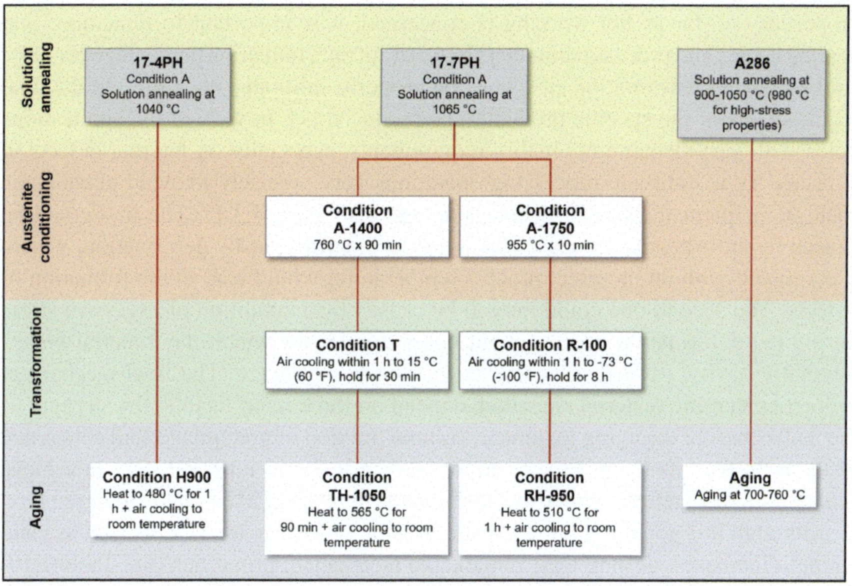


**Fig. 1.11** Possible heat treatment routes for three types of PHSS: martensitic 17-4PH, semi-austenitic 17-7PH, and austenitic A286 (redrawn from [1])

- 17-7PH: also known as AISI 631 (1.4568). Strengthening is given by the precipitation of $\gamma'$ and $\beta$ phases ($Ni_3Al$ and NiAl, respectively), in addition to the martensitic transformation obtained as a consequence of the multiple-stage heat treatment employed for this steel (see Fig. 1.11).
- A286: also known as AISI 660 (1.4980). The precipitation strengthening is given by the formation of the $\gamma'$ phase (see Appendix B).

Worthy of mention is the case of the austenitic A286. On one hand, its fully austenitic structure at any temperature from solidification downward allows it to show high toughness over a wide range of temperatures from − 200 °C to more than 600 °C. In addition to this, the alloy is characterized by excellent creep and stress rupture strength. On the other hand, the fact that the precipitation of the reinforcing $Ni_3$(Al, Ti) phase occurs at temperatures around 700 °C makes it possible to use the alloy at fairly high temperatures with limited risk of overaging. For these reasons, A286 is often considered a Fe-based superalloy, as, ideally, it can be considered the progenitor of this family of alloys as well as a precursor to Co- and Ni-based superalloys [1].

The most important technological process to which PHSS undergo is certainly the combination of heat treatments necessary to obtain the desired structure and properties. Thanks to Fig. 1.11, an example of a possible heat treatment route for one member of each of the three groups can be observed. As far as processing possibilities are concerned, these steels can be subjected to hot working (e.g., hot rolling or forging), but also to cold working which leads to further increase in mechanical properties. As far as hot working is concerned, it is important to remember that heating cycles and process parameters (e.g., strain rate, reduction percentage per pass, cooling during deformation etc.) must be properly calibrated according to the final requirements of the specific product to be produced [1]. In view of the applications fields with very demanding quality requirements, especially on internal defects, it is necessary to calibrate processing operations very precisely to avoid phenomena such as incipient melting or decohesion (see also Sect. 4.3.1). The first could be caused by an excessive increase in temperature during plastic deformation, which, if combined with an incorrect process temperature, would lead to the formation of defects. The second one could instead be caused by a condition of excessive stress during the deformation process, both hot and cold, for which the material breaks internally having exceeded locally its maximum resistance. The final mechanical properties that the material can attain depend on three main factors: the strength of the matrix before the aging treatment, volume fraction of precipitates and coherency of them. In fact, as it can be seen from the data in Table 1.10, PHSSs can achieve remarkable mechanical properties, easily exceeding 1300 MPa yield strength [1].

Although it is generally known that corrosion resistance tends to worsen as yield strength increases, despite their remarkable mechanical properties (see Table 1.10) PHSSs have very good corrosion resistance characteristics. For example, although aged 17-4PH has a higher yield strength than the martensitic AISI 410, its corrosion resistance is much better. Furthermore, PHSSs are characterized by a better corrosion resistance when compared with traditional high-strength steels thanks to their

**Table 1.10** Mechanical properties of some PHSS with related aging temperature [1]

| Type | EN | UNS | Aging temperature (°C) | Tensile strength (MPa) | Yield strength (MPa) | Elongation (%) | Hardness |
|---|---|---|---|---|---|---|---|
| 13-8Mo | 1.4534 | S13800 | 510 | 1520 | 1410 | 6.0–10.0 | 45 HRC min |
| 15-5PH | 1.4545 | S15500 | 480 | 1310 | 1170 | – | – |
| 17-4PH | 1.4542 | S17400 | 480 | 1310 | 1170 | 5.0–10.0 | 40 HRC min |
| Custom 450 | – | S45000 | 480 | 1240 | 1170 | 3.0–5.0 | 40 HRC min |
| Custom 455 | 1.4543 | S45500 | 510 | 1530 | 1450 | ≤ 4.0 | 44 HRC min |
| 15-7Mo | 1.4574 | S15700 | 480 | 1650 | 1590 | 1.0 | 46 HRC min |
| 17-7PH | 1.4568 | S17700 | 510 | 1450 | 1310 | 1.0–6.0 | 43 HRC min |
| AM355 | – | S35500 | 540 | 1170 | 1030 | 12.0 | 37 HRC min |
| A286 | 1.4980 | S66286 | 730 | 860–965 | 655 | 4.0–15.0 | 24 HRC min |

"stainless nature". PHSS are prone to HE (hydrogen embrittlement), especially in alloys with increasing yield strength [1]. From the market associated with them, it is possible to say that the most common type is that of martensitic PHSSs. They are mainly produced in the form of forgings, bars and hot-rolled coils, but they can also be produced in sheets and strips thanks to cold-rolling processes. Obviously, the flatness obtained will be slightly worser than that of non-PH grades. On the other hand, semiaustenitics are more employed in production of components that need extensive forming prior to aging treatment and are more suitable to be produced in the form of strips, sheets and wires. Finally, austenitics are the last numerous types. They are used in applications at lower or higher temperatures than the other types (where these are less performing), in applications where their nonmagnetic nature is required and where higher coefficient of thermal expansion could be an advantage [17]. PH stainless steels are widely used in aerospace (e.g., turbine wheels, blades, fasteners, springs, landing gear and fuselage components), power generation (e.g., blades and fasteners), automotive (e.g., injection system components), and oil and gas industry [1].

### *1.2.6 Duplex Stainless Steels*

Among the five major families into which stainless steels have been divided, duplex stainless steels (DSSs) are the youngest. Their name comes from the simultaneous, and clearly separate, presence of ferrite and austenite in their structure (see Fig. 1.12a). This combination allows them to take advantage of the characteristics of both phases and possess sometimes better characteristics than the two individual microconstituents.

Two-phase (ferritic-austenitic) stainless steels have been the subject of study since the 1920s. In fact, DSSs were first developed at that time, but have remained aside virtually unused until much more recent times. AVESTA Steelworks (Sweden) produced a steel that could be defined as a duplex in 1932. There are several reasons that contributed to the late development of duplex stainless steels, first of all the optimization of the alloy system. This involves, in the first instance, compositional design and control, understanding of the types and mechanisms of secondary phase formation, design of manufacturing processes (e.g., steelmaking, hot and cold working). What is stated above was found to be subordinate to two events, both related to an important alloying element: nitrogen. The first of the two was the advent of the AOD process, which allowed a better control of the N content, while the second one concerns the understanding of the thermodynamics of the system thanks to an approach based on computational thermodynamics [17, 25, 26]. Several advantages can be ascribed to DSSs:

- Higher strength than austenitics and higher toughness than ferritics.
- Better resistance against HE than ferritics.

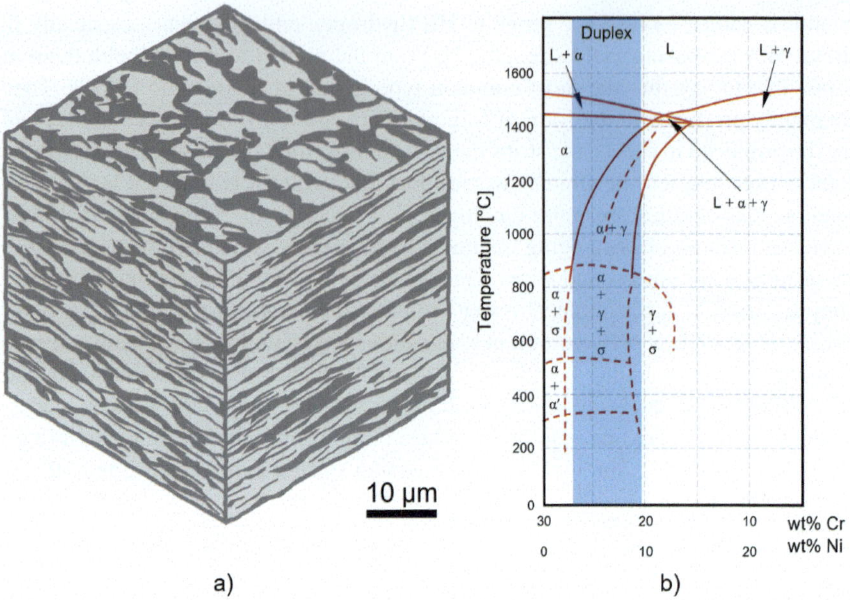


**Fig. 1.12** **a** Three-dimensional reconstruction of the microstructure of a rolled duplex stainless steel with evidence of the two phases highlighted in different shades of grey: austenite and ferrite; **b** quasi-binary section of the Fe–Cr–Ni phase diagram at 70% Fe (redrawn from [25])

- Increased resistance against general, intergranular, pitting, crevice corrosion and SCC.
- Better thermal conductivity than austenitic stainless steels.

On the other hand, some drawbacks have to be obviously taken into account:

- Complex chemistry and advanced knowledge required for their production and processing.
- The complex chemical analysis leads to complexity in both transformation behavior and phase precipitation.
- Secondary phases (carbides, nitrides, intermetallics) have deleterious effects on corrosion resistance and mechanical properties.

In this context, a classification of DSSs into four families takes place according to the latest developments and to the PREN value: lean duplex $\leq 35 <$ duplex $< 40 \leq$ superduplex $< 45 \leq$ hyperduplex. The last two types will be discussed in Sect. 1.2.7. The workhorse of the duplex family is undoubtedly the 2205 grade (1.4462). It is considered a material capable of bridging the corrosion resistance gap between traditional austenitic and ferritic grades and the superaustenitic family (this part will be explored further later in upcoming sections with the introduction of the PREN index, i.e., the pitting resistance equivalent number) [17, 25, 26].

DSSs are considered to have high corrosion resistance due to the minimum Cr value of around 20% (see Table 1.11). Thanks to a relatively low Ni content they are more economical than austenitic grades, when the comparison is made considering the same corrosion resistance.

Unfortunately, due to their complex chemical analysis, they cannot be employed at high temperature because of the formation of deleterious phases, and to the variation of the balance between $\alpha$ and $\gamma$ (see Fig. 1.14). Obviously, the existence of deleterious compounds is affected by the phase in which the formation process takes place (i.e., ferrite or austenite). For what concerns ferrite, the two main undesirable phases are $\alpha'$ and $\sigma$ (see Sect. 1.2.2 and Appendix B for better description of their detrimental effects in ferrite). On the other hand, as far as austenite is concerned, again $\sigma$ is the main undesirable phase, especially for high-Mo grades. As already mentioned, alloying with N is beneficial in terms of limiting $\sigma$ formation and this is one of the reasons why the use of N instead of Mo is employed in leaner alloys. In addition to the $\sigma$ and $\alpha'$ phases, R, $\chi$, and $\pi$ can also form in DSSs (see Appendix B). However, due to the fact that $\sigma$ has severe consequences, forms in larger amounts and, most importantly, forms earlier than the others (see Fig. 1.13), interest in the other phases is more research-oriented than practical. As explained in Appendix B, duplex grades can also be subjected to carbides and nitrides formation. For example, the $Cr_2N$ nitride can form upon welding and its presence results in Cr depletion. Therefore, a decrease in the corrosion resistance of the alloy takes place [17].

It is worth mentioning that DSSs solidify by forming ferrite as the first phase, while the volume fraction of austenite starts to increase as temperature decreases. This point is very important because it is fundamental in the calibration of hot forming processes and heat treatment. Considering the Fe–C binary phase diagram, $\gamma$ is stable at higher temperatures than $\alpha$ and this could lead to the misconception that an increase in temperature causes an increase in volume fraction of austenite. The correct system to take as reference in this case is the ternary Fe–Cr–Ni from which, considering a quasi-binary section at 70% Fe given in Fig. 1.12b, it is possible to derive that the stable phase at high temperature is $\alpha$. Another possible solidification mechanism is the so-called *flip-flop* solidification, but this occurs only in the region near the eutectic point (see Fig. 1.12b) and for increased amounts of $\gamma$-stabilizing elements. From these considerations, in combination with the phase balance also resulting from the precipitation of the other compounds, Fig. 1.14 shows how the phase ratio for the 2205 DSS varies with temperature. This representation is particularly useful with regard to manufacturing as it provides information on both heat treatment and processing temperatures [25, 26].

Generally, the process temperature for hot working lays between 1100 and 1230 °C, depending on the specific grade and process. If the process temperature is too low, the deformation would tend to accumulate within the weaker phase ($\alpha$) and could result in its cracking in the deformed region [27]. Obviously, Fig. 1.14 should be considered similarly to precipitation diagrams (see, for instance, Fig. 1.13), since it is clear that DSSs at low temperature do not consist of $\sigma$ phase and austenite. The residence time at a certain temperature is not given in Fig. 1.14, so it is clear that with cooling rates high enough to quickly cross the critical intervals showed in Fig. 1.13,

**Table 1.11** Chemical composition of selected duplex stainless steels [18, 27]

| Generation | Type | UNS | C | Cr | Ni | Mo | N | Mn | Others |
|---|---|---|---|---|---|---|---|---|---|
| 1 | – | S32900 | 0.08 | 23.0–28.0 | 2.5–5.0 | 1.0–2.0 | – | – | – |
| 1 | – | S31500 | 0.03 | 18.0–19.0 | 4.3–5.2 | 2.5–3.0 | 0.05–0.10 | – | – |
| 2 | Lean | S32101 | 0.04 | 21.0–22.0 | 1.35–1.70 | 0.1–0.8 | 0.20–0.25 | 4.0–6.0 | – |
| 2 | Lean | S32304 | 0.03 | 21.5–24.5 | 3.0–5.5 | 0.05–0.6 | 0.05–0.20 | 2.5 | Cu = 0.05–0.60 |
| 2 | Mo-containing lean | S32003 | 0.03 | 19.5–22.5 | 3.0–4.0 | 1.5–2.0 | 0.15–0.20 | 2.0 | – |
| 2 | Mo-containing lean | S82441 | 0.03 | 23.0–25.0 | 3.0–4.5 | 1.0–2.0 | 0.2–0.3 | 2.5–4.0 | Cu = 0.1–0.8 |
| 2 | 25% Cr | S31200 | 0.03 | 24.0–26.0 | 5.5–6.5 | 1.2–2.0 | 0.14–0.20 | 2.0 | – |
| 2 | 25% Cr | S31260 | 0.03 | 24.0–26.0 | 5.5–7.5 | 2.5–3.5 | 0.1–0.3 | 1.0 | Cu = 0.2–0.8<br>W = 0.1–0.5 |
| 2 | 25% Cr | S32550 | 0.04 | 24.0–27.0 | 4.5–6.5 | 2.9–3.9 | 0.10–0.25 | 1.5 | Cu = 1.5–2.5 |
| 2 | Standard | S31803 | 0.03 | 21.0–23.0 | 4.5–6.5 | 2.5–3.5 | 0.08–0.20 | 2.0 | – |
| 2 | Standard | S32205 | 0.03 | 22.0–23.0 | 4.5–6.5 | 3.0–3.5 | 0.14–0.20 | 2.0 | – |
| 2 | Standard | S32808 | 0.03 | 27.0–27.9 | 7.0–8.2 | 0.8–1.2 | 0.3–0.4 | 1.1 | W = 2.1–2.5 |
| 2 | Standard | S32950 | 0.03 | 26.0–29.0 | 3.5–5.2 | 1.0–2.5 | 0.15–0.35 | 2.0 | – |

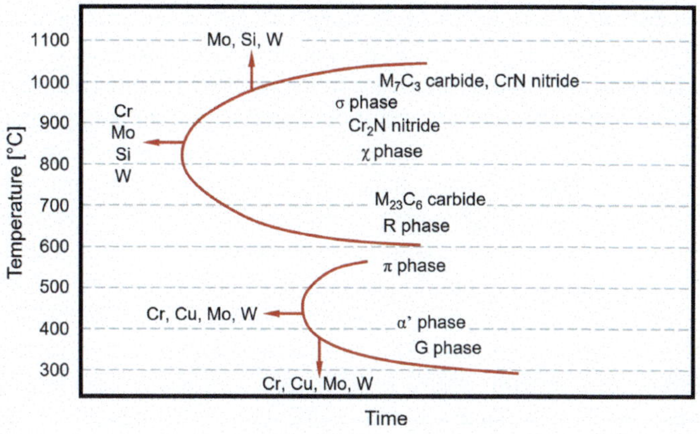


**Fig. 1.13** Qualitative precipitation diagram of duplex stainless steel (redrawn from [26])

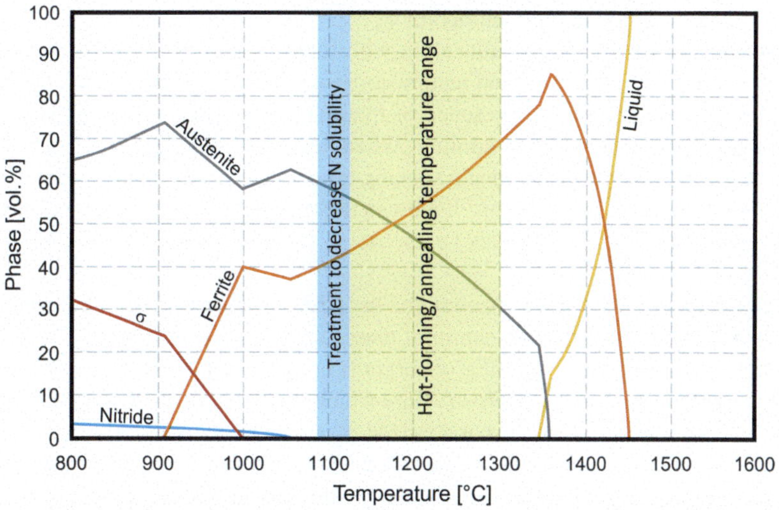


**Fig. 1.14** Calculated phase ratio versus temperature diagram for the standard 2205 (1.4362) duplex stainless steel with schematic indication of heat treatment and hot forming (i.e., forging, rolling) temperature ranges (redrawn from [25])

the volume fraction of α at room temperature would approach that predicted by the Schäffler diagram (see Fig. 1.7) in absence of unwanted second phases [25, 26].

The high-temperature ductility of DSSs highly depends on the process temperature and tends to increase approaching the austenite dissolution temperature. Temperatures for hot working are usually higher than 1100 °C and do not go beyond 1250 °C. However, the exact value depends on the chemical composition and on the desired phase balance at room temperature (which, however, varies as temperature increases as depicted by Fig. 1.14). Generally, ductility of α at high temperatures is better than that of γ, however the situation is strongly impaired by the contemporaneous presence of a second phase. This, in addition to the previous considerations, makes it more complex to calibrate the process and leads to the need for considering other parameters in addition to the chemical composition. For instance, great importance relays on strain rate, type of process (i.e., hot rolling, forging etc.) and specific equipment design (i.e., rolls configuration, dies shape etc.), temperature drop during processing etc. There is usually general agreement in the literature that an increase in temperature is beneficial to hot working. In practice, however, it has been found that this is not always the case. In addition to practical evidence in the production of these steels (e.g., hot rolling or forging), hot rolling laboratory tests have also confirmed this trend. This occurs because an additional parameter must be considered. In fact, it has been shown that ductility above 1100 °C increases as the strain rate decreases. In addition to this, phase balance also has an effect on high-temperature ductility. In fact, the reduction in ferrite volume fraction has much more negative effects than the reduction in temperature. However, as it can be seen from Fig. 1.14, volume fraction of ferrite increases with increasing temperature, so there is no risk of decrease in ferrite volume fraction at high temperature [28].

After hot forming, solution annealing with rapid cooling is always required to modify mechanical properties and restore corrosion resistance characteristics. Both these features depend on the presence of internal stresses and on the presence of precipitate phases formed upon cooling. When heat treating DSSs, great care must be taken to calibrate temperature and soaking time. In fact, at the solution annealing temperature the material is so soft that it can deform under its own weight and this could lead to problems for complex geometries or for products such as wire and wire rod coils.

It is possible to say that DSSs combine the best of both universes, i.e., austenite and ferrite. Due to the biphasic nature of these steels, plastic deformation resulting from hot working (e.g., hot rolling or forging) or cold working (e.g., cold rolling) produce an elongation of the microstructure which is parallel to the major strain axis, thus leading the material to be characterized by a certain degree of anisotropy (see Fig. 1.12a). As far as toughness is concerned, DSSs are characterized by the simultaneous presence of regions with *fcc* and *bcc* structure and have, therefore, a low-temperature impact strength that is intermediate between ferritics and austenitics, but this depends on the specific grade, chemical composition and microstructure balance. In fact, it is worth noting that DSSs may show very good impact strength at − 100 °C, up to 100 J. This property can be improved thanks to a fine-grained microstructure and is negatively affected by cold work [17]. In order to give practical evidence of mechanical properties of DSSs, Table 1.12 gives the most relevant values of selected grades.

In addition to excellent strength and toughness values, DSSs possess outstanding corrosion resistance, also against SCC and corrosion fatigue. To give an example, the lean 2304 (EN 1.4362, see Table 1.11) has more or less the same corrosion resistance of AISI 316L, but with double the yield strength. The fact of having two phases coexisting in the same material has interesting consequences on the resistance to corrosion as a function of the chemical analysis. In fact, the ferritic phase

**Table 1.12** Mechanical properties of selected duplex stainless steels

| Generation | Type | Trade or common name | EN | UNS | Tensile strength (MPa) | Yield strength (MPa) | Elongation (%) | HRC | Charpy V-notch at − 40 °C (J) |
|---|---|---|---|---|---|---|---|---|---|
| 1 | – | AISI 329 | 1.4460 | S32900 | 620 | 485 | 15.0 | 28 | – |
| 1 | – | 3RE60 | 1.4424 | S31500 | 630 | 440 | 30.0 | 30.5 | – |
| 2 | Lean | 2101 | 1.4162 | S32101 | 650 | 450 | 25.0* | 32 | 40 |
| 2 | Lean | 2304 | 1.4362 | S32304 | 600 | 400 | 25.0* | 31 | 40 |
| 2 | 25% Cr | 44LN | – | S31200 | 690 | 450 | 25.0* | 31 | – |
| 2 | 25% Cr | DP-3 | | S31260 | 690 | 485 | 20.0* | 31 | – |
| 2 | 25% Cr | Ferralium 255 | 1.4507 | S32550 | 760 | 550 | 15.0* | 32 | – |
| 2 | Standard | 2205** | 1.4462 | S31803 | 620 | 450 | 25.0* | 31 | – |
| 2 | Standard | 2205** | 1.4462 | S32205 | 640 | 460 | 25.0* | 32 | 40 |
| 2 | Standard | 7-Mo PLUS | – | S32950 | 690 | 485 | 15.0* | 32 | – |

* refers to $A_5$ elongation; ** refers to EN 1.4462 in the old (S31803) and new (S32205) version [17, 18]

by its nature tends to contain a greater amount of Cr than the austenitic one. As a consequence of this, the pitting resistance is greater in ferrite, since this attitude is proportional to the Cr content. In the previous sections, the beneficial effect of Mo addition, which is also an α-stabilizer and preferentially partitioned into ferrite, was highlighted. It might be thought that its addition would exasperate the difference in corrosion resistance between α and γ, but this does not occur totally thanks to the presence of N. Nitrogen is concentrated almost entirely in austenite, greatly reducing the activity of Cr, which may therefore be present in greater than usual amounts within γ. This leads to a further stabilization of austenite, to a more constant phase ratio as temperature increases (with obvious limitations as seen in Fig. 1.14), and leads to increased pitting resistance of the γ phase. These features have been made possible thanks to the knowledge of the thermodynamics of the alloy system and the way alloying elements partition in the material. In this context, as already pointed out, computational thermodynamics has played a key role. Furthermore, duplex stainless steels have an advantage over austenitics in corrosion resistance in specific environments. In fact, when the environment contains NaOH, HCl, $H_3PO_4$, $H_2SO_4$ and organic acids they have a better response, resisting better to the aggressive condition. This behavior is maintained even when the aggressiveness of the environment is emphasized by the presence of other contaminants. Resistance to localized corrosion, pitting in particular, is highly dependent on phase balance, chemical composition and specific alloy type (i.e., wrought or cast). However, this resistance can be quantified by the PREN index (pitting resistance equivalent number), which will be analyzed in Sect. 3.3.11 with specific regard to superaustenitics [17].

Thanks to their remarkable characteristics in terms of mechanical properties and corrosion resistance, duplex stainless steels can be employed in several applications where their features can be exploited. One example is the pulp and paper industry, where DSSs are used for the fabrication of black liquor storage tanks, bleaching equipment and digestors. Another application field is the production of components for desalination plants. In addition to the previously mentioned possibilities, DSSs find broad use in the oil and gas industry thanks to their superior pitting resistance (e.g., piping systems, separators, parts of pumps). Finally, DSSs are also employed for architectural purposes and in the food and drug industry [27].

### *1.2.7 Super and Hyper Grades*

Previous paragraphs have highlighted the existence of five major families of stainless steels: ferritics, martensitics, austenitics, PH and duplex. Thanks to the great variety of chemical analysis made possible by the Fe-based system, within each family it is possible to find many members and this leads to the existence of more than 150 standard grades. This is not difficult to imagine since up to this section the chemical analyses of as many as 75 standard grades have already been provided. Furthermore, there are also non-standard grades developed by the manufacturers to meet special customer needs (e.g., machinability, corrosion resistance, deformability, etc.). Often, these steels do not have standard or unified designations. In many cases they have only one designation but no correspondence in the different designation systems (e.g., EN designation without the AISI counterpart) and therefore often take a name developed by the company that produces them. Therefore, to the previously cited numbers, a large quantity of non-standard grades that differ in the content of specific elements should be added. The universe of stainless steels is very wide indeed. Further details about this discussion will not covered, however, within the five families there are particular grades that due to some particular characteristics have deserved the prefix *super*, and in some cases even *hyper*. These types are the superferritic, supermartensitic, superaustenitic, superduplex and hyperduplex stainless steels. The prefixes used hint at a reason why these alloys are somehow better performing than the "standard" versions. The "super performance" in this case refers primarily to the ability of these materials to be much more resistant to aqueous corrosion (and its specific forms) than previously-developed alloys. With the exception of the superaustenitics, which will be the main topic of the entire textbook, some basic information on the other four types mentioned will be given in the following, starting from superferritics.

Superferritic stainless steels are a type of ferritic alloys that contain more than 25% Cr and also have Mo in their chemical analysis. In some cases, this class of ferritic stainless steels is also referred to as high performance ferritic stainless steels [29]. Research activities aimed at increasing Cr above 25% with Mo addition and decreasing C led to the development of this group. Thanks to this further increase in Cr, addition of Ni avoiding the problem of $\alpha$ destabilization was possible, thus leading to positive effects on corrosion resistance (Ni improves resistance to crevice and pitting corrosion) [10]. E-Brite® was the first commercially produced superferritic stainless steel (1970, Allied Vacuum Metals) and premium techniques such as VIM (vacuum induction melting), EBM (electron beam melting) and ESR (electroslag remelting) were employed in its production in order to minimize C- and N-related detrimental effects on the alloy. A few years later, 29Cr–4Mo was also developed (DuPont), but despite the superior performance in high chloride environments, high production costs limited the applications of these innovative alloys [30].

Subsequently to 29Cr–4Mo, AL 29-4C® was developed, in practice the stabilized version of 29Cr–4Mo [17]. Until the late 1970s, additional new grades were sequentially developed, the chemical analyses of which are given in Table 1.13. As already explained in Sect. 1.2.2, the management of the melting and refining process is fundamental in order to obtain low levels of C and, consequently, avoid exceeding with Cr for compensating the possible precipitation of Cr carbides. In case of superferritics, Cr content is intentionally high, as is Mo content relatively so, while C is kept at low values (see Table 1.13). For this reason, AISI 446 (see Table 1.3) despite having a Cr content greater than 25% is not always classified as superferritic because it belongs to a first generation of alloys where the C content was not so low and where Mo is not present [20]. Due to their chemical composition showing high Cr and Mo values, superferritic stainless steels are prone to $\sigma$ and $\chi$ phase precipitation. Other characteristics, such as being susceptible to 475 °C embrittlement, are similar to those of the other ferritics. Superferritic stainless steels are attractive thanks to their good mechanical properties (e.g., S44735 has a yield strength of 440 MPa, a UTS of 560 MPa and 20% elongation to rupture together with 95 HRB), high thermal conductivity and moderate cost [30]. Superferritic stainless steels are characterized by outstanding corrosion resistance. Therefore, they are used for applications that require high corrosion and oxidation resistance. Furthermore, they show a PREN value of about 40. This allows them to be compared with many superaustenitic and superduplex grades for what concerns corrosion resistance in aqueous environments. Unfortunately, differently from superaustenitics and superduplex, these steels cannot be readily welded in thick sections due to alterations in the HAZ (heat-affected zone) that might lead to deleterious consequences (e.g., grain growth, embrittlement). These materials have found successful applications in replacing, for example, AISI 316L in situations where SCC is an issue (ferritics are practically immune to SCC) [17, 20, 31]. For all the abovementioned positive features, the use of superferritic stainless steels has strongly increased since 1999. Thanks to their ability to guarantee high chloride and acid resistance, some of the possible applications are: piping systems, heat exchangers for Cl-containing aqueous solutions (including seawater), condensers, components in gas-fired heating systems and ventilation systems and thin strips [20, 30, 31].

Supermartensitic stainless steels (hereafter also SMSSs) are a particular class martensitic stainless steels based on the Fe–Cr–Ni–Mo system and characterized by an extremely low content of interstitial elements (C and N). Due to this characteristic, the tetragonal distortion associated with the martensitic transformation (see Sect. 1.2.3) is limited, the microstructure after tempering consists usually in lath martensite and the resulting alloy is characterized by a relatively low hardness.

**Table 1.13** Chemical composition of selected superferritic stainless steels [17, 30]

| Trade or common name | EN | UNS | C | Cr | Mo | Mn | Ni | Others |
|---|---|---|---|---|---|---|---|---|
| AISI 448 FS10 29-4-2 | – | S44800 | ≤ 0.01 | 28.0–30.0 | 3.50–4.20 | ≤ 0.30 | 2.0–2.5 | C + N ≤ 0.025 Si ≤ 0.20 |
| – | 1.4575 | – | ≤ 0.015 | 26–30 | 1.80–2.50 | ≤ 1.0 | 3.0–4.5 | Nb ≥ 12 × %C Nb ≤ 1.2 |
| E-Brite® 26-1 XM-27 | – | S44627 | ≤ 0.01 | 25.0–27.5 | 0.75–1.25 | 0.40 | 0.50 | Cu = 0.2 N = 0.015 Nb = 0.05–0.20 Si = 0.4 |
| SeaCure® 26-3-3 | – | S44660 | 0.025 | 25.0–27.0 | 2.5–3.5 | 1.0 | 1.5–3.5 | N = 0.035 Ti + Nb = 0.2 + 4 × (C + N) min, 0.80 max |
| Monit® | – | S44635 | 0.025 | 24.5–26.0 | 3.5–4.5 | 1.0 | 3.5–4.5 | N = 0.035 Ti + Nb = 0.2 + 4 × (C + N) min, 0.80 max |
| 29Cr–4Mo | – | S44700 | 0.01 | 29.0 | 3.9 | – | 0.15 | – |
| AL 29-4-C® | – | S44735 | 0.015 | 29.0 | 3.75 | 1.0 | 0.4 | Ti + Nb = 0.2 + 4 × (C + N) min, 0.80 max |

There is a non-negligible amount of Ni in these steels, which serves to make them transformable and to stabilize $\gamma$, which can be then found as retained austenite in the annealed structure, and transformed into reverted austenite upon tempering. Reverted austenite is not a harmful phenomenon in this case because it has some positive effects: increase in toughness and ductility at the acceptable expense of UTS, yield strength and hardness [29]. Taking two standard martensitic stainless steels such as AISI 410 and AISI 420 as reference, it could be said that SMSSs are the result of a two-step modification of the chemical analysis. The first step consists in a reduction of C with addition of Ni and Mo, thus obtaining a softer martensite. Starting from the result of this first modification, the next one would concern further addition of Ni and Mo and reduction of Cr and interstitial elements (i.e., C and N). As a consequence of this simple explanation, it is possible to say that SMSSs are somehow a further development of soft martensitics with very low levels of interstitials [31]. Their chemical composition usually includes %C $\leq$ 0.02, Cr around 13%, Ni between 4 and 6% and Mo concentrations between 0.5% and 2.5% [32]. However, this is a relatively young class of stainless steels that is still the subject of study and development, so it is difficult to find unified UNS designations or an EN number. An example of a SMSS that has been given a designation is UNS S41525 (also known as 13CrS or Fe–0.01C–12Cr–6Ni–2.5Mo–Ti). Often, trade names such as Sumitomo Super 13Cr are given, but unfortunately it is usually difficult to find precise specifications. Some of the most common designations are those that explicitly refer to the chemical composition, such as 13Cr–5Ni–1Mo–0.025Nb–0.09V–0.06N or Fe–0.017C–13Cr–6.2Ni–2.4Mo–0.6Mn–0.3Si [31]. For all these reasons, unlike the other families, a table with specific chemical analysis will not be provided. From an alloy design point of view, with these materials some interesting goals have been achieved with respect to standard martensitic grades (e.g., AISI 420 or AISI 440C):

- Increase of Cr content in solid solution (better corrosion resistance).
- Reduction of %C (better weldability and corrosion resistance).
- Presence of Ni to maintain martensitic structure without $\delta$ ferrite (formation is suppressed thanks to Ni) and Mo to improve corrosion resistance, especially against localized phenomena [32, 33].

This sub-typology of martensitics was developed in the last quarter of the twentieth century primarily for offshore applications and for the oil and gas industry for components such as flowlines, downhole tubulars and line pipes. Concerning temperature, their application field lays in the range between -50 °C and 250 °C, depending on chemical composition and heat treatment. For these reasons, SMSSs find application where they appear economically attractive such as the case of specific components in the off-shore/oil and gas sector where, for instance, the corrosion resistance of C steel is insufficient, and where the employment of DSSs for that specific use would be excessive. Hence, they are placed in an economic niche between carbon and low-alloy steels, typically employed together with corrosion inhibitors, and highly alloyed DSSs [31].

As already mentioned in Sect. 1.2.6, superduplex grades are those showing a PREN greater than 40, while hyperduplex are characterized by further improvement of the corrosion resistance and are capable to reach PREN values in excess of 45. Superduplex stainless steels are characterized by very good general corrosion resistance in several environments, high pitting resistance even in direct contact with seawater and at relatively high temperature. Furthermore, thanks to their relatively high ferrite content, SDSSs have high resistance to SCC when the phenomenon is induced by the presence of chlorides [31]. One example of this family of alloys is UNS S32750 (trade name F53, 1.4410). As shown in Table 1.14, this alloy has an average Cr content of 25%, Ni around 7%, average Mo content of 4% and even more controlled additions of N than 2205 (see Table 1.12).

Superduplex stainless steels (SDSSs) have a chemical composition that makes them very prone to deleterious phases formation (e.g., $\sigma$ phase). $\sigma$ and other deleterious compounds (see Appendix B) can form quickly in the temperature range between 400 °C and 800 °C during welding operations, or after heat treatment when

**Table 1.14** Chemical composition of selected superduplex (SDSS) and hyperduplex (HDSS) stainless steels together with the typical PREN range value [27]

| Type | Trade or common name | EN | UNS | C | Cr | Ni | Mo | N | Mn | Others | PREN |
|---|---|---|---|---|---|---|---|---|---|---|---|
| SDSS | 255 | 1.4507 | S32550 | 0.04 | 24.0–27.0 | 4.4–6.5 | 2.9–3.9 | 0.10–0.25 | 1.5 | Cu = 1.5–2.5 | 38–41 |
| SDSS | 2507 F53 | 1.4410 | S32750 | 0.03 | 24.0–26.0 | 6.0–8.0 | 3.0–5.0 | 0.24–0.32 | 1.2 | Cu = 0.50 | 40–43 |
| SDSS | – | 1.4501 | S32760 | 0.03 | 24.0–26.0 | 6.0–8.0 | 3.0–4.0 | 0.20–0.30 | 1.0 | Cu = 0.5–1.0 W = 0.5–1.0 | 40–43 |
| SDSS | – | 1.4477 | S32906 | 0.03 | 28.0–30.0 | 5.8–7.5 | 1.5–2.6 | 0.3–0.4 | 0.8–1.5 | Cu = 0.8 | 41–43 |
| HDSS | – | – | S32707 | 0.03 | 26.0–29.0 | 5.5–9.5 | 4.0–5.0 | 0.3–0.5 | 1.5 | Cu = 1.0 | 49–50 |
| HDSS | – | – | S33207 | 0.03 | 29.0–33.0 | 6.0–9.0 | 3.0–5.0 | 0.4–0.6 | 1.5 | Cu = 1.0 | 52–53 |

proper cooling is not employed. This latter phenomenon leads to negative consequences for what concerns mechanical properties (i.e., ductility and toughness) and corrosion resistance. For this reason, SDSSs are chosen only when their superior corrosion resistance combined with excellent mechanical properties leave no choice to the designer and therefore justify the higher associated cost and greater manufacturing complexity [31]. In principle, the fabrication of SDSSs does not present any particular difficulties, provided to have established operating practices. As for the operations following melting and refining, these steels can be cast in ingots or using continuous casting machines. In this latter case, the process parameters (casting speed, cooling etc.) must be well calibrated to avoid defects in the solidified product (e.g., edge cracks). Regarding hot forming and heat treatment operations, the considerations that can be made are similar to those already expressed for DSSs (see Sect. 1.2.6 and Fig. 1.14). However, in practice it has been seen that these steels tend to have narrower processing temperature windows. Excesses, even if very small, above or below the ideal process temperature hardly leave room for the production of a defect-free component. For what concerns mechanical properties, SDSSs show higher strength than standard austenitic grades and are about 10% stronger than standard 2205 grade. The widest field of application for SDSSs is the oil and gas industry (e.g., manifolds, flow lines, risers, seawater systems), particularly for components that have to work with the most severe corrosion and temperature conditions [31].

Hyperduplex grades are characterized by higher Cr and Mo contents than those typical of SDSSs. Hyperduplex stainless steels (HDSSs) such as UNS S32707 (see Table 1.14) have very high PREN values and have the highest CPT (critical pitting temperature) along with the best mechanical properties among all currently existing duplex grades. The driving force behind the development of steels with better performance than superduplex, both in terms of corrosion resistance and mechanical strength, derives in large part from the oil and gas sector's need to reach reservoirs at ever greater depths. This type of applications is characterized by high temperature, high pressure and extremely corrosive environment where even SDSSs find it difficult to be applied. As it can be inferred from their chemical analysis (see Table 1.14), HDSSs are highly alloyed, and obviously a challenge related to this feature concerns the possibility of developing alloys where the risk of deleterious phase formation is minimized (e.g., $\sigma$, $\chi$, $Cr_2N$; see Appendix B). Again, as seen in previous sections for other stainless steel types, computational thermodynamics has been used in the alloy design and has played a key role in the development of these materials. For what concerns mechanical properties, HDSSs show excellent values. Just to give an idea without going too much into details in this context, a HDSS can reach yield strength values as high as three times those of a standard AISI 316L. HDSSs are used in extreme environments such as hot tropical seawater in particular for components or assemblies characterized by the presence of crevices, ultra-deep seawater, refineries and challenging oil drilling plants. One possible application is the realization of subsea umbilicals, which are used to connect the control station of offshore platforms to the wellhead on the seabed. These components are used to provide necessary control signals and to inject chemicals to subsea oil and gas wells [27, 34].

### 1.2.8 Phases in Stainless Steels

According to what has been stated within previous sections, stainless steels feature a wide variety of possible grades and chemical analyses. This great breadth of chemical compositions, both in terms of the spices involved and their relative amount, leads to the possibility of forming a large number of compounds within the alloy. In fact, it is known that the number of phases that are able to coexist in an alloy is proportional to the number of alloying elements it contains. Consequently, in view of the information regarding the chemical composition of stainless steels, which can be relatively simple as in the case of martensitics, but can also be very complex as in the case of SDSSs; it is not surprising at all that numerous phases can be found in this family of alloys.

By greatly simplifying the discussion of this topic, it is possible to say that the formation of compounds with a different chemical composition, or stoichiometry, than that of the matrix in which they are formed is due to the thermodynamics that characterizes the specific alloy system. This is essentially the basis for the study of phase diagrams (see Fig. 1.6) in which transformation lines define the stability range of one phase with respect to another, highlighting also the presence of solubility limits. It is easy to see that the discriminant for the existence of a particular phase is its greater stability compared to others that can be found in the same system and this depends on the temperature and chemical composition. Of course, the analysis of binary systems is quite simple, but as mentioned in the previous sections, as the number of alloying elements increases the difficulties in the interpretation of the thermodynamic behavior of the alloy increase considerably. For this reason, it is necessary to use, in addition to intense experimental activity, also software for computational thermodynamics in order to have predictions on phenomena of this type. In fact, the presence of more alloying elements alters the stability ranges of one phase relative to the other and it becomes necessary to employ even more complex representations. One example in this regard is the evolution of the diagram in Fig. 1.6 brought back in Fig. 1.15 in the moment when Ni is added to the Fe–Cr system, thus leading to the ternary Fe–Cr–Ni alloy system. From the comparison of the two figures, it is possible to see that the σ phase that forms in the binary system is still present in the ternary one, however the extent of its stability field is also influenced by the presence of Ni. Moreover, this time it is possible to find it accompanied not only by α, but also by the γ phase. The deleterious nature of the σ phase has already been discussed, is presented in Appendix B, and will be discussed further in subsequent sections. However, this specific phase will now be used to expose some information that should be intended as a reference for the formation of any possible second phase of deleterious nature. In case the phase diagram states that σ can exist for certain combinations of temperature and chemical composition, methods do have to exist to be able to avoid its precipitation or to limit it. One possibility may be the addition of specific alloying elements that limit or retard its formation (e.g., N). Alternatively, it is necessary to rapidly cross the temperature range in which precipitation can occur using appropriate cooling trajectories from high temperature. The latter concept is particularly evident when considering Fig. 1.13 where, to avoid

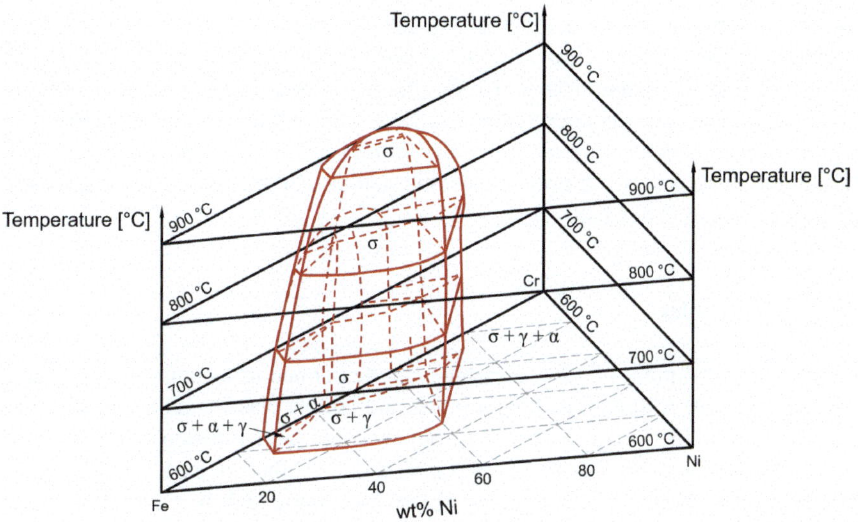


**Fig. 1.15** Three-dimensional representation of the Fe–Cr–Ni equilibrium diagram in the temperature range between 600 and 900 °C (redrawn from [10])

entering the stability field of deleterious second phases, it is necessary to cool the alloy from high temperature while avoiding intercepting the related transformation lines.

The discussion turns totally opposite in case of talking about hardening phases (also known as reinforcing phases), such as the intermetallics introduced in Sect. 1.2.5 with regard to PHSSs. In this particular case, the material is heated on purpose to the temperature field in which precipitation of these phases takes place and is held at that temperature for a time consistent with the formation of precipitates of appropriate size for the desired degree of reinforcement. The soaking time is a critical parameter as there is increasing reinforcement for increasing volume fraction of second phases and increasing size up to a threshold value. Beyond this critical size the precipitates continue to grow, but the reinforcement obtained starts to fade and the alloy tends to soften. This last condition is defined as *overaging*. An explanation of the mechanisms of precipitation hardening, the interaction between dislocations and precipitates and all the related topics is beyond the scope of this textbook, and the information presented so far was preparatory to the introduction of the phases that can form in stainless steels. A more comprehensive discussion about this topic can be found in the volumes in the reference list and further reading list. On the other hand, what can be declined in greater detail are the specific characteristics of the most common second phases found within stainless steels. To accomplish this task, Appendix B was developed, which collects detailed information on precipitates and second phases typical of these steels. Thanks to the field experience and extensive literature review that lays behind that table, Appendix B gives information on the chemical composition, unit cell and relative parameters, occurrence and main effects of these compounds. The idea behind the development of Appendix B is that the reader may consider it as a sort of guide to the compounds found in stainless steels, to be linked to the parts of the text where they are mentioned. Finally, in order to complete the introductory overview of stainless steels, a brief description of the most important phases that can be identified in stainless steels will be provided hereafter. It is important to note that for superaustenitic stainless steels, the relevant phases will be further analyzed in Chap. 2.

#### 1.2.8.1 Carbides

From a chemical point of view, a carbide can be defined as a compound of C and, at least, a metal. Although making a quick and simple classification may be difficult, one possibility for carbides is to be classified as a function of the type of atomic bond that is established between C and the other element. According to this definition, there are four families of carbides: ionic (salt-like carbides), covalent (e.g., SiC), interstitial compounds, transition metal carbides (e.g., TiC) [35]. In addition to this classification, carbides can be categorized in primary and secondary carbides. Primary carbides are formed from the liquid upon solidification and are characterized by a strong inertia to be modified once they are formed. On the other hand, secondary carbides precipitate in the solid state and are more sensitive to heat treatment, hence being subjected to alteration once they have been generated in the material. The existence of carbides is inextricably linked to the presence of C in solid solution within the alloy. It is possible to state that in most cases C is a detrimental impurity for stainless steels from a corrosion-resistance point of view due to its high tendency of bonding with Cr. This, of course, excludes the hardening function in steels due to the martensitic transformation and the related supersaturation degree of the resulting metastable structure (i.e., martensite; see also Sect. 1.2.3); the strengthening of austenite at high temperatures and the secondary hardening of some martensitic grades. Carbides that can be found in stainless steels are: $M_{23}C_6$, MC, $M_7C_3$, $M_3C$ and $M_6C$ and their precipitation occurs at different rates and threshold C concentrations between austenite and ferrite. In the latter, precipitation occurs at GBs and is very rapid and cannot practically be avoided by quenching unless with concentrations between 20 and 50 ppm (very rapid cooling from high temperature is still mandatory). Concerning austenite, C is practically insoluble in it at room temperature, however, quenching from high temperature can freeze C values of about 0.1% in supersaturation (see the case of AISI 301, which composition is given in Table 1.7). In austenitic stainless steels the precipitation of carbides occurs first at GBs and starts from about 900 °C downward for steels with higher C content. Starting temperature for precipitation decreases with decreasing C percentage down to the limit of 500 °C (see also Fig. 1.9 for the precipitation of the $M_{23}C_6$ carbide as a function of the C concentration). Below this latter value, in practice, precipitation stops due to the excessive slowness of C diffusion [17]. Some alloying elements can affect the solubility of C in the alloy matrix decreasing its value and thus promoting carbides precipitation. This is the case, for example, of Ni and Mo. In addition to what was stated above, the carbides used for alloy stabilization (i.e., MC carbides such as TiC and NbC) bring beneficial effects to the material, as declared in previous sections. This type of carbide, which is thermodynamically more stable than $M_{23}C_6$, is formed in place of this latter avoiding the phenomenon of Cr depletion from the matrix and, thus, preventing sensitization. More detailed information about the specific type of carbide can be found in Appendix B.

#### 1.2.8.2 Intermetallics

The term intermetallic (or intermetallic compound) is used to define a class of solid-state ordered compounds that are formed as a result of the combination between two or more elements of metallic nature, including post-transition metals (e.g., Al, Ga, In, Tl) and metalloids (e.g., As, Te). An important and interesting feature of intermetallics is that they have different composition and properties from that of the constituent elements. The interest in intermetallic compounds is dual in nature. On one hand these compounds are studied due to their reinforcing action or deleterious nature (in terms of mechanical properties and corrosion resistance). On the other hand, they can be used on their own and not as secondary phases for particular applications, for instance at high temperature [1, 36]. In the traditional meaning of the definition of intermetallic, interstitial compounds (i.e., carbides and nitrides) are not included. However, interstitial intermetallic compounds such as cementite ($Fe_3C$), which have a metal in their chemical composition, are included because they have stoichiometry and properties similar to those of the compounds defined above [36]. There are several possibilities for their classification, one might be to categorize these compounds according to their composition. In this case two types can be identified: stoichiometric intermetallics which have fixed composition (e.g., $Fe_3C$, $Ni_3Al$) or nonstoichiometric intermetallic compounds. In this latter case a range of compositions characterizes the compound and often the term intermediate solid solution is used for their nomenclature [36]. Another possibility for the classification of intermetallics is the effect they have on the alloy: deleterious or reinforcing. For what concerns the former type, they are normally hard and brittle and can adversely affect the properties of the material by making it brittle when formed along grain boundaries (in particular for what concerns TCP phase, which will be discussed in Sect. 2.4.6). In addition to this, another problem arising from the formation of this type of intermetallics is the depletion of important alloying elements from the region that surrounds the intermetallic itself, as for example the case of compounds containing Cr and Mo (deleterious effect similar to that related to sensitization caused by carbides). Intermetallics formed by diffusion of substitutional-type alloying elements are characterized by a precipitation that is generally slower than that of carbides, which instead can rely on the interstitial nature of C. Steels that are particularly rich in alloying elements, however, may exhibit faster precipitation, even in the range of minutes. In addition to this, there are phenomena that can promote the diffusion of substitutional species. It is the case of deformation that, promoting substitutional diffusion, is able to contribute to the acceleration of the formation of intermetallics.

Reinforcing intermetallics are phases characterized by coherent precipitation and have a precise lattice misfit with a value below a certain threshold. Simplifying the discussion, the presence of reinforcing precipitates leads to a different behavior of the dislocations in the alloy, whose movement is hindered due to the presence of the reinforcing phase. The most important reinforcing intermetallics in stainless steels are $\gamma'$-$Ni_3(Al, Ti)$, NiAl and CRPs (copper-rich precipitates) [17]. Further information on specific intermetallics has already been given in previous sections with reference to specific families of stainless steels. Detailed information on nomenclature, unit cell, chemical composition and other specific features of each of the selected types of intermetallics can be found in Appendix B.

#### 1.2.8.3 Nitrides, Carbonitrides, Borides and Inclusions

Nitrogen is the uncombined element present in the greatest quantity in the world accessible to mankind, and this is not difficult to be figured out since, for example, about the 78% of the volume of the Earth's atmosphere is made up of N. Moreover, N is able to combine with almost all elements in the periodic table and for many of them it is possible to observe several stoichiometries. Nitrides are nitrogen compounds where this element is found in the form of the $N^{3-}$ ion and are often classified into four types: salt-like (e.g., $Be_3N_2$, $Ca_3N_2$ and $Li_3N$), covalent (e.g., $P_3N_5$ and $S_2N_2$), diamond-like that are isoelectric with graphite (e.g., AlN, BN, GaN); and metallic. These latter compounds are also called interstitials and represent the largest group with general formulae MN, $M_2N$, $M_4N$ (M is the metal). Common features of metallic nitrides are high hardness, chemical inertia, refractoriness and several similarities with borides and carbides [35, 37]. For what specifically concerns stainless steels, nitrogen has a lower tendency to form compounds with Cr than C, for this reason it has a dual nature in that it is considered beneficial in austenite, but harmful in ferrite as in this latter phase it has practically zero solubility resulting in high tendency to form nitrides in the material. At high N levels, such as the case of duplex stainless steels and superaustenitics, formation of $Cr_2N$ may occur. This can happen in case the alloy is heated up to temperatures in the solution annealing temperature range, where the solubility of N is higher in the alloy. Another nitride that can form is CrN, which is a primary nitride structurally similar to MC carbide. This latter phase can form, for instance, in the HAZ of welded joints [17]. Further and specific information about four types of nitrides that can form in stainless steels can be found in Appendix B. As with intermetallics and carbides, information on chemical composition, unit cell etc., will be provided to the reader.

Carbonitrides, as the name of these compounds suggests, are phases consisting of C, N and a third metallic element. A concept therefore similar to what happens with the carbides presented in Sect. 1.2.8.1. With regard to carbonitrides, stainless steels are prone to the formation of such compounds in Nb- or Ti-stabilized grades. In addition, Ti carbonitrides can also form in those alloys that have sufficient Ti for reasons related to the precipitation of reinforcing phases (e.g., the austenitic PH A286 with the formation of the $\gamma'$ precipitate). However, titanium carbonitrides have effects that can be neglected in austenitic grades. On the other hand, with regard to Ti- or Ni-stabilized ferritic grades, the carbonitrides present in these alloys are initially formed as nitrides during solidification. The nitrides enrich themselves in C when the temperature falls below 1050 °C. Due to this aspect, their role in jeopardizing corrosion resistance is marginal if the annealing treatment of the final product is properly carried out. However, if the annealing temperature is too high, the decomposition of carbonitrides could bring enough C in solid solution to promote the formation of Cr carbides on cooling, thus causing subsequent corrosion resistance problems. In addition to this, Ti and Nb carbonitrides can act as preferential nucleation sites for corrosion phenomena and brittle fracture in the moment when the material is attacked by strong acids [38].

As mentioned in Appendix A, B additions in stainless steels leads to improvements in mechanical properties and, with specific reference to austenitics, in hot workability and creep resistance. Beneficial effects of B additions are therefore to be considered referred to the condition in which this element is in solid solution. Borides are compounds formed between B and an element characterized by a lower electronegativity. This family of substances comprises more than 200 binary compounds to which numerous nonstoichiometric phases of variable composition must be added. Metal-rich borides are characterized by high hardness, chemical inertia, refractory nature and high electrical conductivity, often higher than that of the parent metals [35]. Considering steels, in case the B content is too high, this could lead to the formation of a eutectic compound, i.e., $(Cr, Fe)_2B$. This boride has relatively low melting temperature (1150–1225 °C) which could give problems of *hot-shortness* during welding and hot working operations. Stainless steels characterized by B content greater than 30 ppm may exhibit the precipitation of two orthorhombic borides: $M_2B$ (M = Cr, Fe) and $M_3B$ (M = Cr, Fe, Mo). The latter type is found less frequently and only in alloys with high Mo content, while the former is found more frequently and tends to precipitate mainly at GBs. Furthermore, $M_2B$ boride is often found to form at the incoherent phase boundary of primary carbides. Fortunately, given the small amounts of B in stainless steels, the volume fraction of borides that can be formed is small and does not significantly affect the alloy properties [39]. However, for a seek of completeness in addressing the topic of second phases in stainless steels, some information about borides is offered in Appendix B, with the same structure proposed for the previously mentioned compounds.

Finally, to complete the general picture offered by this section, it is worth to mention also other compounds that can be found in steels, and therefore also in stainless steels: inclusions. Inclusions can be classified into three types: type I formed in the melt, type II formed during solidification process, and type III formed in the solid. Concerning these compounds, a detailed description will not be provided and the reader can refer to textbooks in the reference list. However, it is worth remembering that although in most cases inclusions are deleterious, there are cases where they are formed deliberately to achieve specific alloy properties. For example, sulfides and oxysulfides can be beneficial for machinability because they may act as chip-breakers and solid-state lubricants and are formed in the melt according to specific steelmaking practices. In all other cases, inclusions are physically removed in the stages before casting while the steel is still liquid [17].

## 1.3 Stainless Steels: Nomenclature and Applications

As seen in previous sections, there is a very wide variety of stainless steels due to the many possibilities in terms of chemical composition. Within the tables showing chemical analysis and mechanical properties, different designations have been given, including the most common trade names. As it is evident from the above tables, there are different designations depending on the entity (i.e., institute, society etc.) that developed them, also depending on geographical location (e.g., Europe, United States, North America etc.). Excluding trade names, within the world of stainless steels the most used designations are essentially three: one based on the EN European standards, one based on the AISI standard and one defined by the Unified Numbering System (UNS) [1]. The designation according Euronorm is carried out in accordance with EN 10027-1 and 10027-2. The EN 10027-1 is divided into two parts, one assigning the name according to use characteristics and properties, the second according to chemical analysis. With reference to the latter one, it provides an alphanumeric nomenclature that for stainless steels begins with the letter X followed by a digit corresponding to the percentage of C multiplied by 100. After this first digit the symbols of the main alloying elements are shown followed by their respective average percentage concentrations (e.g., X5CrNi 18 10 corresponds to the austenitic AISI 304). EN 10027-2 provides a numerical designation that is also taken by Annex C of EN 10088-1: 2014 and provides a first digit (1 for steels, 2 for Ni alloys etc.) followed by a dot and four digits divided into two pairs, the so-called *Werkstoffnummer* (WN, German word for substance number indicating the material number). The first pair refers to the material group (4x for stainless steels), while the second pair refers to the specific type of alloy. Therefore, the following groups can be recognized based on this designation:

- Corrosion resistant alloys:
  - 1.40xx for steels with %Ni $<$ 2.5%, without Mo and without special alloying elements.
  - 1.41xx for steels with %Ni $<$ 2.5%, with Mo and without special alloying elements.
  - 1.43xx for steels with %Ni $\geq$ 2.5%, without Mo and without special alloying elements.
  - 1.44xx for steels with %Ni $\geq$ 2.5%, with Mo and without special alloying elements.
  - 1.45xx and 1.46xx for special grades containing relevant amounts of elements such as Cu, Nb or Ti.
- Heat resistant alloys:

  - 1.47xx for steels with %Ni < 2.5%.
  - 1.48xx for steels with %Ni ≥ 2.5%.

- Creep resistant alloys
  - 1.49xx.

The AISI (American Iron and Steel Institute) designation is a type of designation used extensively in the commercial field and features the AISI noun followed by three numbers to identify the type of steel. Grades whose first digit is 2 are austenitic Cr–Mn–Ni stainless steels, such as AISI 201. In case the designation begins with 3 it refers to Cr–Ni austenitics (e.g., AISI 304) and 4 refers to martensitic or ferritic grades (e.g., AISI 420 or AISI 430, respectively). Some particular grades, such as PHSSs, are designated with 6 as first digit (e.g., AISI 660, AISI 630 and AISI 631). Particular features such as low carbon or relevant nitrogen content are highlighted with letters after numbers, i.e., L and N, respectively (e.g., AISI 304L or AISI 304N). Concerning the UNS designation system, this standard is particularly accepted in North America and can be found in many textbooks and scientific publications [1]. The code used is alphanumeric and includes a first letter (S for stainless steels) followed by five digits. The first three often refer to the corresponding AISI (older), while the last two digits indicate particular alloy variations. For example, AISI 304 corresponds to the UNS S30400, while AISI 304L is UNS S30403. Standard designations are an essential tool in the materials market and therefore also in that of stainless steels. This is precisely the reason why it was important to introduce the designation before to talk about market and applications.

As already and partially mentioned in the previous sections of this introductory chapter, there are a variety of applications for stainless steels. Given the breadth of possibilities offered by these materials, in order to expose and understand the characteristics of the uses that make stainless steels a necessary resource, it is important to be able to group and summarize the applications. As is often the case in this field, it is not possible to have a univocal classification of the applications as this is dependent on the level of detail used and, not least, on the moment in time in which this classification is made. The latter point is particularly true due to the continuous development and optimization of materials, as well as the continuous research for new applications. In this chapter, therefore, a classification into application areas as depicted by Fig. 1.16 is provided, thus defining 8 industrial fields in which stainless steels are used. In addition to what is declared in Fig. 1.16, some corollary information will be provided below in order to conclude the introductory part and complete the information necessary for an optimal understanding of the upcoming chapters.

Starting from the aerospace application field, it is worth saying that use of stainless steels in the aerospace industry has seen many changes over time, in particular a decrease for certain parts in favor of superalloys. In any case, the service conditions are very different and depend on the function of the component. Perhaps, the most important requirement in aerospace is the weight minimization and this finds expression in the use of light alloys (e.g., Al- and, with lesser extent, Mg-based alloys) and

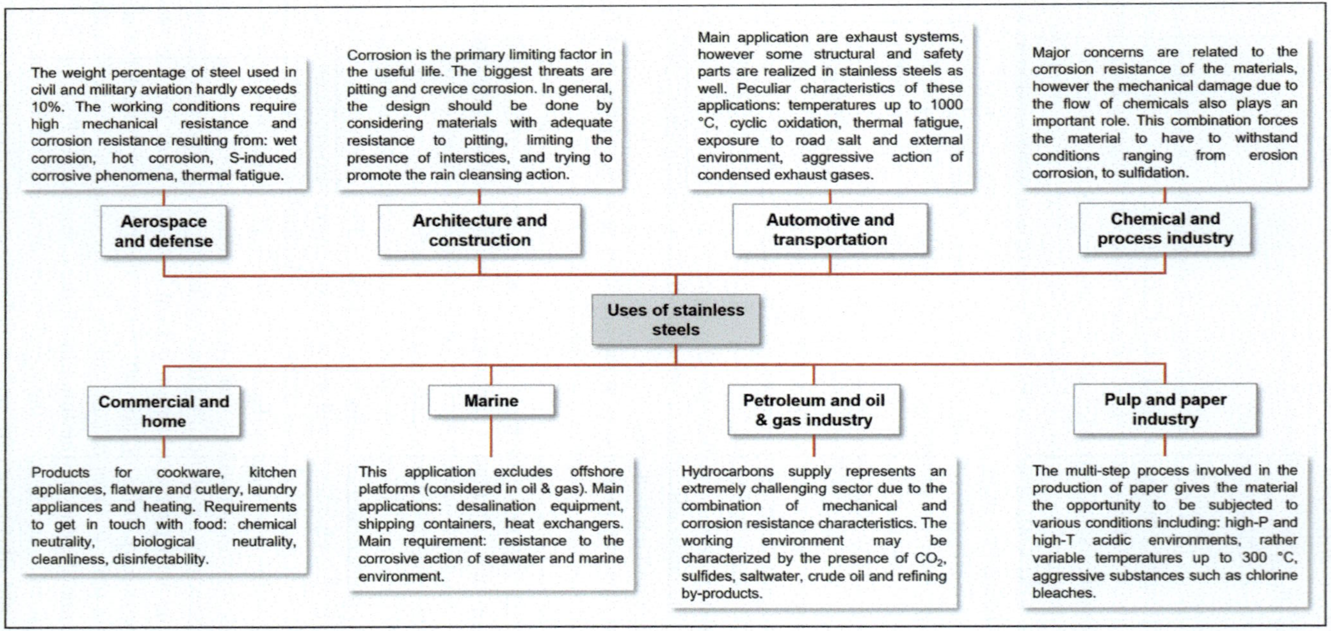


**Fig. 1.16** Overview of industrial fields in which stainless steels are used. For each sector, the main characteristics and requirements that the material must meet for use are listed. Further details are given in main text

composite materials for aircraft structures. Steels, as well as superalloys, have a relatively high density and are therefore subjected to more limited use. Steels are only employed when the designer cannot choose otherwise due to the requirements in terms of tensile strength, wear resistance, corrosion resistance, stiffness and relative cost. Furthermore, the use of steels is often reserved for extremely critical components, both for what concerns safety and performances. For these reasons, the weight percentage of steel used, for example, in civil and military aviation hardly exceeds 10%. In addition to mechanical resistance requirements, the aerospace sector is also characterized by particular corrosive conditions that cover a very wide range of environments and phenomena: from wet corrosion to hot corrosion and sulfur-induced corrosive processes. High-temperature corrosive processes are mostly located inside GTEs and also here stainless steels find applications for the manufacture of various components from afterburner parts to exhaust components and medium-temperature engine structural parts. With regard to austenitic stainless steels, in particular cases, application temperatures approaching 900–1100 °C can be reached. An example is AISI 321, which is able to guarantee a good combination of oxidation resistance at high temperature and adequate mechanical properties over a fairly wide temperature range (400–820 °C). Martensitic stainless steels are mainly used in the production of components that must have good tribological features combined with corrosion resistance. Probably the best known, and most effective use, is in the production of ball bearings made of AISI 440C. The family that is still most widely used in aerospace is certainly that of PH. This is particularly evident when thinking about the fact that A286 (AISI 660) is considered the forerunner of Ni-based superalloys and that for a long time it has been used in the manufacture of turbine wheels and shafts before being supplanted by more high-performance high-temperature superalloys. Despite this, PHSSs still have a very large market in the aerospace industry and are used to manufacture various components, including parts of missiles[1].

As declared in Fig. 1.16, the most limiting factor in the useful life of any architectural artifact exposed to the environment is corrosion. Stainless steel is a markedly suitable material for use in architecture and construction thanks to its aesthetic beauty and to its versatility. As already mentioned in previous sections, this latter feature derives from the countless chemistries that these alloys can show. Consequently, different grades will lead to products with different aesthetic appearance, resistance (mechanical and to corrosion), availability and, obviously, cost [17]. Just to mention two examples of architectural use of stainless steels, it is possible to refer to two buildings with iconic design: the Chrysler Building in New York and the Walt Disney Concert Hall in Los Angeles. The first was completed in 1930, at a time when the stainless steel production process was far from optimization and where the absence of the AOD process weighed heavily on the optimization of chemical compositions. Nonetheless, even today, nearly 100 years after its completion, it still stands gleaming on the New York skyline in an environment characterized by urban pollution and proximity to the sea cost. The Walt Disney Concert Hall is of more recent construction, it was completed in 2003 and is characterized by a particular shape, definitely different from that of the Chrysler Building. The curved outer surface of the building is made of austenitic stainless steel panels. Since the two main corrosive threats are pitting and crevice corrosion, materials selection must consider the use of alloys that avoid/limit the occurrence of pitting at the lowest possible cost, while the design must be carefully optimized to avoid presence of interstices that might promote crevice. While an extreme simplification of the selection rules involves the use of AISI 304 and AISI 430 for indoor applications, AISI 304 for outdoor applications in environments without chlorides, and AISI 316 in the case of proximity to the coast; an effective selection criterion is given by the PREN value (the calculation will be explained Sect. 3.3.11). As a consequence of this, the selection should be made accordingly: steels with higher PREN value for applications where the risk of pitting is higher. An additional factor to take into consideration while using stainless steels in exterior infrastructures is the surface finish. This aspect has a twofold implication: on one hand the aesthetics and the way the surface will affect the perception of the structure, on the other hand it is known how the surface finish affects the corrosion resistance (i.e., higher surface roughness implies lower corrosion resistance). Like any manufactured product, stainless steel structures for outdoor use require maintenance. However, there is one ally that is independent from human intervention: cleansing action of rain. Rain is not able to totally clean surfaces, however it is able to solubilize many compounds and consequently dilute any harmful substances and physically remove solid deposits. As a function of this cleansing action, certain design considerations are well worth adhering to:

- Areas that may be exposed to splashing/splattering and sheltered areas should be designed for easy cleaning.
- Avoid, as far as possible, any design that creates drainage patterns or uneven flows that lead to uncleansed areas.
- Avoid, as far as possible, any design that promotes the accumulation of dirt (e.g., horizontal surfaces, recesses) [17].

In addition to the abovementioned information about architectural applications of stainless steels, one possible use that might seem unusual, and perhaps unjustified at first glance, is that of concrete reinforcing bar. In fact, it is known that steel inside concrete is in a basic environment that is not aggressive to the alloy. However, structures in coastal areas such as bridges may be subjected to seawater penetration into the concrete over time. In case the internal rebar undergo corrosion, the increase in volume associated with the formation of corrosion products would lead to concrete spallation and possible structural failure. The use of stainless steel will not bring immunity to pitting corrosion, but it will limit generalized corrosion by remedying the situation described above. The most commonly used stainless steel types are AISI 316 and duplex grades such as 2304 (1.4362) and 2205 (1.4462). Duplex grades, in particular, are particularly well suited for this use because they provide excellent mechanical properties, corrosion resistance and SCC resistance at an acceptable cost.

Considering now the automotive and transport sector, among the most important requirements there is certainly an adequate durability which should be compatible or, better, exceed the expected life span of the vehicle (cars, buses, trucks etc.). Although early in the automotive history stainless steels were used more for ornamental purposes (e.g., wheel covers) than anything else, over time the use became more functional and less decorative. Nowadays, components made of stainless steels are highly-engineered and, important for many recent developments, steel producers have been working closely with automotive manufacturers for a long time. This has created a positive improvement loop in which automotive manufacturers demand better performing materials, raise and develop quality standards. As a result, steel manufacturers respond by improving production processes, chemical analysis control, inclusion content control as well as reduce the amount and size of defects tolerated. The components for the realization of which stainless steels are mainly used are exhaust systems, however these materials find also application for the production of structural components and safety parts such as airbag and seat belt components, engine parts such as valves and gaskets, various tubing, hose clamps and filters. As far as the exhaust system is concerned, it usually consists of three main components, connected by piping: a manifold, a catalytic converter and a muffler (see Fig. 1.17). For the realization of these components, it is necessary to use stainless steels with specific characteristics of corrosion resistance, oxidation resistance, creep, thermal fatigue and formability that are depicted directly in Fig. 1.17. Stainless steels are also used for structural components. For example, cold-worked AISI 301 is the standard choice for safety components such as seat belt anchors and hose clamps. Furthermore, when crash worthiness is the major issue, austenitic stainless steels such as AISI 201 and AISI 301 become an excellent choice for structural applications. Ti-alloys, Al-alloys and heat-treated alloy steels could also be chosen for these types of applications. However, their lower formability, durability and higher cost make the abovementioned austenitics a good choice. For what concerns austenitic stainless steels it is possible to say that they have excellent energy-absorption-to-weight and strength-to-weight characteristics (for this specific use), making them excellent structural materials for automotive applications. In addition to this, their superior corrosion resistance can greatly limit the use of coatings which, for car manufacturers, are a great source of cost and investments, both for raw materials and equipment. Additional applications for stainless steels include trailers for corrosive substances or food transportation, space frames for buses and tank wrappers [17].

Stainless steel has become a standard choice for products dealing with food, health care and laundry. This has made possible thanks to the combination of durability in terms of corrosion resistance, strength and ability to be sterilized, disinfected and cleaned. Certainly, with specific regard to food, the specifications required and the way the material interacts with food are of extreme interest. As a result of this, talking in a general way, a material that must be in contact with food must necessarily possess four characteristics, whose details are summarized in Fig. 1.18. The fact that stainless steel possesses the features depicted in Fig. 1.18 makes it an excellent material for a variety of applications in the commercial and home field. The most common of them are indicated in Fig. 1.19.

In the past, it was common to consider stainless steels as materials not suitable for use in marine applications and this idea remained unchanged until about the 1970s.

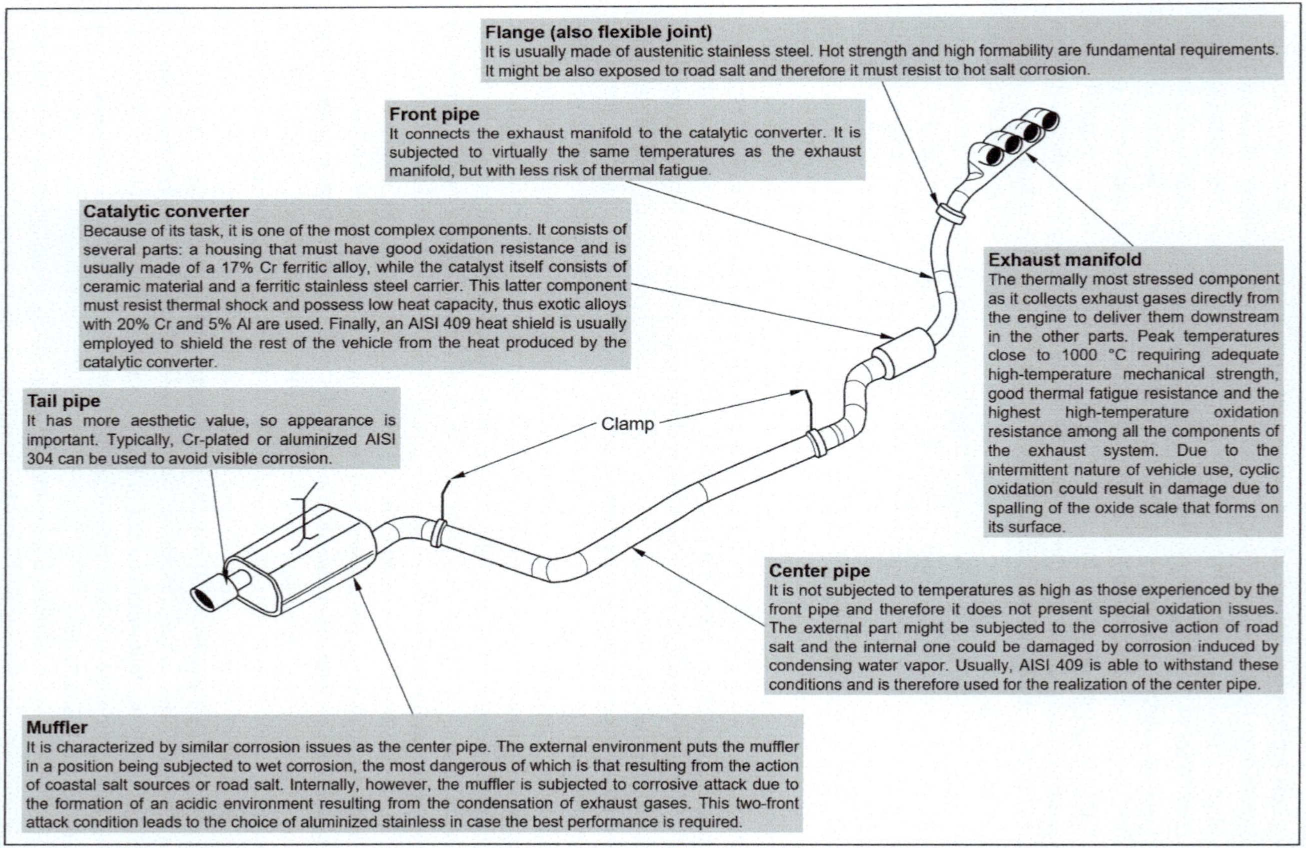


**Fig. 1.17** Typical possible structure of an exhaust system with indication of main constituent parts and details useful for material selection (redrawn from [40])

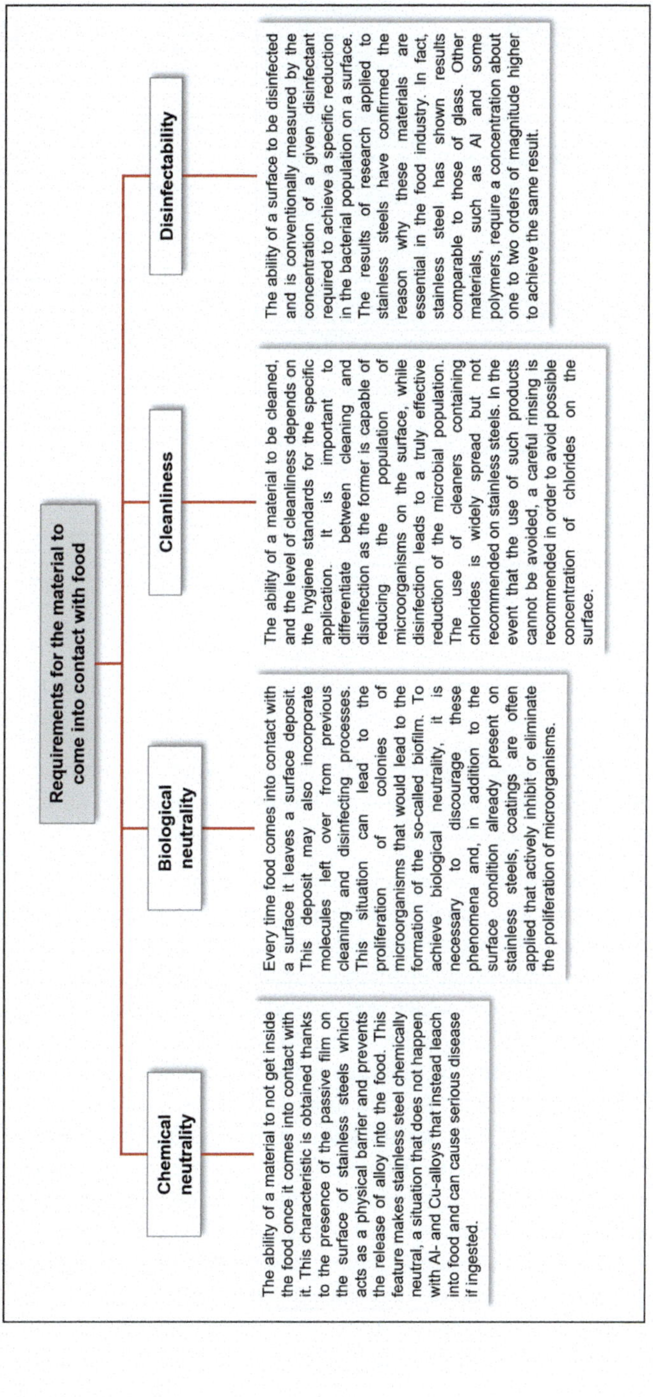


**Fig. 1.18** Requirements for materials to come into contact with food and related brief explanation [17]

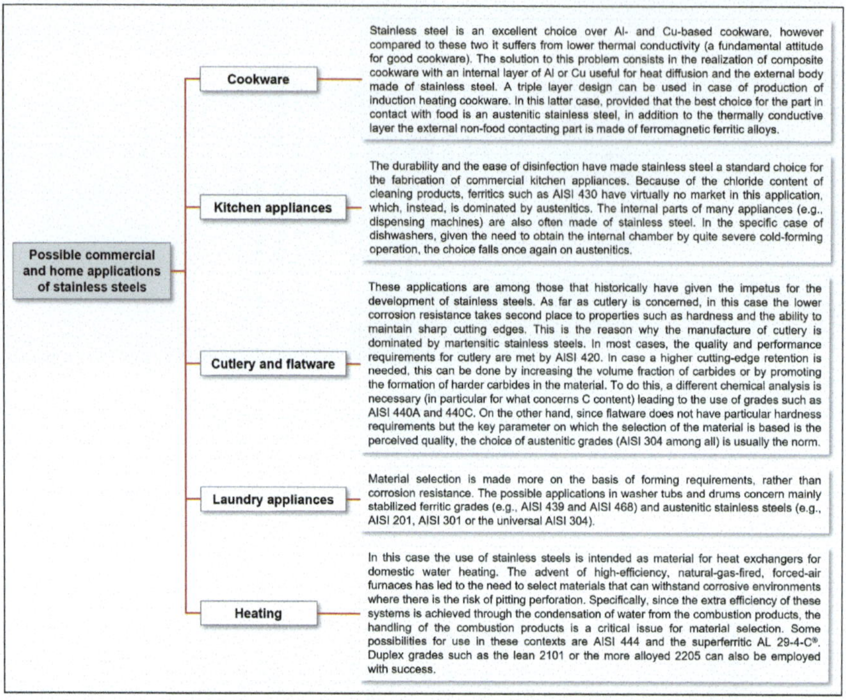


**Fig. 1.19** Common application of stainless steels in the commercial and home field [17]

Then, thanks to the remarkable progress in alloy design and in the development of new types of stainless steels, this kind of thinking has become obsolete. Developments have led to the use of stainless steels for various applications such as:

- Desalination equipment: the commercial availability of desalination process dates back to the 1950s and was initially based solely on a process of evaporation and recondensation of salt-free water using a process called multi-stage flash (MSF). Subsequently a low-temperature distillation process called MED (multiple-effect distillation) was introduced to which was also added the reverse osmosis process (RO), the latter based on the use of polymeric membranes to carry out the separation. For these applications the first really effective stainless steels were superferritics (use limited to tubing) and the more expensive superaustenitic grades, which instead can be used with any thickness. The introduction of duplex grades in the market allowed them to be materials for the beginning of the success story of stainless steels applied to desalination. In terms of material requirements for these applications, the 2205 duplex grade (PREN = 35) is sufficient for seawater handling up to 20 °C. If seawater at higher temperature, or with higher salinity, needs to be processed, the use of more alloyed steels such as 2207 (superduplex grade with PREN = 40–43) is recommended. On the other hand, less alloyed grades such as AISI 304, AISI 316, 2101 (duplex) or AISI 439 are usually a good choice, depending on the combination of temperature and salinity of the output water.

- Shipping containers: this application sees the possibility of transporting the most varied goods, from chemicals to food and beverage to liquid natural gas. Since 2000, marine chemical tankers have become the application with the greatest use of duplex stainless steels. This is primarily due to the fact that duplex grades can provide the widest range of corrosion resistance along with higher yield strength. This latter detail allows the design of the container on the basis of the yield strength, with an obvious advantage in terms of lower weight compared to austenitics (an issue of extreme importance for the reduction of costs of transporting goods by sea). For cryogenic applications, however, the choice still falls on austenitics, given the absence of a DBTT.
- Heat exchangers: for these applications the material must have resistance to corrosion by seawater or even by brackish water. For thin piping the choice falls on superferritics, while applications with thicker walls usually imply the choice between duplex and superduplex grades or superaustenitics [17].

Petroleum industry and stainless steel development have always influenced each other since the early days of both fields. If on one hand new steels could find successful applications in this industry, on the other hand the ever-increasing requirements drove the development of new alloys. Some of the typical characteristics of the petroleum and oil & gas applications have already been provided within the paragraphs on SDSSs and HDSSs in Sect. 1.2.7. However, it is worth noting that future hydrocarbons supplies are expected to be characterized by amounts of $CO_2$, sulfides (e.g., hydrogen sulfide) and saltwater that will make corrosion issues a major priority and the greatest driving force for materials selection. In addition to this, the search for new reservoirs goes to greater depths characterized also by higher temperature. Stainless steels become at this point practically necessary materials, especially because of the potential risks deriving from the presence of the so-called wet $CO_2$, which is extremely corrosive for carbon steels and alloy steels. Because of the abovementioned situation, in some specific environments/applications, stainless steels are required above a certain threshold value of $CO_2$, regardless of the hydrogen sulfide content. Depending on an increasing level of $CO_2$, the choice falls first on 13Cr-type martensitic alloys, then on duplex alloys for increasing levels of $CO_2$ and hydrogen sulfide up to about 250 °C (in this latter case, alloys with 25% Cr are preferred). Finally, for $H_2S$ partial pressure values greater than 1 atm it is necessary to shift from stainless steels to another family of metallic alloys: Ni-based alloys. Applications such as line pipe and flow lines are the typical field of use for duplex stainless steels. For line pipes specifically, the requirements for corrosion resistance and mechanical properties are in line with those of this family, which can also guarantee another fundamental attribute for this specific use: weldability. For subsea applications, DSSs with 22% Cr are often cathodically protected, condition that is not necessary for grades with 25% Cr. As already mentioned in Sect. 1.2.7, umbilical tubing is characterized by operating conditions that may present a need to resist high pressure-generated stresses, corrosion in various forms (including SCC) and increasing temperature. Therefore, the use of SDSSs and HDSSs is a standard choice. Platforms are also an extended application area for stainless steels. This application presents an interesting combination of high maintenance costs, very aggressive corrosive environment and strong limitations in the weight, that must be kept as low as possible. In this context, the components that can be made of stainless steel are many: flanges, fittings, piping, pumps etc. Also in this case, it is common to use DSSs with 22% Cr and cathodic protection, or DSSs with 25% Cr without protection. Finally, for refinery applications, corrosion resistance is definitely the main parameter to consider in materials selection. Crude oil, as it is well known, can be an extremely corrosive fluid, but great attention must also be paid to resistance to the chemicals used for refining and to the by-products resulting from the process. One of such substances is polythionic acid ($S_x(SO_3H)_2$ with $x > 2$), which can form accidentally when air moisture reacts with the products of sulfide-related corrosion. The corrosive attack intergranular and therefore the use of alloys susceptible to this corrosive phenomenon is not recommended. In addition to this, in some cases, temperatures can also be reached that are able to promote the formation of embrittling phases.

The chemical industry represents an interesting application field for stainless steels, it is sufficient to think about the fact that it is with the production of nitric acid that stainless steel had its first application in this industrial field. It is also interesting to think that the materials in this field must not only contain or allow the flow of aggressive chemicals, they must also be able to withstand the manufacture conditions of such substances. In this context, new processes are being developed all the time, involving ever-changing environmental conditions for which suitable materials must be defined. The main role in the selection of materials for the chemical industry is played by corrosion resistance, leaving mechanical strength in second place on the podium, although it has its own importance that can influence costs. Furthermore, there is also the influence of temperature which, in this field, has a very wide range of values. This implies the possibility to reach remarkable temperatures where the corrosion resistance is expressed by the oxidation resistance and the mechanical strength by the creep resistance. In addition to this, the material could be asked to work in the so-called single- or dual-environment systems. In the first case, the chemical substance contained or transported in the storage container/vessel is the only source of corrosion, while in the second case the corrosive action of the environment outside the vessel (e.g., a sulfuric acid container exposed to a marine environment) is added to the condition of single-environment system. The main forms of corrosion that materials must resist in applications in the chemical industry can be summarized as:

- Erosion corrosion.
- General corrosion.
- Intergranular corrosion.
- Localized corrosion such as pitting and crevice corrosion.
- Stress corrosion cracking.
- Oxidation.
- Sulfidation.

The first five in the list are associated with liquid-dependent corrosion phenomena, while the last two refer to gas-phase corrosion (which could also be originated by other gases besides those listed) [17].

As is well known, the raw material used for the manufacture of paper is wood. Through a multi-step process that involves the use of aggressive chemicals and rather variable temperatures (up to 300 °C), the raw material is transformed into the finished product: paper. The typical conditions of industrial papermaking needed the use of stainless steels with higher performances than austenitics, and so duplex stainless steels have become the standard choice for these applications. Without going into the specifics of the phases of paper production, it is interesting to analyze some of the steps to allow the reader to understand what conditions, even opposing ones, the material has to face. One of the first stages involves the dissolution of the lignin-connecting wood fibers at relatively high-temperature and high-pressure acidic environment characterized by a pH between 2.0 and 4.0. This process is named digestion and is carried out inside large vertical tanks named digestors. A good option in terms of materials selection in this case is the 2101 lean duplex (see Tables 1.11 and 1.12). In the following stages the substance contained in the digestor (i.e., the pulp suspension) is transferred, also at high velocity, and in this case erosion-corrosion becomes a major issue. In the following steps (washing and screening, delignification, bleaching), the requirements for the material are still related to the resistance to previously mentioned conditions to which the resistance against the substances used during the bleaching phase has to be added. Bleaching can be done by ozone/peroxide bleaches or even by chlorine bleaches. In this latter case, since the process takes place in closed environments, chloride levels can rise to such an extent that it is necessary to use materials with very high resistance such as superaustenitics or superduplex grades. In case bleaching is carried out in the absence of chlorides (TCF process), the corrosivity of the environment decreases and less alloyed grades, such as the lean duplex 2101 or 2304, can be employed. As far as the recycled paper industry is concerned, the criteria for materials selection remain the same and so, once again, duplex stainless steels are one of the most suitable choices [17].

## References


1. Gialanella, S., Malandruccolo, A.: Aerospace alloys. Springer (2020)
2. Ashby, M.: Materials Selection in Mechanical Design, 4th ed. Butterworth-Heinemann (2011)
3. Nicodemi, W.: Metallurgia. Zanichelli (2007)
4. Hosford, W.F.: Iron and Steel. Cambridge University Press, Cambridge (2012)
5. Ayto, J.: Word ORIGINS—The Secret Histories of English Words from A to Z. A & C Black, London (2005)
6. Hoad, T.F.: The Concise Oxford Dictionary of English Etymology. Oxford University Press (1996)
7. Cobb, H.M.: The History of Stainless Steels. ASM International, Materials Park, Ohio (2010)
8. Stodart, J., Faraday, M.: Experiments on the alloys of steel, made with a view to its improvement. Phil. Mag. **56**(257), 26–35 (1820)

9. Stodart, J., Faraday, M.: On the alloys of steel. Philos. Trans. R. Soc. Lond. **112**, 253–270 (1822)
10. Totten, G.E.: Steel Heat Treatment—Metallurgy and Technologies, 2nd ed. CRC Press (2007)
11. Olsson, C.-O.A., Landolt, D.: Passive films on stainless steels—chemistry, structure and growth. Electrochimica Acta **48**, 1093–1104 (2003)
12. Pedeferri, P.: Corrosion Science and Engineering. Springer (2018)
13. McCafferty, E.: Introduction to Corrosion Science. Springer Verlag (2010)
14. McCafferty, E.: Surface Chemistry of Aqueous Corrosion Processes. Springer (2015)
15. Calinski, C., Strehblow, H.: ISS depth profiles of the passive layer on Fe/Cr alloys. J. Electrochem. Soc. **136**(5), 1328 (1989)
16. Massalski, T.B., et al.: Binary Alloys Phase Diagrams, Vol. 2. ASM International, Materials Park, Ohio (1986)
17. McGuire, M.F.: Stainless Steels for Design Engineers. ASM International, Materials Park, Ohio (2008)
18. ASM International: ASM Handbook Volume 1—Properties and Selection: Irons, Steels and High Performance Alloys. ASM International, Materials Park, Ohio (1993)
19. Kain, V.: Stress corrosion cracking (SCC) in stainless steels. In: Raja, V., Shoji, T. (Eds.), Stress Corrosion Cracking: Theory and Practice, pp. 199–244. Woodhead Publishing Limited (2011)
20. Campbell, F.: Elements of Metallurgy and Engineering Alloys. ASM International, Materials Park, Ohio (2008)
21. Isalgué, A., et al.: Study of the spinodal decomposition of an Fe-28Cr-2Mo-4Ni-Nb alloy by small-angle neutron scattering. J. Mater. Sci. **25**(12), 4977–4980 (1990)
22. Schweitzer, P.A.: Metallic Materials—Physical, Mechanical and Corrosion Properties. Marcel Dekker Inc (2003)
23. Capelli, F.: L'acciaio INOX. Centro Inox Servizi Srl (2021)
24. Bhadeshia, H.K.D.H., Honeycombe, R.W.K.: Steels: Microstructure and Properties. Butterworth-Heinemann (2017)
25. Knyazeva, M., Pohl, M.: Duplex steels. Part I: Genesis, formation, structure. Metall. Microstruct. Anal. **2**(2), 113–121 (2013)
26. Knyazeva, M., Pohl, M.: Duplex steels. Part II: Carbides and nitrides. Metall. Microstruct. Anal. **2**(5), 343–351 (2013)
27. IMOA: Practical Guidelines for the Fabrication of Duplex Stainless Steels, 3rd ed. International Molybdenum Association (IMOA), London (2014)
28. Alvarez-Armas, I., Degallaix-Moreuil, S.: Duplex Stainless Steels. Wiley—iSTE (2009)
29. Braga Moura, L., et al.: Computational thermodynamic analysis of secondary phases in super ferritic stainless steels. J. Mater. Res. Technol. **2**(3), 282–287 (2013)
30. Janikowski, D.S.: Super-ferritic stainless steels rediscovered. Paper presented at the Stainless Steel World Conference and Expo, KCI Publishing, Maastricht, the Netherlands, 8–10 Nov 2005
31. Farrar, J.C.M.: The Alloy Tree: A Guide to Low-Alloy Steels, Stainless Steels and Nickel-Base Alloys. Woodhead Publishing Limited (2004)
32. Ma, X., et al.: Microstructure and properties of 13Cr5Ni1Mo0.025Nb0.09V0.06N super martensitic stainless steel. Mater. Sci. Eng. A. **539**, 271–279 (2012)
33. De Sanctis, M., et al.: Study of 13Cr-4Ni-(Mo) (F6NM) steel grade heat treatment for maximum hardness control in industrial heats. Metals **7**(9), 351 (2017)
34. Chail, G., Kangas, P.: Super and hyper duplex stainless steels: structures, properties and applications. Procedia Structural Integrity **2**, 1755–1762 (2016)
35. Greenwood, N.N., Earnshaw, A.: Chemistry of the Elements, 2nd ed. Butterworth-Heinemann (1997)
36. Askeland, D.R., Wright, W.J.: The Science and Engineering of Materials, 7th ed. Cengage Learning (2016)
37. Miessler, G.L., et al.: Inorganic Chemistry. Pearson (2013)
38. Kovach, C.W.: High-Performance Stainless Steels. The Nickel Institute (2000)

39. Padilha, A.F., Rios, P.R.: Decomposition of austenite in austenitic stainless steels. ISIJ Int. **42**(4), 325–337 (2002)
40. Renugadevi, R., et al.: Modeling and analysis of damping effect in exhaust system using Ansys. Int. J. Aerosp. Mech. Eng. **3**(5), 38–43 (2016)

## *Further Reading*

ASM International: ASM Specialty Handbook: Stainless Steels. ASM International, Materials Park, Ohio (1994)

Cahn, R., Haasen, P.: Physical Metallurgy, 4th ed. North Holland (1996)

Kobayashi, S., Takasugi, T.: Mapping of 475 °C embrittlement in ferritic Fe-Cr-Al alloys. Scripta Mater. **63**(11), 1104–1107 (2010)

Massalski, T.B., et al.: Binary Alloys Phase Diagrams, Vol. 1. ASM International, Materials Park, Ohio (1986)

McCafferty, E.: General relations regarding graph theory and the passivity of binary alloys. J. Electrochem. Soc. **150**(5), B238 (2000)

McCafferty, E.: Graph theory and the passivity of binary alloys. Corros. Sci. **42**(11), 1993–2000 (2000)

McCafferty, E.: Graph theory and the passivity of iron-chromium binary alloys. Electrochem. Solid-State Lett. **3**, 28 (2000)

Porter, D., Easterling, K.: Phase Transformations in Metals and Alloys. CRC Press (2009)

Smallman, R., Ngan, A.H.W.: Modern Physical Metallurgy, 8th ed. Butterworth-Heinemann (2014)

Tylecote, R.F.: A History of Metallurgy. Maney Publishing (2002)

# Chapter 2
# Superaustenitic Stainless Steels: Metallurgy and Chemistry

**Abstract** Superaustenitic stainless steels are one of the families of high-performance stainless steels, the so-called "super" grades. Although they share the face-centered cubic lattice with standard austenitics, these alloys have a more complex chemical composition and are able to provide a unique combination of outstanding corrosion resistance and high mechanical properties. The chemical composition is certainly the basis from which the analysis of the characteristics of superaustenitic stainless steels has to start. In fact, it allows them to be distinguished from each other and classified into two different subgroups: 6Mo superaustenitics and high-N superaustenitics. The chemical composition of each alloy defines it in all its characteristics; it also plays a fundamental role in defining its possible uses while also conditioning its market share. In addition to this, the chemical composition also determines the fundamental metallurgical aspects related to the alloy. Among these, key ones appear to be those related to the solidification process, the evolution of the microstructure and the possibility of secondary phases formation. Complex compositions rich in certain alloying elements such those of the super-austenitics necessarily have to deal with the phenomenon of segregation and related consequences. The study of the thermodynamics and kinetics associated with these materials becomes critical for understanding the types of secondary phases that can form. Finally, understanding the nature of secondary phases is critical to address the consequences of their presence and strategies for their elimination, as well as the possibilities for preventing their formation.

**Keywords** Chemical composition · Solidification · Segregation · Secondary phases · Intermetallics · Carbides · Nitrides



A. Malandruccolo, *Superaustenitic Stainless Steels*, Engineering Materials,
https://doi.org/10.1007/978-3-031-68744-0_2

## 2.1 Introduction and Origin of Superaustenitic Stainless Steels

As already mentioned in Sect. 1.2.7, some particular grades deriving from the five families into which stainless steels are classified have the prefix *super* that identifies them. This also applies to a particular group of austenitic stainless steels that are called superaustenitic (SASSs, i.e., superaustenitic stainless steels). SASSs belong to a group that can be classified as high-performance stainless steels. These alloys differ from the so-called "standard" types because of their resistance to corrosion and in particular to the phenomena of localized corrosion (i.e., crevice and pitting) and SCC, especially in environments containing aggressive species such as the chloride ion. These performances derive in the first instance from the chemical composition that characterizes these steels regarding the high levels of some key alloying elements: Cr, Mo, N and Ni. As explained in Chap. 1, the birth of stainless steels is a relatively recent event in the history of the modern era (these materials are just over 100 years old), and for the superaustenitics an even more recent period is involved. In fact, in order to be able to effectively use the abovementioned four chemical species as alloying elements, it was necessary to wait until the advent of specific processing technologies starting around the middle of the twentieth century. Among all of them, the most important ones are VOD (vacuum oxygen decarburization) and AOD (argon oxygen decarburization), which will be briefly presented in Chap. 4 and among which AOD is certainly the one that has had the greatest impact [1]. In the past, during the 70s and into the 80s of the twentieth century, much attention was given to this recent family of stainless alloys. The basis for the development of SASSs was the introduction, in 1951, of a new alloy by the Carpenter Steel Company (now Carpenter Technology Corporation): Stainless 20. Subsequent refinements in both the production process and chemical analysis, mainly aimed at producing a wrought material, led to the development of the 20Cb-3 alloy in 1965. This alloy is to all intents and purposes a steel since the major constituent is Fe, however due to its high Ni content, it has been classified with UNS nomenclature of a Ni-based alloy: N08020. In addition to this, it is referred to as superaustenitic because by plotting its chemical analysis in the Schäffler diagram (see Fig. 1.7), the position is much higher than the austenite-ferrite boundary region. Although it is an alloy that became popular in the chemical industry as a cost-effective alternative to Ni alloys to replace AISI 316, its pitting resistance is only marginally better than that of AISI 316L. As a result, performance in the marine environment were not considered sufficiently improved [2]. It is common practice to define superaustenitic stainless steels as those austenitic grades possessing a PREN value greater than 40. This particular characteristic, which will be discussed in detail in Chap. 3, is obtained thanks to the particular chemistry of these steels that typically implies high Mo content (usually 6%, or in excess of this value) [2, 3]. In order to provide the reader with a better guide in this introductory part, in addition to the historical development notes just reported, reference will be now made to Fig. 2.1 depicting the evolutionary path of development of SASSs starting from AISI 304. AISI 304 is naturally taken as a reference as it is always classified as the standard

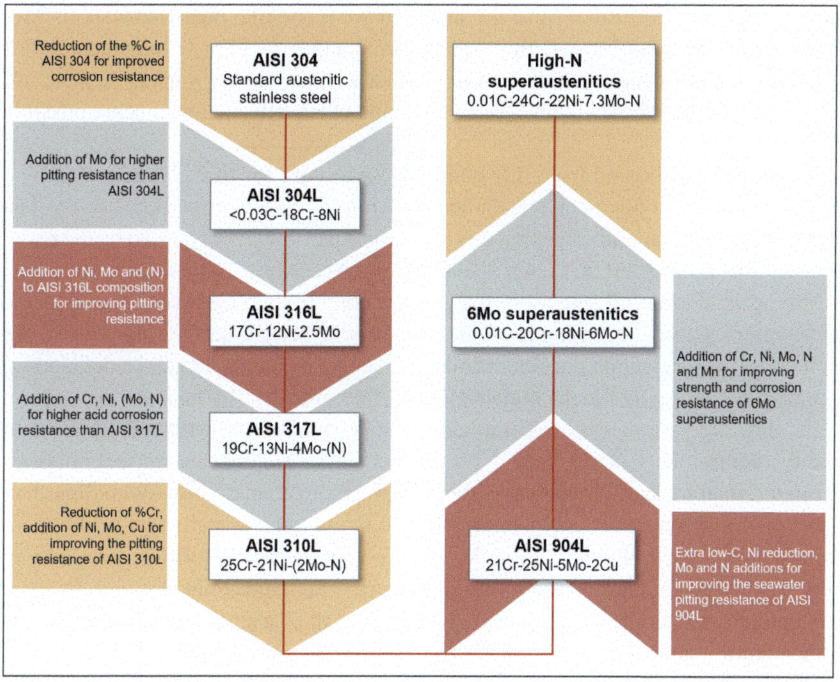


**Fig. 2.1** Evolutionary path of development of superaustenitics starting from AISI 304. Next to each type of steel, the fundamental steps to allow its evolution to the next type are given. Hence, the reduction of %C in AISI 304 to increase corrosion resistance leads to AISI 304L [4]

austenitic grade and, therefore, the ideal starting point for comparison of any further variant of this type of stainless steel.

As can be seen from Fig. 2.1, compositional changes to the standard AISI 304 austenitic grade lead, through C reduction, Mo and Ni addition, to AISI 316L. Further modifications, in particular in the content of elements such as Cr, Ni, Mo and N lead, thanks also to the addition of Cu, to AISI 904L. Addition of elements such as Ni and Nb would lead this "family tree" to evolve into corrosion-resistant Ni-based alloys. Remaining in the universe of austenitic stainless steels, the lowering of C to extra-low levels (i.e., ELC, extra-low carbon) with a slight reduction of Ni, addition of Mo and N leads to the development of the branch of SASSs defined as 6Mo and characterized by better resistance to seawater pitting corrosion. The last evolutionary step involves the increase in the percentage of Cr, Ni, Mo, N and the addition of Mn to obtain increased mechanical and corrosion resistance properties (against pitting and crevice, in particular). Hence, the class of high-N superaustenitic stainless steels is defined. This latter group, similarly to what happens with HDSSs with regard to the duplex family, probably represents the development limit for austenitic grades for what concerns the chemistry modifications to achieve better corrosion resistance. Actually, a further improvement in corrosion resistance would make it necessary to further raise the levels of some specific elements such as Mo. However, an increase beyond the typical percentages of high-N superaustenitics would no longer be manageable for the Fe matrix, thus making it necessary the use of Ni-based alloys where higher levels of Mo and other critical elements are possible [4, 5].

As it will be discussed from this point on, SASSs are for sure characterized by a very wide range of both mechanical and corrosion resistance properties. For this last feature it is worth remembering that this property is particularly exploited by the resistance against pitting and SCC (see also Chap. 3). According to these characteristics, for the optimal material selection, the service conditions (i.e., severity of chloride environment as well as strength requirements) will determine the final choice for what concerns fabrication considerations. As a consequence of the more demanding requirements and consequent higher chemical complexity that characterizes these steels, they are certainly more complex in terms of metallurgy, manufacturing and quality requirements (see Chap. 4 for further details). Anyway, the peculiarities just mentioned make their use possible in applications much more demanding than other types and where the corrosion resistance required is typically that of the more expensive Ni-based alloys. Although in some respects SASSs differ greatly from their "standard" counterparts (e.g., AISI 304, AISI 316), there are common features between the two. In the first instance, SASSs consist of a single *fcc* structure when in the solution annealed condition. They are not able to be hardened by heat treatment and are also non-magnetic (further details about austenite will be provided in Sect. 2.2). As for all austenitic stainless steels, yield strength is no particularly high, but higher than that of standard types thanks to the solution hardening effect of some alloying elements. Furthermore, the work hardening tendency is high as is the tensile strength attainable (see Chap.3). As it will be explained in Chap. 4, they are capable of good ductility, formability and toughness especially at low temperature (absence of ductile-to-brittle transition). A substantial difference with respect to standard types is that due to the chemical composition SASSs have a higher corrosion resistance, but they are also subject to rapid formation of secondary phases at high temperature. These phases are mainly deleterious in terms of both mechanical and corrosion resistance properties. This creates substantial limitations on their use, in particular they are rarely employed above 500 °C and specific care must be taken in operations involving a heating cycle (e.g., welding and plastic deformation operations such as rolling and forging) [1].

One of the most interesting aspects of SASSs is that they are manufactured starting from the same scrap and low-cost alloys, as well as the same equipment, as more common, less-performing grades such as AISI 316L (more details will be given in Chap. 4). Although more specific fabrication practices than standard grades must be employed for their production, these steels can be supplied at a lower cost than Ni-based alloys. Clearly, as it can be deduced from the examination of the specific chemical analysis of SASSs and the explanations that will be given in the following, there are different grades in this family and, equally obviously, there are different producers around the world. This last aspect is also expressed by the fact that each manufacturer has developed over time a customized production process based on different experiences, different production capabilities (i.e., equipment available and know-how) and often with different applications as target use of its material. However, despite what manufacturers would like, and apart from the use of very specific techniques (e.g., air melting versus vacuum melting or remelting practices), the real differences for the same grade produced by different manufacturers are not that incident for the various applications and the choice is made on the basis of commercial considerations, availability and supply, technical support and price [6].

In order to better understand all the key aspects of SASSs that will be presented in Chaps. 3 and 4, information of fundamental importance for understanding metallurgy and chemistry of these materials will be presented in the upcoming sections. A first part will focus on the austenite phase and its characteristics. Once the nature of the phase from which these materials take their name is understood, the chemical composition of the most common and relevant types will be exposed. After this, specific reference will be made to the solidification process and on the possible related issues. Finally, the secondary phases that can be found in these steels and that result from the relationship between the chemical composition and the process parameters (i.e., temperature, deformation etc.) will be discussed.

## 2.2 Austenite

For what concerns the Fe–C system, austenite is defined as an interstitial solid solution of C in Fe, whose structure is *fcc*. Metallographic analysis shows austenite as a homogeneous solid with straight grain boundaries and often presence of twins, as shown by Fig. 2.2. Austenite was given its name in honor of the English metallurgist Sir William Chandler Roberts-Austen.

As explained in Sect. 1.2, it is important to note that in the Fe–C binary system, austenite is a phase that is not stable at room temperature. In order to make it possible for austenite to exist at room temperature, it is necessary to modify the chemical composition of steel by adding appropriate alloying elements capable of extending towards lower temperatures the field of existence of austenite (see also Appendix A). In case of austenitic stainless steels, this task is accomplished by the so-called $\gamma$-stabilizing elements, the most important of which can be considered, in this specific context, Ni. Nickel, both as an alloying element for stainless steels and as a base for a very wide variety of alloys, is particularly interesting because it is characterized by *fcc* structure from solidification down to any temperature. Of course, it is always worth remembering that in addition to Ni, other metals also show *fcc* structure at room temperature: Ag, Al, Au, Cu, Pb and Pt [8]. As set forth within Sect. 1.1, steel has held, and still holds, a very high importance in human society and is often used as a benchmark material for the evaluation of all other materials in terms of characteristics such as mechanical and resistance properties. Probably, also as a consequence of this fact, in metallurgy it is common to hear the adjective *austenitic* referred to many structures characterized by a *fcc*-type lattice. For this reason, it is common to hear about the fact that Ni and its alloys have an austenitic structure at any temperature,

**Fig. 2.2** Optical micrograph of the austenitic microstructure of UNS S32615. This grade is less common than AISI 304 and belongs to the family of austenitic stainless steels. It is characterized by high Si content that allows it to resist against corrosion in strongly oxidizing acid environments, in particular in presence of sulfuric and nitric acids. Its average chemical composition features 0.04% C, 18% Cr, 20.5% Ni, 5.5% Si, 0.9% Mo, 1.5% Mn and 2% Cu. Chemical etching: hydrogen peroxide + HCl (ASTM E 407 No. 22) (chemical composition expressed in wt% after [7], picture courtesy of Omeco S.r.l.)

despite the fact that in the *fcc* lattice of Ni there is no C as the definition of austenite would require.

In comparison with the other stable interstitial solid solutions formed in the Fe–C system (i.e., ferrite, with *bcc* structure), austenite is characterized by a higher solid-state solubility for carbon. The reason behind this phenomenon lies in the fact that the octahedral sites, within which C can stay as an interstitial, in the *fcc* lattice are much larger than those of the *bcc*. When looking at the unit cell of austenite, the *fcc* lattice is characterized by atoms occupying the corners of the structure and the center of the faces. For the *fcc* unit cell, the close-packed planes are the diagonal ones inside the cube and are denoted as the {111} planes. On the {111} planes the atoms that make up the lattice touch along the <110> directions. As a result of this, it is possible to say that the *fcc* unit cell is characterized by having both close-packed directions <110> and close-packed planes {111}. As can be easily inferred from Fig. 2.3, the six atoms in the center of the faces contribute by 1/2 to the unit cell, while the eight in the corners contribute by 1/8. Consequently, the calculation of atoms per unit cell results in each *fcc* unit cell having four atoms, twice as many as the *bcc* unit cell [8].

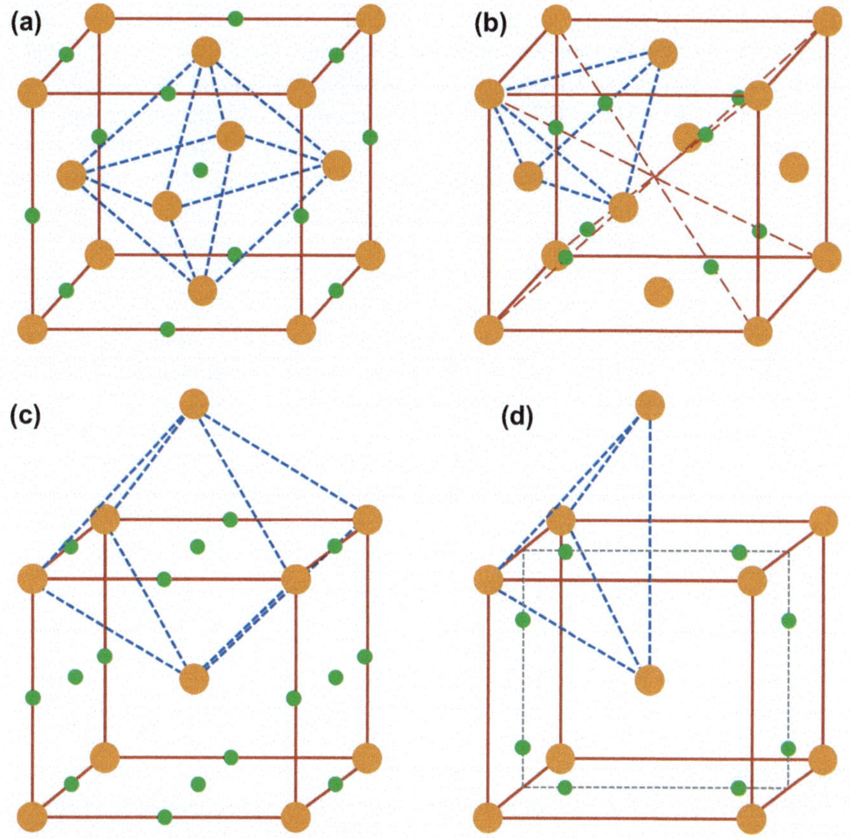


**Fig. 2.3** Interstitial sites in *fcc* and *bcc* lattices: **a** octahedral sites in *fcc*; **b** tetrahedral sites in *fcc*; **c** octahedral sites in *bcc*; **d** tetrahedral sites in *bcc*. Please note that the representations are not proportional to the data in Table 2.1 and that their function is purely explanatory. The reader, to get a real idea of the proportion between the dimensions of the interstitial sites, should refer to the data in Table 2.1 and to the main text (redrawn from [8])

Clearly, the type of crystal structure will have influence on various characteristics of the material that possesses it. In particular, as far as the present context and the upcoming chapters and sections are concerned, it is possible to give a first indication of three areas in which the crystal structure has strong influence:

- Alloying possibilities: the lattice type influences the possibility of forming alloys because the solid-state solubility between two elements is also a function of their crystalline structure. This results in increased solid solubility between elements that show the same lattice (e.g., the isomorphous binary Cu–Ni system, both elements with *fcc* lattice). In case of elements with different crystal lattices, the solubility in the solid state has more limitations.

- Diffusion: the possibility for atoms to migrate within a crystal lattice varies depending on how densely the lattice is filled. In this context, with reference to the APF (atomic packing factor) it is possible to compare the *bcc* (APF = 0.68) with the *fcc* (APF = 0.74) lattices. As the name of this parameter suggests, a higher APF value is an indication of a greater degree of spatial filling of the unit cell by the atoms. With specific reference to steels and C diffusion within the lattice, this phenomenon will occur faster in the *bcc* structure than in the *fcc* one, as the latter has a higher APF (for the calculation of APF, the reader may refer to Haidemenopoulos [8]). Another phenomenon of great relevance to the potential uses of materials at high temperature is strongly influenced by solid-state diffusion: creep. Creep involves diffusion of atoms and vacancies within the material as a result of the application of stresses at high-temperature. A material characterized by slower diffusion in the solid state will show a better creep resistance. For this reason, materials with *fcc* lattice have higher creep resistance than *bcc* ones (e.g., Ni-based superalloys used in the hottest parts of GTEs) [5, 8].
- Plastic deformation: plastic deformation in metallic materials is made possible by the glide of crystallographic defects (i.e., dislocations) on close-packed planes. As already mentioned, *fcc* lattice has close-packed planes, the {111} planes. For this reason, metals and alloys exhibiting *fcc* lattice are positively prone to plastic deformation processes.

With reference to what has been expressed, since the APF for both the two cubic lattices mentioned is below the unit value, even if they are close-packed structures, this implies the possibility of having enough space to accommodate chemical spices other than the one that constitutes the lattice itself. The phenomenon just described allows the formation of solid solutions that can be of two types: interstitial or substitutional. α ferrite and γ ferrite (austenite) are both interstitial solid solutions of C in α-Fe and γ-Fe, respectively. Obviously, not all atoms are able to occupy the free space within the lattice. This void space between the atoms is called interstice, and an atom must have small atomic radius (as a rule, less than 1 Å) to be able to occupy it. As a result of this latter criterion, it is possible to identify the most important chemical spices that have interstitial behavior: B ($r_a$ = 0.97 Å), C ($r_a$ = 0.77 Å), H ($r_a$ = 0.46 Å), N ($r_a$ = 0.71 Å) and O ($r_a$ = 0.60 Å). The ability of a lattice to host interstitial atoms, i.e., the solid solubility for interstitial solid solutions, is dependent on the size of the interstitial sites in the lattice itself. Again, with reference to the *fcc* and *bcc* lattices, both are characterized by the presence of two types of interstitial sites: octahedral and tetrahedral. Octahedral sites are surrounded by 6 atoms, forming, in fact, an octahedron. On the other hand, tetrahedral sites are smaller and surrounded by 4 atoms, thus leading to the formation of a tetrahedron (see Fig. 2.3). Regarding the *fcc* lattice, octahedral sites are larger in size than tetrahedral sites. In the *bcc* lattice, on the contrary, the octahedral sites are smaller than the tetrahedral ones, as stated in Table 2.1 [8].

The size of the interstitial sites takes on fundamental importance for the formation of interstitial solid solutions. This is particularly evident in the Fe–C system, which is why it has been chosen to explain it for providing the reader with important

**Table 2.1** Size, CN (coordination number) and differences between interstitial positions in *fcc* and *bcc* unit cell types

| Unit cell type | Interstitial position | CN | $r_{max}$ | Dimension difference with respect to *fcc* |
|---|---|---|---|---|
| *fcc* | Tetrahedral | 4 | 0.225R | Taken as reference |
| *fcc* | Octahedral | 6 | 0.414R | |
| *bcc* | Tetrahedral | 4 | 0.291R | + 29.33% with respect to the tetrahedral *fcc* sites |
| *bcc* | Octahedral | 6 | 0.154R | – 168.83% with respect to the octahedral *fcc* sites |

R is the atomic radius of the hosting atoms, $r_{max}$ is the maximum radius of the interstitial position expressed as a function of R. The dimension difference with respect to the *fcc* unit cell expresses the percentage increase or decrease of the space available in the interstitial position with reference to the *fcc* unit cell. This latter comparison is intended to highlight the differences between the sites of the two unit cells and to get rid of the better attitude of the *fcc* cell for hosting interstitial spices in octahedral sites [8]

information related to the characteristics of austenite. In fact, γ-Fe has octahedral sites that can host a 0.52 Å radius atom, while tetrahedral sites could fit only a 0.28 Å radius atom. As for C, it is obvious that the best possibility will be to be hosted in the octahedral sites, leading, however, to a minimal lattice strain due to the size difference. On the other hand, as far as α-Fe is concerned, octahedral sites could fit an atom with radius equal to 0.19 Å, while tetrahedral sites could host atoms with radius 0.36 Å. In this latter case, the only chance for C to form interstitial solid solution in α-Fe is to occupy the tetrahedral sites. However, compared to the octahedral sites of austenite, the difference between the space needed and the space available for the atom is greater. As a consequence of this, it becomes obvious at this point how it is more difficult for C to form interstitial solid solutions in the *bcc* lattice of α-Fe. Furthermore, when this happens (see α-ferrite in the Fe–C binary system), the formation of interstitial solid solutions is accompanied by large lattice strains. Regarding the phenomenon of solid solubility, it is worth clarifying that α-Fe despite an APF equal to 0.68 (γ-Fe has a value of 0.74) has a lower solid-state solubility for C in comparison to γ-Fe. The explanation lies in the fact that while there is more empty space available, it is distributed in interstitial sites that are smaller in size. By greatly simplifying, it is possible to state that in austenite the void space is lower due to a higher APF, but due to the larger size of the interstitial sites, the solubility of C in the lattice results greater. This phenomenon is reflected in the fact that for the simple Fe–C system, the maximum solubility of C in γ-Fe is 2.11 wt%, while in α-Fe the value is practically 100 times lower [8].

As far as substitutional solid solutions are concerned, solute atoms will not be located in interstices, but they substitute those of the solvent. As for the solid solutions of interstitial type, also for the substitutional ones there are different factors that influence the possibility to form such solutions. This means that there will be limits to the possibility of certain combinations of elements to form substitutional solid solutions. The conditions required for complete solid solubility when dealing with substitutional compounds are described by the so-called Hume-Rothery rules, which can be summarized as follows [5, 8]:

- Difference in atomic size (expressed, for instance as atomic radius or diameter) in the $\pm$ 15% range. A difference that lays outside this range will lead to limited solid solubility.
- Same crystal structure of the chemical spices involved.
- Same valence of the chemical spices involved.
- Small differences in electronegativity between the values associated with the involved chemical spices.

In this context, this topic will not be explored further, however it is worth saying that the difference in size also leads to lattice strains causing local nonuniformity in the crystal lattice. This latter phenomenon is responsible for a reinforcing mechanism called solid solution hardening (or solid solution strengthening) resulting from greater difficulty for dislocations to move within the lattice. This phenomenon is a direct consequence of the non-uniformity of the lattice. Consequently, this inevitably requires the application of higher stresses to allow the dislocations movement. Hence, an increase in the stress required for dislocation movement results in reinforcement for the alloy as plastic deformation will only occur for higher levels of applied external stress.

## 2.3 Types, Chemical Composition and General Features

A detailed description of the chemical composition of members of the SASSs family is useful for a better understanding of as many aspects of these materials as possible. In this context, Table 2.2 provides the reader with the chemical composition of the superaustenitic grades and will be used as reference from this point on. Chapter 1 and Appendix A are structured to provide the reader with all the basic knowledge necessary to understand the discussion and Fig. 2.1 will be recalled and used as starting point. Regarding SASSs, the part to refer to in Fig. 2.1 is definitely the one on the right which, from the bottom, starts with AISI 904L.

AISI 904L is often considered a SASS, although due to its chemical composition and high-temperature deformation behavior it lies somewhere between austenitic stainless steels and Ni alloys, as suggested by its UNS nomenclature beginning with the letter N. To be effectively precise, AISI 904L is not really a superaustenitic stainless steel if specific reference is made to the definition linked to the PREN value. However, in practice (even by the steel manufacturers) it is considered as a SASSs both for the fact of having a PREN close to 40, and because it was the starting point for the development of the 6Mo series (see Fig. 2.1). Hence, it can be defined as the precursor of modern SASSs. For these reasons, in the specific context of this textbook, while stating that according to the SASS definition AISI 904L should not be part of

**Table 2.2** Classification and chemical composition of superaustenitic stainless steels

| Group | Most common name | WN | UNS | C max | Cr | Ni | Mo | Mn | N | Others |
|---|---|---|---|---|---|---|---|---|---|---|
| High-performance austenitic stainless steels | AISI 904L | 1.4539 | N08904 | 0.02 | 19.0–23.0 | 23.0–28.0 | 4.0–5.0 | 2.0 | – | Cu = 1.0–2.0 |
| | AISI 904LN | – | – | 0.02 | 19.9–21.0 | 24.0–26.0 | 4.0–5.0 | – | 0.04–0.15 | Cu = 1.0–2.0 |
| 6Mo superaustenitics | 20Mo-6 | – | N08026 | 0.03 | 22.0–26.0 | 33.0–37.0 | 5.0–6.7 | – | 0.1–0.16 | Cu = 2.0–4.0 |
| | 1925 hMo | 1.4529 | N08925 | 0.02 | 19.0–21.0 | 24.0–26.0 | 6.0–7.5 | 1.0 | 0.1–0.2 | Cu = 0.8–1.5 |
| | NAS 254N | – | S32053 | 0.03 | 23.0 | 25.0 | 5.5 | – | 0.20 | – |
| | 25-6Mo 1925 hMo | 1.4529 | N08926 | 0.02 | 19.0–21.0 | 24.0–26.0 | 6.0–7.0 | 1.0 | 0.15–0.25 | Cu = 0.5–1.5 |
| | Uranus SB8 | 1.4537 | N08932 | 0.02 | 24.0–26.0 | 24.0–26.0 | 4.7–5.7 | – | 0.17–0.25 | Cu = 1.0–2.0 |
| | 254 SMO | 1.4547 | S31254 | 0.02 | 19.5–20.5 | 17.5–18.5 | 6.0–6.5 | 1.0 | 0.18–0.22 | Cu = 0.5–1.0 |
| | Nicrofer 3127 hMo | 1.4562 | N08031 | 0.02 | 26.0–28.0 | 30.0–32.0 | 6.0–7.0 | – | 0.15–0.25 | Cu = 1.0–1.4 |
| | AL-6XN® | – | N08367 | 0.03 | 20.0–22.0 | 23.5–25.5 | 6.0–7.0 | – | 0.18–0.25 | Cu = 0.75 |
| High-N superaustenitics | Uranus B66 | 1.4659 | S31266 | 0.03 | 23.0–25.0 | 21.0–24.0 | 5.0–7.0 | 3.0 | 0.35–0.60 | Cu = 0.5–3.0 W = 1.0–3.0 |

(continued)

**Table 2.2** (continued)

| Group | Most common name | WN | UNS | C max | Cr | Ni | Mo | Mn | N | Others |
|---|---|---|---|---|---|---|---|---|---|---|
| | 654 SMO | 1.4652 | S32654 | 0.02 | 24.0–26.0 | 21.0–23.0 | 7.0–8.0 | 2.0–4.0 | 0.45–0.55 | Cu = 0.3–0.6 |
| | Nirosta® 4565S | 1.4565 | S34565 | 0.03 | 23.0–26.0 | 16.0–19.0 | 3.5–5.0 | 3.5–6.5 | 0.4–0.6 | Nb < 0.15 |

Values are expressed in wt%. Please note that although N08925 and N08926 have the same WN, they are two different steels that are the latter an evolution of the former. Moreover, N08926 has two common trade names used with virtually the same frequency. Therefore, it is the only one with two trade names within the table and in the subsequent ones [1–4, 9, 10] this family, this steel will still be analyzed because the usual market and manufacturers/producers' classification makes it a high-performance material from a corrosion resistance point of view (in fact, it is classified as a high-performance austenitic stainless steel) [1, 4]. This alloy was developed in the second half of the twentieth century to meet the corrosion resistance requirements of industrial fields that were subjected to rapid development such as the chemical and pulp and paper ones. Thanks to a PREN value higher than 35, this steel offers excellent corrosion resistance performance against pitting and crevice in chloride environments. Furthermore, due to the presence of Cu in its chemical analysis, the resistance against general corrosion in many acid environments with reducing agents is also remarkable, provided, however, that these do not contain nitric acid. AISI 904L is also characterized by excellent resistance to $H_3PO_4$ and is also suitable for uses involving the handling of $H_2SO_4$. Furthermore, the Ni content allows excellent resistance to SCC (not at high temperature) and the low C percentage limits intergranular corrosion (additional information on corrosion resistance provided in Chap. 3). AISI 904L is always supplied in the solution annealed condition and shows a stable fully austenitic microstructure [2, 4]. Regarding processability, with particular regard to welding, although there are electrodes and consumables made of this alloy, it is common practice to use Ni-based alloys (e.g., Alloy 625) for welded parts that must come into contact with environments where severe chloride conditions are there [4]. As already mentioned, AISI 904L is mainly used in the chemical and pulp and paper industries for the realization of piping systems, process vessels and tanks, agitators, valves for use in fertilizer, phosphoric, sulfuric and acetic acid plants, rotors and cast pumps. Applications can be also found in the marine environment, but are more limited in offshore platforms and oil and gas equipment. Further possibilities regard the use in bleaching systems, heat exchangers, piping systems and pollution control equipment [2, 4]. One very interesting application, although under many aspects the material is more performing than needed, is in watchmaking. In fact, Rolex® uses a variant of AISI 904L with a chemical analysis specifically developed for the brand under the name Oystersteel. This alloy is used for the manufacture of cases and watch bracelets and certainly possesses corrosion resistance and mechanical properties that are more than suitable for this specific use. This is, however, a special use case where performance requirements are not the driving force in material selection. The use of a more expensive material, highly-alloyed and therefore *precious*, probably derives from the desire to give further value to the product; to be able to guarantee practically infinite durability. Not only that, another reason why could be to make it perceive that watches are built with the highest quality materials and technology.

### 2.3.1 6Mo Superaustenitics

From AISI 904L, the next development step, or natural evolution, was that to further increase the Mo content up to 6%, trying however to limit/delay the consequent possible easy formation of the deleterious $\sigma$ phase with N addition. This concept came to fruition with the birth of the family of 6Mo superaustenitics, of which 254 SMO and Al-6XN® were among the first representatives [2]. It is important to emphasize from now on that the group of 6Mo superaustenitics does not consist of a single chemical composition, but there are several proprietary compositions (e.g., 1.4529 and 1.4547) that fall within a more generic analysis that, by the way, includes Cr content in the 20–24% range and Ni content between 18 and 25%. As shown in Fig. 2.1, this class of alloys is an evolution (the first) of AISI 904L and thanks to the PREN value they are capable to achieve, they have been named *superaustenitics*. They are an important family of alloys developed for the first time in the second half of the twentieth century in the Scandinavian regions in order to obtain materials able to withstand increasingly aggressive corrosive conditions of chlorine and chlorine dioxide pulp bleaching and of seawater at near-ambient temperature [1, 4]. In Chap. 1 the positive effect of an increase in the percentage of Mo in stainless steels on corrosion resistance was discussed. In this concern, these steels are designed to have an even greater resistance to corrosion than AISI 904L, specifically to pitting and crevice corrosion. This is achieved mainly thanks to an increase in %Mo up to a typical average value of around 6%. However, as explained in Appendix 1 and in Chap. 1, Mo is a α-stabilizing element and therefore it is not positive for austenite stabilization. Furthermore, high Mo contents might promote segregation, thus reducing corrosion resistance. As always in these cases, in order to avoid austenite destabilization, the α-stabilizing action of Mo is counterbalanced by an increase in the γ-stabilizing elements such as Ni and N (this latter with fairly high values, around 0.2%). Moreover, Ni, besides stabilizing austenite, has a synergistic action for what concerns corrosion resistance and contributes to improve the SCC resistance in chloride environments, as well as good general resistance against acids [1]. In addition to this, it is also worth remembering that Mo can lead to the formation of deleterious phases in the steel such as σ, χ and Laves phases (see Appendix B, Chap. 1 and upcoming sections). In order to avoid, or at least limit, the formation of such compounds it will be necessary to act, among all, on the balance of the chemical analysis considering the elements that delay or suppress the formation of deleterious phases and to avoid service temperatures, heat treatments or cooling processes at temperatures where such phases could be formed. N specifically contributes to delaying formation of deleterious phases, increases resistance against localized corrosion; and provides a substantial boost to mechanical properties, raising them to higher values than standard austenitics (see also Chap. 3). Nevertheless, as it will be seen in Chap. 3, these steels possess good properties but remain less performing than duplex grades of equal corrosion resistance. This latter aspect results in the need to use larger sections and therefore penalize the product from point of view of weight [1, 4]. As far as processability is concerned, with specific reference to welding processes, the high Mo content imposes limitations on the use of consumables made of the same alloy. In fact, the segregation tendency of Mo (with all the harmful consequences from the mechanical and corrosion resistance performance point of view) requires the use of electrodes or consumables made of corrosion resistant Ni-based alloys. As usual with austenitic grades, members of the 6Mo family are supplied in the solution annealed condition characterized by a fully austenitic structure and maximum corrosion resistance [4].

As stated at the beginning of this section, the 6Mo superaustenitics were originally designed to resist corrosion in pulp-bleach plants and in seawater and this was made possible by relatively high levels of Cr, Ni, Mo and N [1]. These steels have a PREN value above 43 when the multiplication factor for N is 16 (see also Table 3. 5). As it will be seen Chap. 3, there are various equation for calculating PREN as a function of the chemical composition and type of stainless steel and there still remains some disagreement about the nitrogen coefficient for highly-alloyed stainless steels containing N [1, 4]. 6Mo superaustenitics are used in environments and applications where the risk of general, pitting, crevice and SCC is high, in particular in the chemical and pulp and paper industry for applications in chemical and power plants, pulp-bleach plants, heat exchangers and piping systems handling seawater. Some examples of components that are generally made from these alloys are: agitators, tanks and digester vessels, piping systems, rotors, cast pumps and valves. There are also applications in seawater environments, e.g., firewater systems and process pipework in the oil and gas sector for such applications where seawater and hydrogen sulfide contamination are present. Although for many oil and gas applications DSSs and SDSSs are preferred, due to the service experience in the Scandinavian paper industry, 6Mo superaustenitics are used in the North Sea oil and gas industry by the Norwegian sector of this industry (duplex grades are more popular in the UK sector). This case represents an example of the use of a group of alloys that, because of the reliability demonstrated in a specific sector and because of the confidence with them, is still preferred to materials that could potentially be more performing, but where confidence and experience are not yet so crystallized. Such situations are still very common in industry and represent a difficult hurdle for technicians and engineers involved in the materials selection process. This happens because in addition to technical considerations, trust and confidence with a group of materials becomes a non-negligible selection parameter [1, 4]. As stated previously, there are several alloys within the 6Mo superaustenitic group, partly due to the existence of proprietary compositions of the companies that developed them. 254 SMO, AL-6X®, AL-6XN® and 25-6Mo are the four most important and well-known grades in this group and more information will be provided on them.

#### 2.3.1.1 254 SMO

254 SMO is a SASS used in a variety of applications where corrosion resistance is a key requirement for components fabrication, for which the materials selection and design process identified this alloy as the optimal choice. From an alloy development point of view, 254 SMO evolves from standard austenitics with, however, the addition of significant amounts of N. N has a strong austenitizing action and is also cheaper than Ni as alloying element. Furthermore, N is an essential alloying element for high-Mo alloys because it delays the formation of deleterious Cr–Mo-rich phases

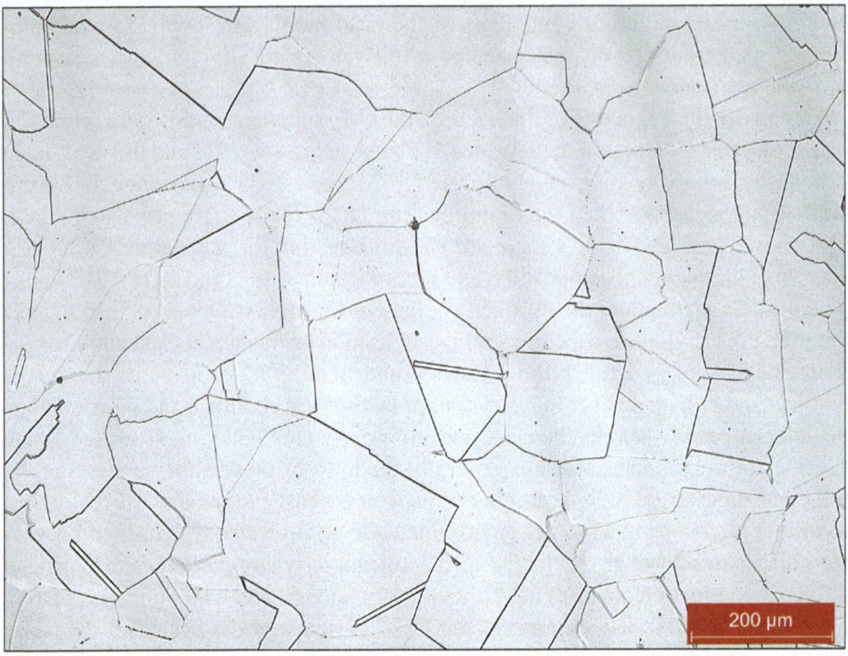


**Fig. 2.4** Optical micrograph of the austenitic structure of 254 SMO. Chemical etching: electrolytic with nitric acid 60% (ASTM E 407 No. 219) (courtesy of Omeco S.r.l.)

(e.g., σ) as it will be extensively discussed in upcoming sections. The microstructure upon metallographic analysis is provided in Fig. 2.4. [1].

The chemical composition particularly rich in alloying elements makes this alloy characterized by a poor machinability. In fact, compared with standard austenitics, 254 SMO is characterized by solid solution hardening and higher strain hardening attitude. This, in addition to being reflected in greater mechanical properties, results in machining difficulties due to work hardening of the contact zones between the tool and the material itself. Furthermore, unlike some other austenitics, there are no high-S variants (e.g., AISI 303) or versions with optimized distribution of micro inclusions of 254 SMO. As a result of this, there is no room for improving machinability through such practices (see also Sect. 4.5). On the other hand, as far as weldability is concerned, the characteristics and prescriptions are those common to all members of the superaustenitic family (i.e., use of specific processes, attempt to limit heat input to avoid precipitation of deleterious second phases, use of high-alloy filler material etc.). From a corrosion resistance point of view, this alloy was designed for maximum resistance to crevice corrosion and pitting. The PREN value well above 40 effectively defines it as a superaustenitic stainless steel and makes its use in seawater environment possible (see Sect. 3.3.12). Its mechanical and corrosion resistance characteristics make it a suitable choice where standard austenitics cannot withstand the combination of mechanical requirements and corrosive environment and where duplex, superduplex or hyperduplex stainless steels prove to be direct competitors. Some examples of possible applications are: pulp mill and bleach plants, high-chloride environments (e.g., brackish water), and other high-chloride process streams [4].

#### 2.3.1.2 Al-6X®

Unlike 254 SMO, AL-6X® has as starting point for its development the family of Ni alloys. Thus, this SASS uses Ni to stabilize austenite and does not involve voluntary alloying with nitrogen. The introduction on the market of a version containing N with better mechanical properties, but still retaining the original Ni content, marked the decay of this first material and took place in 1984 leading to an alloy that was named AL-6XN® [1].

#### 2.3.1.3 Al-6XN®-N08367

Al-6XN® was originally developed for uses where resistance to seawater is mandatory. Nonetheless, its chemical composition has allowed it to show resistance to a great variety of corrosive environments. Its UNS designation is N08367, despite the fact that the major constituent is Fe (see 20Cb-3 alloy in Sect. 2.1 for the explanation of the UNS nomenclature typical of Ni alloys, but assigned to a steel). Its most known name, Al-6XN®, is a registered trademark of Allegheny Ludlum Corporation [4]. Unlike AISI 904L, the Cu content is very low (practically a trace element), while the high content of both Ni and Mo contributes to the excellent corrosion resistance against crevice and pitting in oxidizing chloride solutions. Furthermore, Ni and Mo also provide the alloy with improved resistance to chloride SCC [4].

This alloy is a cost-effective alternative to more expensive Ni-based alloys in applications where the less expensive AISI 316L cannot guarantee adequate values of strength and corrosion resistance. In fact, Al-6XN® is characterized by excellent formability, weldability (the low C content makes it effectively an L grade), strength and corrosion resistance. Obviously, the most interesting resistance characteristics for this alloy are those related to corrosion. For these reasons the material is also approved by the ASME (American Society of Mechanical Engineers) for use up to 427 °C in unfired pressure vessels (ASME Boiler and Pressure Vessel Code, Sect. 11.8, Division 1). The alloy is also approved for uses in contact with food [4]. Al-6XN® is characterized by excellent resistance to crevice attack, pitting and SCC in high-chloride environments. Furthermore, it has very good corrosion resistance against reducing solutions, seawater and oxidizing chlorides. Al-6XN® can be also employed in handling of acetic, formic, nitric, phosphoric and sulfuric acids at various concentrations and temperatures (see Table 2.3) [2]. Last but not least, this alloy finds successful application in the realization of components used in the chemical industry (condensers, heat exchanges, pipelines, power plant flue-gas scrubbers,

**Table 2.3** Compatibility of Al-6XN® with various chemicals and related maximum allowable temperature for which data have been collected

| Chemical substance | Concentration (%) | Maximum temperature (°C) |
|---|---|---|
| Acetic acid | 20 | 99 |
| Acetic acid | 80 | 103 |
| Formic acid | 45 | 104 |
| Formic acid | 50 | 104 |
| Nitric acid | 10 | 90 |
| Nitric acid | 65 | 116 |
| Oxalic acid | 10 | 99 |
| Phosphoric acid | 20 | 99 |
| Phosphoric acid | 85 | 76 |
| Sulfamic acid | 10 | 99 |
| Sulfuric acid | 10 | * |
| Sulfuric acid | 60 | 50 |
| Sufluric acid | 95 | 30 |
| Sodium bisulfate | 10 | 99 |
| Sodium hydroxide | 50 | 99 |

Incompatibility between the chemical substance and the alloy is expressed by *. In case of compatibility, the corrosion rate associated with the interaction with the material is lower than 20 mpy [2]

process vessels), in the food industry (food processing equipment), as well as in nuclear plants (service-water piping) [4].

#### 2.3.1.4 25-6Mo

The alloy designated with WN 1.4529 has several trade names, according to the producer: the name according to the UNS standard is N08926, it is often referred to as Alloy 926, but it is also called Nicrofer 1925 hMo, Cronifer 1925 hMo or 25-6Mo (here two of the major manufacturers define the most common trade names). In the first cases, the name refers to Cr and then Ni content, in the second to Ni and Mo content. The version currently on the market is an evolution of UNS N08925 and differs from the latter in some values in chemical analysis (see Table 2.2). Alloy 25-6Mo is characterized by outstanding resistance in environments containing chlorides and other halides, as well as non-oxidizing acids such as $H_2SO_4$ and $H_3PO_4$. Furthermore, 25-6Mo is characterized by high strain hardening coefficient in comparison with standard austenitic grades such as AISI 316, and this is mainly due to its chemical composition rich in Mo and N. As a consequence of this, cold plastic deformation processes such as cold upsetting or cold heading might result in cracking (see further details in Sect. 4.3). In addition to this, it is important to underline that any cold deformation process with consequent strain hardening can affect corrosion resistance, especially in chloride- and sulfide-containing environments. Again, as a result of the significant strain hardening this alloy can show, its machinability is poor when compared with standard austenitic grades. Its remarkable mechanical properties also make it possible to have relatively high design values, thus allowing the use of thinner resistant sections compared to other materials, such as AISI 316L. Thanks to its chemical composition, in particular the 6% Mo content, 25-6Mo has high PREN value and is resistant also to MIC (microbiologically-induced corrosion), typical of brackish and wastewater environments. This is made possible by the fact that molybdenum gives the alloy protection from Mn-bearing, S-bearing and reducing types of bacteria in general. Again, considering PREN, this steel is able to have better performance than 2205 duplex grade (PREN = 36). In addition to the already listed corrosion resistance features, 25-6Mo is characterized by the fact of having a corrosion rate lower than 1 mpy (mils per year, 0.0254 mm per year) in saturated NaCl environments and pH in the range of 6–8. Furthermore, in more aggressive oxidizing conditions involving sodium chlorate, 25-6Mo still retains a corrosion rate lower than 1 mpy without being subjected to pitting corrosion, even at temperatures above boiling [4]. 25-6Mo can be employed in the fabrication of various components such as pump shafts, heat exchangers, storage tanks, flanges, bolting, equipment for pulp bleaching, naval applications, components in contact with sterilizing solutions and other products for which the corrosion resistance offered by standard grades (e.g., AISI 316) and special ones (e.g., AISI 904L) is not enough.

#### 2.3.1.5 Nicrofer 3127hMo

Nicrofer 3127 hMo is also known by the trade name VDM® Alloy 31 and is a superaustenitic stainless steel belonging to the group of 6Mo SASSs. The latter nomenclature specifically refers to the average chemical composition of this alloy with regard to Ni (31 wt%) and Cr (27 wt%), as it can be seen by looking at Table 2.2. Furthermore, Cu is added to the alloy in order to improve corrosion resistance in many environments. Another trade name by which this alloy is known is URANUS 31. As with many members of this stainless steel family, the driving force behind its development was the need to produce alloys that could bridge the gap between austenitic stainless steels and Ni-based alloys in terms of corrosion resistance. Nicrofer 3127 hMo also differs from standard austenitics such as AISI 316L because of higher mechanical properties, mainly obtained thanks to its higher alloying elements content. What makes it particularly distinctive, however, is its resistance to corrosion, in particular to localized forms (i.e., crevice and pitting). It has been observed that the material has outstanding corrosion resistance in acidic and alkaline halide-containing media, as well as in oxidizing media and in boiling azeotropic nitric acid. Furthermore, Nicrofer 3127 hMo displays excellent corrosion and erosion resistance in phosphoric acid media, as well as against corrosive attack in sulfuric acid-containing environments, even with high concentrations. As with all members of the family of SASSs, Nicrofer 3127 hMo is typically supplied in the solution annealed condition.

The alloy is also characterized by relatively good machinability when compared to other SASSs and by the possibility to be welded using conventional processes such as GTAW (gas tungsten arc welding), GMAV (gas metal arc welding) and plasma welding. Autogenous welding is possible, but to maximize the weld strength and corrosion resistance of the joint, the use filler materials that are austenitic in nature, but more highly alloyed, is recommended. Nicrofer 3127 hMo can be used in various fields to produce a reasonably wide variety of components, some examples are: components for FGD (flue-gas desulfurization) systems, components in plants for the production of $H_3PO_4$ via wet digestion process, pickling plants, systems for evaporation and crystallization of salts, refineries and mineral oil production plants, pipes handling seawater or brackish water, chillers and condensers [11, 12].

### *2.3.2 High-N Superaustenitics*

There is a further subgroup of SASSs, the high-N superaustenitics (see Fig. 2.1) which includes a grade that can be considered as the highest level of performance development ever achieved among all austenitic stainless steels: 654 SMO. Members of this group exhibit high mechanical strength combined with outstanding resistance to localized corrosion [1]. High-N superaustenitics are also known as high Mo superaustenitics or *superalloyed*, high-N austenitic stainless steels. They are one of the most recent developments as they saw the light of day towards the end of the twentieth century. The driving force for their development was essentially the desire to bridge the performance gap (in particular corrosion resistance) between the 6Mo superaustenitics and the more expensive, higher-performance Ni-based alloys. High-N superaustenitics are alloys that contain a greater quantity of alloying elements with respect to 6Mo SASSs, in particular the N content is noteworthy and reaches 0.5%. Within this group there are few chemistries, among which that of Alloy 654 (654 SMO) with an even higher Mo content than 6% and with the highest PREN value among all austenitic stainless steels. Another member of this group is UNS S31266, where Mo is partly replaced by W and where there is also addition of Cu to improve resistance against acids. A third member of this family is UNS S34565, which differs from 654 SMO by having a lower Ni content and a higher Mn content. Steels belonging to this group are designed to have better corrosion resistance (i.e., general, pitting, crevice and SCC) than 6Mo SASSs. They are also credited with excellent corrosion resistance in halide-containing solutions with performances that are comparable to those of some Ni-based alloys [4]. In fact, it is worth remembering that high-N SASSs are able to show good resistance to SCC in the boiling 45% $MgCl_2$ test. In addition, their good resistance against localized corrosion is retained also under very severe crevice conditions and at temperatures even significantly higher than the ambient one (see also Sect. 3.3) [1]. As already mentioned, they possess the absolutely highest PREN values among austenitic stainless steels (even greater than 50). Again, due to the already mentioned controversy over the exact method of calculating this index, if N multiplication factor were 30 instead of 16, the PREN value would rise over 60 [4]. As explained in Appendix A, N has the ability to increase the mechanical properties of steel. In the case of the high-N group, unlike 6Mo SASSs, the even higher content of this interstitial element leads to a noticeable increase in tensile properties. Specifically, yield strength values up to around 30% higher than 6Mo class can be achieved, which puts high-N SASSs on the same strength level as 22% Cr duplex grades. This makes them potentially attractive for use in weight critical applications where high corrosion resistance is required [4]. Thanks to these extraordinary characteristics for Fe-based alloys, high-N SASSs are able to approach the best of Ni-based alloys in terms of localized corrosion resistance contemporary combined with high mechanical properties [1]. As far as welding processes are concerned, these alloys present the same critical issues as 6Mo SASSs, therefore electrodes and welding consumables tend to be supplied in Ni-based alloys with high Mo content [4]. The specific applications of these materials are those where few other alloys can be used successfully and where the higher cost is still justified by the benefits associated with their use. Thanks to their superior corrosion resistance, high-N SASSs can be used in chemical industry applications where it is necessary to handle halide-containing solutions, both neutral or acid [4]. Furthermore, this group of SASSs has been specifically designed to have the potential to solve problems related to crevice corrosion happening in systems handling seawater at high temperature, in gasketed joints and in several systems operating in corrosive environments at high pressure [1]. Other possible applications are those related to use in environments with severe crevice corrosion conditions such as seawater-cooled plate heat exchangers, bleach plants, power station FGD systems and flanges in seawater piping systems [4].

#### 2.3.2.1 654 SMO

654 SMO (WN 1.4652) is a high-N SASS also known as Alloy 654 or UNS S632654. This steel has been developed by Avesta Sheffield and brought to market starting from 1992 and produced through conventional AOD process in 70-ton commercial heats. In addition to this, 654 SMO can also be continuously cast. As anticipated, its corrosion resistance properties are improved over 6Mo SASSs and superduplex grades [13, 14]. 654 SMO can be regarded as a very viable and cost-effective alternative to Alloy 625, which is often chosen as standard for special applications in contact with highly corrosive agents (e.g., piping, tubing, parts of pumps and plate heat exchangers in industries such as the paper industry, the chemical industry or the petrochemical industry). Before the end of 2021, it was possible to quantify a typical cost reduction, for the realization of the same component, between 30 and 40%. As a result of international events that began in the first quarter of 2022, and the resulting geopolitical and trade situation worldwide, the fluctuating price and availability of raw materials make it difficult to offer an updated estimation. Prior to the market *shake-up* that began in late 2021, taking AISI 316L as a benchmark and giving it a unit price value, 654 SMO averaged 6.5 times more expensive. Alloy 625, however, averaged 10.5 times the price of AISI 316L. As a consequence of this, it

is easy to understand the reason for its possible choice based on purely economic considerations.

The chemical composition of Alloy 654 features high contents of Cr, Mo and N, as depicted in Table 2.2. In particular, Mo and N values are among the highest, if not the highest, found in steels used in modern technology. In comparison, Mn is maintained at relatively lower values. As explained in Appendix A, one of the functions of Mn is to increase the solubility of N in liquid steel, but this aspect takes a back seat when considering that Mn could induce problems related to the possibility of formation of deleterious intermetallic phases, given the high Cr and Mo contents typical of this steel [14]. 654 SMO can be considered as the highest performing member of the superaustenitic family together with UNS S31266, both in terms of mechanical properties and corrosion resistance. In fact, as far as the first aspect is concerned, its mechanical strength is about double that of AISI 316L [2]. Compared with many Ni alloys of equal corrosion resistance, 654 SMO has often better mechanical properties, and this, combined with a lower density (more details in Chap. 3) allows for fabrications with smaller cross-sections, lower weight, and equal strength. Despite high mechanical properties, also conferred by solid solution hardening operated by N, 654 SMO maintains high elongation. This last feature allows it to be suitable also for plastic deformation processes involving stretch-forming operations. In addition, as it is typical for austenitics, the impact toughness at low temperatures is retained at high values [13, 14].

654 SMO is also readily weldable. In most cases, to achieve the best possible result, especially in terms of corrosion resistance of the welded joint, a Ni-based filler with high Mo content is preferred. Corrosion resistance is affected by the welding process in any case, however this steel, even after the permanent joining process, retains better corrosion resistance than unwelded superduplex and 6Mo superaustenitic grades. The high Mo content (higher than 7%) allows this material to have a corrosion resistance comparable to that of Ni-based alloys. Indications of use place Alloy 654 on the same level as Alloy C276 (UNS N10276, a Ni-based alloy). As a result of this latter feature, this grade is often called the "world's most stainless steel grade" [2]. In addition to the direct comparison with Ni alloys, which can be considered quite close relatives of SASSs, 654 SMO can also be considered an interesting alternative to Ti alloys. In fact, in some applications where the material must come in contact with fluoride ions, 654 SMO has better resistance to corrosion than Ti alloys. This is because, this type of compound is particularly damaging to Ti alloys.

As it will be explained in more detail in Sect. 3.3, this steel has the highest PREN value, which puts it in the position of having, at least theoretically, the highest resistance to localized forms of corrosion. This places it at a higher level than the SDSSs and the 6Mo superaustenitics. In addition to this, some experimental evidence also showed that 654 SMO was more resistant to crevice corrosion in seawater for conditions where Alloy C276 had failed [14]. The outstanding result in terms of corrosion resistance is based on tests conducted in bleach plants, filtered seawater and other aggressive chloride environments. Consequently, it is possible to say that 654 SMO has the highest resistance to localized corrosion among all the SASSs.

With regard to applications involving the handling of high-chloride environments, this alloy is intended as a competitor to Ti and Ti alloys [2]. For all the abovementioned reasons, 654 SMO is capable to replace the more expensive Ni-based alloys, as well as Ti alloys, in many applications where high corrosion resistance is mandatory, but allowing a cost saving due to the lower Ni content. Due to its undisputed qualities, 654 SMO finds use in applications such as: components and systems used for the construction of offshore platforms (structures to be employed in marine environment and seawater in general), components for the nuclear power generation (e.g., condenser tubing), components in waste incineration plants, construction of chemical and oil refining equipment, FGD systems and devices, plate heat exchangers exposed to seawater and underwater structures and equipment, products for the chemical processing industry and for the pipeline industry, pulp bleach industry (although in this case Alloy C276 might be preferred to it), solvent recovery, systems and components in pharmaceutical plants exposed to aggressive gases [14].

#### 2.3.2.2 UNS S31266

UNS S31266 (WN 1.4659) is a high-N superaustenitic stainless steel that is also known with the trade name Uranus B66 (or simply UR™ 66). Its chemical composition is quite similar to that of 654 SMO, as shown in Table 2.2. However, compared to the other member of the same family, one alloying element makes a non-negligible difference between the two: tungsten. As explained in Appendix A, W is a α-stabilizing, carbide-forming element that is used to increase the mechanical properties of steels. Although not that common as an alloying element in stainless steels, W is also found in WN 1.4501 SDSS (see Table 1.14). Obviously, the chemical analysis of the alloy ensures that the α-stabilizing action of W is counterbalanced by the austenitizing action of the ensemble of the other alloying elements. At this point, in addition to significantly increasing the mechanical properties of the alloy, W performs another important task. In fact, the synergistic action of W with N leads to a non-negligible decrease in the formation kinetics of σ phase, while it seems that other phases can be stabilized by its presence (e.g., Laves phases). It was found that, this feature emerges when comparing this alloy with its high-Mo and high-N counterparts, but without W. The chemical composition of this steel, in addition to the characteristics just mentioned, also gives it excellent corrosion resistance, markedly better than the 6Mo SASSs group when tested according to the common standards in this concern (e.g., ASTM G48D, ASTM G48E). For example, the resistance of Uranus B66 against corrosion in pure $H_3PO_4$ is maintained up to the boiling temperature regardless the concentration of the acid. The alloy also exhibits remarkable corrosion resistance in HCl and $H_2SO_4$, as well as in seawater. Regarding the calculation of the PREN value associated with this alloy, please refer to Sect. 3. 3. In fact, addition to the already mentioned calculation differences regarding the consideration of N, in this specific case the presence of a non-negligible amount of W has an impact on the result. In terms of mechanical properties, the values associated to this alloy in terms of strength are similar to the other two members of the family, i.e., 654 SMO and UNS S34565, with much higher values, especially the yield strength, than conventional austenitics (see Table 3.3). In terms of formability, the material can be processed both by means of hot forming, and cold forming processes. However, due to the high Mo content, work hardening is much higher than in standard austenitics that also contain Mo, such as AISI 316 and AISI 316L. With regard to other technological processes, such as welding, as with other members of the SASS family, the fully austenitic and highly-alloyed microstructure could be susceptible to hot cracking phenomena or deleterious phase transformations. For these reasons, processes using heat sources with high power density and low heat transfer away from the joint are definitely preferable. Techniques such as GTAW, GMAW can be successfully employed. In case of filler-free welding, a post-weld solution annealing treatment is recommended. On the other hand, in the case filler material is employed, Ni alloys should be used, as for the other SASSs. Uranus B66 is used effectively in a variety of industrial fields such as: oil and gas industry, pulp and paper industry for the realization of bleaching equipment, chemical industry, pollution control sector (marine scrubbers, FGD ducts and adsorbers), geothermal energy production, seawater applications and hydrometallurgy.

#### 2.3.2.3 UNS S34565

UNS S34565 is a high-N superaustenitic stainless steel and, similarly to 654 SMO, is characterized by the combination of high strength and outstanding corrosion resistance in a wide range of corrosive environments. Another name by which this alloy is known in the market is NIROSTA® 4565S. The path that led to its development started from the need to have a material suitable for applications such as high-chloride environments, hence seawater in particular. As indicated in Table 2.2, this steel has a high N content allowing it for outstanding crevice and pitting corrosion resistance. In addition to the benefits already mentioned, the high N content is also responsible for delaying the precipitation of deleterious second phases upon cooling from the solution annealing temperature, or after welding operations. In this latter context, the corrosion resistance of welded joints certainly takes advantage from such feature. In addition to this, good results are also shown from a mechanical point of view since a drop of the ductility of the material is manifested after times approaching 1 h at temperatures of about 800 °C (clearly incompatible with the duration of a welding process that, for obtaining good quality, must have short duration and minimization of the heat transferred away from the welded joint). Thanks to the combination of its resistance properties, S34565 finds application in equipment used where the higher cost is justified by the requirements for use. Some examples are the transport and storage of chemicals, the pulp and paper industry, the production and transport of hydrocarbons in on- and off-shore facilities, FGD systems, chemical industry in general, seawater desalination plants and other technologies used for environmental purposes [10].

## 2.4 Metallurgical Considerations and Secondary Phases in Superaustenitic Stainless Steels

In order to better understand the upcoming chapters on the properties (Chap. 3) and the technological processes to which superaustenitic stainless steels may be subjected (Chap. 4), it is important to dwell on some aspects related to the metallurgy of these alloys. Furthermore, the considerations outlined hereafter will be helpful in better understanding the peculiarities of the alloys described within Sect. 2.3. Taking into consideration the general information given within Sect. 1.2 and Appendices A and B, the first step in the analysis of relevant metallurgical considerations is surely a closer examination of the most important aspects related to the solidification process of this family of alloys. In this context, in order to understand the discussion related to microstructure formation and segregation phenomena, a brief summary of the main concepts related to nucleation, growth and segregation will also be provided. In fact, many studies have shown that SASSs are prone to solute segregative phenomena during solidification after processes such as casting and welding. As a result of this, the formation of intermetallics at the end of the solidification process becomes a rather likely phenomenon, and this would lead to the jeopardization of several properties, including the hot formability of the alloy [15]. Immediately after, the most important aspects of the multicomponent alloy systems to which SASSs can be related to will be considered. These two initial steps take on particular and fundamental importance when considering the fact that they constitute the necessary basis for better understanding and analyzing the next sections devoted to processes like casting, heat treatment and welding. In addition to this, given the complexity of the chemical composition of these steels (see Table 2.2), a fundamental step will be to provide a description of the possible solidification path of SASSs. In order to accomplish this task, three relevant examples derived from research literature will be analyzed and taken as reference for the explanation. Finally, once the phenomena resulting from solidification and temperature changes have been described, a description of the secondary phases that can form within these alloys will be provided. This will be aided by the information given in Appendix B, which will be contextualized with reference to the SASSs and whose most relevant aspects will be explored in more detail for subsequent sections pertaining to the production and processing of these alloys.

### *2.4.1 Solidification, Related Phenomena and Relevant Multicomponent Alloy Systems*

Solidification of metals and their alloys is a process that can be defined as a change of state phenomenon (i.e., from liquid to solid) associated with the emission of a certain amount of energy in form of heat and with the formation of a primary crystalline structure on the way of physically- and chemically-driven crystallization process.

The term solidification is also often substituted by *freezing*. This latter nomenclature is certainly useful in describing the change at the atomic level from a state characterized by greater atomic motion and less order, to one where atomic motion is reduced and the degree of order is greater (reference to crystalline solids; for amorphous solids the specific terminology changes, as does the description of the phenomenon of transition from liquid to solid state). Anyway, in order to avoid confusion to the reader, it should be remembered that liquids are intermediate between solids and gases in that they do not exhibit complete disorder like the latter, but neither do they exhibit high degree of order like solids. This is also confirmed by some studies employing X-rays carried out on liquids and indicating the existence of some short-range order. The liquid-to-solid transition is associated with a decrease in enthalpy and the difference between the energy in the solid and liquid state is the latent heat of fusion (often indicated as $\Delta H_f$), which is released by the material to the surroundings [16]. Solidification, as well as melting, is a process associated with a certain change in free energy ($\Delta G$) that is also dependent on the change in entropy and on the temperature. Energy considerations are fundamental in the study of solidification phenomena in that, among other aspects, the equilibrium temperature of solidification/melting is obtained by setting the change in free energy to zero and making the ratio of the enthalpy change to the entropy change. Once the temperature at which the solidification process is energetically allowed to begin is defined, the first step in the process is the formation of the very first stable solid in according to a phenomenon named nucleation. From the perspective of the theoretical analysis, nucleation is described as homogeneous or heterogeneous. The first occurs under ideal conditions with formation of the first stable solid nucleus (also known as *embryo*) randomly throughout the liquid and without contact with other interfaces or surfaces. On the other hand, the heterogeneous one, is associated with real processes and involves the formation of solid nuclei using preexisting solid surfaces as preferential nucleation sites. For the full understanding of heterogeneous solidification, homogeneous nucleation must be first described. To do this, reference could be made to a model in which the solid nucleus is represented by a tiny spherical particle, i.e., the embryo, of radius $r$ placed in the midst of the liquid at a certain temperature below the equilibrium temperature for solidification/melting. The necessary condition for nucleation is dictated by a negative variation in the free energy, the equation for which is given below:

$$\Delta G = A\gamma_{LS} - V\Delta G_V = 4\pi r^2\gamma_{LS} - \frac{4}{3}\pi r^3\Delta G_V \tag{2.1}$$

where

- $\Delta G$ is the free energy variation associated with the formation of the embryo.
- $r$ is the radius of the embryo.
- $A$ is the surface area of the sphere associated with the embryo and equal to $4\pi r^2$.
- $\gamma_{LS}$ is the surface tension at the solid–liquid interface.
- $V$ is the volume of the sphere associated with the embryo and equal to $\frac{4}{3}\pi r^3$.
- $\Delta G_V$ is the change of free energy per volume.

**Fig. 2.5** Free energy change during nucleation. The change of free energy, ΔG, increases with embryo size up to a critical radius, r*. The critical free energy for nucleation is ΔG *(redrawn from [16])

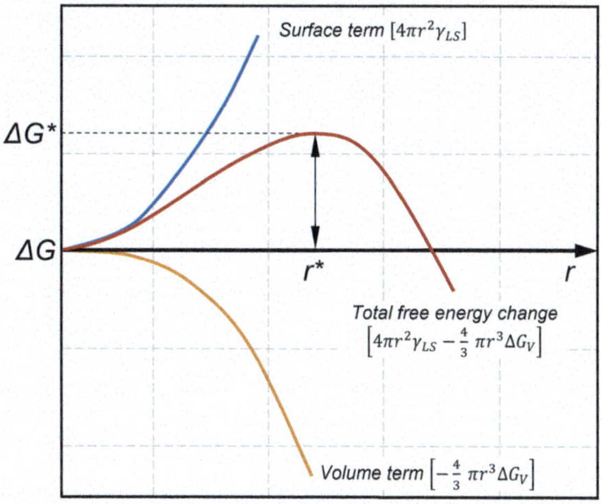


The trend of the function depicted by Eq. 2.1 is given in Fig. 2.5, where it is evident that the total change in free energy has a decreasing trend beyond a certain value associated with the radius of the solid nucleus. Given the necessary condition of negative free energy change just mentioned above, stable nucleation can occur only when the radius of the solid embryo is greater than a value called the critical radius and denoted by $r^*$ [16]. This is because, at very small radii, the growth of solids increases the energy content of the system in agreement with Eq. 2.1 and Fig. 2.5. However, in the moment when the radius of the formed nuclei exceeds the value beyond which the energy of the system decreases with further growth, i.e., $r^*$; all embryos with $r > r^*$ would not redissolve in the liquid, remain stably solid and grow spontaneously [16].

The critical radius can be intended as a barrier beyond which homogeneous nucleation can occur and for its determination, the derivative of Eq. 2.1 in $dr$ has to be set equal to zero, thus obtaining Eq. 2.2.

$$\frac{\Delta G}{dr} = 8\pi r\gamma_{LS} - 4\pi r^2\Delta G_V = 0 \rightarrow r^* = \frac{2\gamma_{LS}}{\Delta G_V} = \frac{2\gamma_{LS}}{\left(\frac{\Delta T}{T_m}\right)\Delta H_V} \tag{2.2}$$

where

- $\Delta G$ is the free energy variation associated with the formation of the embryo for homogeneous nucleation.
- $r$ is the radius of the embryo.
- $\gamma_{LS}$ is the surface tension at the solid–liquid interface.
- $\Delta G_V$ is the change of free energy per volume. For solidification taking place at the equilibrium temperature, $\Delta G_V = \left(\frac{\Delta T}{T_m}\right)\Delta H_V$.
- $r*$ is the critical radius.
- $T_m$ is the equilibrium melting temperature.

- $\Delta T$ is the undercooling, i.e., the temperature difference from the equilibrium melting temperature at which solidification is intended to occur or is being studied.
- $\Delta H_V$ is the enthalpy variation.

The critical radius is associated with a critical free energy value that is usually referred to as the critical activation energy (see also Fig. 2.5), which can be determined as reported by Eq. 2.3.

$$\Delta G^* = \left(\frac{16}{3}\right)\frac{\pi \gamma_{LS}^3}{\Delta G_V^2} = \left(\frac{16}{3}\right)\frac{\pi \gamma_{LS}^3}{\left[\left(\frac{\Delta T}{T_m}\right)\Delta H_V\right]^2} \quad (2.3)$$

where

- $\Delta G^*$ is critical activation energy for homogeneous nucleation.
- $\gamma_{LS}$ is the surface tension at the solid–liquid interface.
- $\Delta G_V$ is the change of free energy per volume. For solidification taking place at the equilibrium temperature, $\Delta G_V = \left(\frac{\Delta T}{T_m}\right)\Delta H_V$.
- $T_m$ is the equilibrium melting temperature.
- $\Delta T$ is the undercooling, i.e., the temperature difference from the equilibrium melting temperature at which solidification is intended to occur or is being studied.
- $\Delta H_V$ is the enthalpy variation.

As already mentioned, homogeneous nucleation, to which Eq. 2.3 specifically refers, is not the correct phenomenon to describe real solidification processes because in such cases the conditions are far from equilibrium. For real processes, in fact, nucleation does not occur randomly throughout the liquid, but takes place on pre-existing surfaces that act as preferential solidification sites. With reference to Eq. 2.1, for heterogeneous nucleation the situation changes. In fact, although the formation of a new solid requires new surface between it and the pre-existing solid (typically the walls of the mold) and between it and the liquid; the surface between the pre-existing solid and the liquid must be destroyed. This leads to a change in the balance between the surface and volume terms resulting in a decrease in the activation energy for the process to happen. The new value of the critical activation energy for heterogeneous nucleation is therefore given by Eq. 2.4 [16].

$$\Delta G^*_{hetero} = \Delta G^* \frac{(2+\cos\theta)(1-\cos\theta)^2}{4} \quad (2.4)$$

where

- $\Delta G^*$ is critical activation energy for homogeneous nucleation (see Eq. 2.3).
- $\theta$ is the wetting angle (also known as contact angle).

In the specific analysis of liquid solidification, the contact angle can be defined as that measured between the tangent at the liquid–solid interface and the preexisting surface as depicted by Fig. 2.6. The contact angle depends on the surface energy

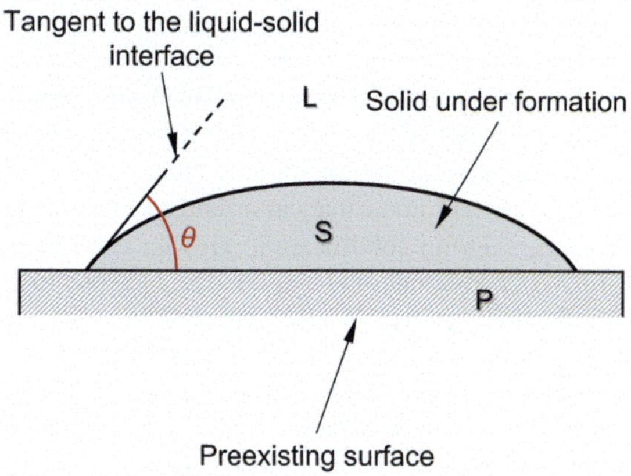


**Fig. 2.6** Determination of the contact angle θ with specific reference to the solidification process of a metal or alloy

values associated with the different interfaces that develop during solidification. As it can be inferred from Fig. 2.6, the interfaces are: liquid–solid, preexisting surface-solid and liquid-preexisting surface. As a result of the interfaces determination, the contact angle can be easily determined from Eq. 2.5.

$$\cos\theta = \frac{\gamma_{LP} - \gamma_{PS}}{\gamma_{LS}} \tag{2.5}$$

where

- $\theta$ is the contact angle.
- $\gamma_{LP}$ is the surface energy of the liquid-preexisting surface interface.
- $\gamma_{PS}$ is the surface energy of the preexisting surface-solid interface.
- $\gamma_{LS}$ is the surface energy of the liquid–solid interface.

By analyzing Eqs. 2.4 and 2.5, it is possible to see that nucleation on the surface turns out to be an energetically favored process when $\gamma_{LP}$ is high, while $\gamma_{PS}$ is low [16]. In case the systems analyzed were metal alloys, the determination of $\Delta G^*$ should take this issue into account and, in the first instance, a redetermination of $\Delta G_V$ should be performed. In fact, the latter value would no longer depend solely on temperature, but also on the composition of the liquid and solid nuclei. As a result of this, for a given liquid composition of the alloy under investigation, the determination of $\Delta G^*$ would require determining the critical values of nucleus composition, as well as size [17, 18].

Further detailed explanation of this topic, however, is beyond the scope of this introductory section (however, issues related to compositional inhomogeneity characteristics between liquid and solid under solidification will be exposed hereafter). Therefore, the reader is invited to refer to the bibliographic references at the end of the chapter for further information. The phenomena described this far served the understanding of what will be set forth in upcoming text and are deliberately limited, despite the fact that the topic is extremely wide and not convenient to be developed in this textbook.

Once stable nucleation is established, the rate of solidification, i.e., the rate at which new solid is formed, is almost totally controlled by the heat-removal rate: higher heat-removal rate will imply higher rate of solidification. High heat-removal rate values will promote nucleation and instead limit the other important phenomenon that occurs during solidification: growth. Growth is the direct consequence of the diffusion phenomena and implies the increase in size of structures formed after solidification (growth is also observed in other contexts, such as the formation of coarse-grained microstructures as a result of exposure to high temperatures or excessive duration of high-T heat treatments). The balance between nucleation and growth will determine features of the microstructure after solidification. The resulting morphology will also be a function of the type of liquid–solid interface that was developed as a result of the combination of solidification rate (also known as freezing rate) and thermal gradient. Metal alloys typically solidify with dendritic morphology, the characteristics of which are influenced by the parameters of the solidification process. The non-homogeneous features of real solidification processes also lead to the possibility for another phenomenon to happen: segregation during solidification. It is important to specify that the segregation phenomenon that will be soon exposed refers to what occurs during solidification since segregation can also occur to surfaces. Segregation is a diffusion-related process and therefore may take place in solids only at high temperatures, where the mobilities of lattice elements are high enough to allow lattice transport to occur. Segregation can be defined as non-uniform compositional distribution (i.e., partitioning) of the elements within the microstructure and can be classified in two forms: equilibrium and non-equilibrium segregation. Segregation is also mentioned for elements that tend to accumulate in the liquid. Equilibrium segregation takes place as a result of inhomogeneities in the solid that promote the formation of sites for which the solute atoms have a lower free energy. These sites are found at defects such as dislocations and stacking faults, phase interfaces, grain boundaries or free surfaces. In these regions there are concentrations of solute atoms that differ from each other and, most importantly, from that of the bulk material. Non-equilibrium segregation, on the other hand, is a function of kinetic events and rate processes. Generally, non-equilibrium segregation would disappear as time tends to infinity and if the diffusive processes are actually allowed to reach complete equilibrium [17, 18].

For the understanding of the aspects related to the solidification of SASSs, and related issues, an overview of the main features of segregation will be provided. For a complete description, again, the reader can refer to the references cited and found in the reference list at the end of this chapter. In order to give as comprehensive a picture as possible and to give useful information for understanding what will be presented, it is possible to start from the following initial assumption: the solidification features of real processes lead to conditions to be different from those that would be predicted by the phase diagram related to the alloy studied, i.e., equilibrium conditions. Since diffusion is also a time-dependent phenomenon in addition to temperature-dependency, compositional homogenization cannot occur unless sufficient time is provided for the process. Furthermore, and unfortunately, diffusion rates are especially low for solids and decrease with decreasing temperature. Thus, during solidification processes, unless invariant transformations are involved, temperature will continuously decrease and diffusion will increasingly face difficulties in maintaining the compositional equilibrium between liquid and solid and, in fact, will fail in the task. As a consequence of this, one possibility could be that diffusion in the solid will not be fast enough to ensure equilibrium, and this will cause continuous solute enrichment in the remaining liquid, as simplified by Fig. 2.7 (in particular, the one represented is one of the most common forms of microsegregation: *coring*). The ultimate result of this phenomenon will be the formation of particularly solute-rich regions between the structures. In case the entire concept was applied to a solidification process characterized by dendritic interface, the interdendritic regions (IDRs) would be those subjected to segregation and thus solute enrichment. As it will be seen later in this chapter, the segregative tendency is different depending on the specific chemical element being referred to (some elements will have higher tendency to segregate in the liquid, some other will behave differently and the definition of a partition coefficient will be mandatory). Finally, a further aspect related to segregation concerns the dimension to which it refers. In other words, when the compositional inhomogeneity extends over a wide range that can vary from a few millimeters to dimensions in the order of an entire ingot, the name macrosegregation is used. To underline its macroscopic nature, it is sufficient to consider that this phenomenon is also influenced by the gravity (density differences between the liquid components of the system will play an important role). If, on the other hand, the segregative phenomenon extends only to microstructural dimensions, microsegregation would be the name associated [17, 18].

Having defined the concepts preparatory to understanding the solidification of SASSs, it is worth declaring at this point that it will not be possible to describe the detailed solidification process for each of the alloys given in Sect. 2.3. This is mainly caused by the fact that the discussion would become too long, deviating from the original goal of giving a general description that can be suitable for each alloy through, instead, detailed description of the specific phenomena that have influence on the process. In fact, given the similarities among these alloys, a few will be selected that, due to their chemical composition and market share, can be used as a reference to derive the behavior of all the others (obviously considering the appropriate corrections resulting from the differences in chemical composition). In summary, what will be exposed is intended to be an adaptable interpretation key to enable the reader to obtain a flexible method of analysis and understanding. Usually, for studying the solidification of binary alloys, it is convenient to start from the analysis of the phase diagram of the system taken as reference, and the transformations upon cooling for a given composition are observed. In case the variation of transformations as a function of chemical composition has to be checked, the procedure just described is repeated by increasing or decreasing the solute concentration, i.e., moving left or right on the horizontal axis (see Fig. 1.6 as reference). Obviously, as already mentioned, for the examination of the solidification process of SASSs it is

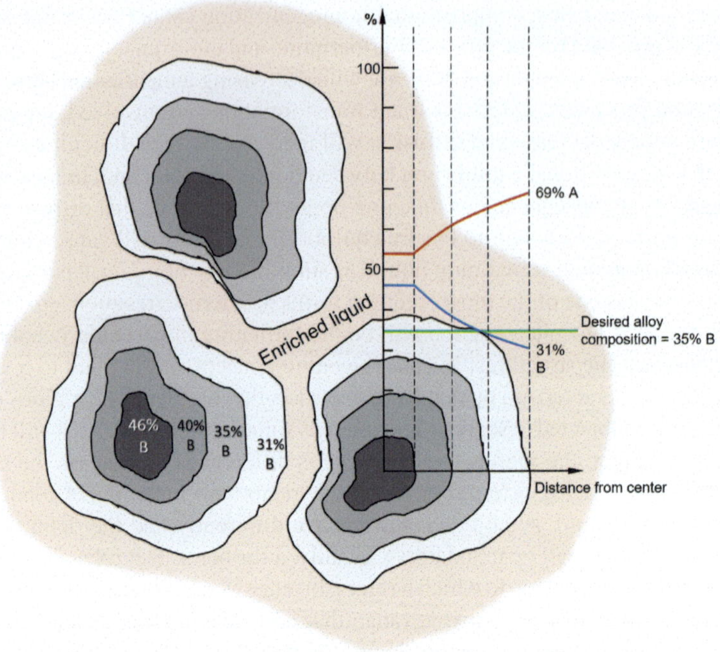


**Fig. 2.7** Schematic representation of the segregation process of a binary alloy leading to inhomogeneous solids and solute enrichment in the last liquid under solidification (the phenomenon is also known as *coring*). A and B are two generic chemical elements that compose the binary alloy AB, whose segregative phenomenon is depicted. Because of the non-equilibrium solidification conditions, the first solid formed will be less A-rich than the last. The trend of alloying elements concentrations as a function of position in the solidified solid is also depicted

not possible to refer to binary systems because the chemical composition is far more complex and counts several elements in addition to the basic ones that define a steel, i.e., Fe and C (see Table 2.2). Moreover, it must be considered that these steels have a high Mo content (4–8%), the presence of which is certainly beneficial for mechanical properties and for corrosion resistance (see Sect. 1.2.1), but still, it is an alloying element that stabilizes α rather than austenite. Then, a high Cr content is also there (19–28%), necessary for attain *stainlessness* and superior corrosion resistance, but acting as α-stabilizing element too (see Appendix A). To counteract the tendency of promoting α ferrite stabilization and to ensure the austenitic structure, Ni (16–32%) and N are there, the latter with values as high as 0.6%. Thanks to this combination of elements, it is possible to ensure an austenitic structure at any temperature. However, it is reasonable to assume that with such a complex ensemble of chemical elements, phase transformations might happen that do not involve the formation of austenite, both with regard to solidification and with reference to possible solid-state reactions. In fact, as reported within Appendix B and as expressed in the previous sections, high Cr and Mo contents can promote the formation of phases such as σ, χ, μ, R and Laves, with greater occurrence and evidence in the literature of the first two.

In view of the information just reported, a convenient starting point might be the analysis of the ternary system that represents the phase transformation of alloys obtained through the combination of the major constituents of SASSs: Fe, Ni and Cr (see Table 2.2). The ternary system just mentioned would imply the need for using the related ternary phase diagram, which, as is well known, is represented with a triangular prism in the moment when both temperature and composition are considered. The complexity behind its interpretation in this three-dimensional form makes it necessary to begin the discussion by considering a sectional diagram, thus cut off vertically at a certain percentage of one of the three elements that constitute the system itself. As depicted by Fig. 2.8, two sections of the ternary diagram obtained at 16% and 20% Ni, respectively, are considered. The percentage of Cr remains variable and that of Fe, clearly, is rebalanced as a function of the sum of the percentages of the other two alloying elements.

Within the two diagrams in Fig. 2.8 the typical range related to the percentage of Cr in SASSs is represented (19–28% according to Table 2.2). By combining the data in Table 2.2 with the sections of the Fe–Cr–Ni ternary diagram depicted in Fig. 2.8, initial hypotheses can be made regarding the solidification of two SASSs: 254 SMO and S34565. The first of the two has a Cr content between 19.5% and 20.5% and Ni in the 17.5–18.5% range (an intermediate value between the two proposed in Fig. 2.8). The second grade, on the other hand, has a Cr percentage between 23 and 26% and a Ni content in the 16–19% range, identical in the minimum value to one of the diagrams in Fig. 2.8 and slightly lower than the other for the maximum value. Looking at the two diagrams in Fig. 2.8, it can be seen that as Ni content increases, the field of simultaneous existence of δ + γ + α widens. On the other hand, the

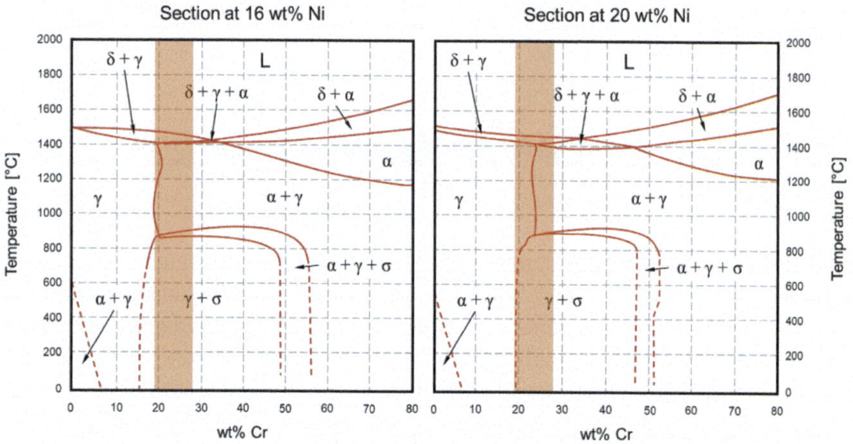


**Fig. 2.8** Section through the Fe–Cr–Ni ternary phase diagram with different Ni concentrations: 16 wt% and 20 wt%, respectively. The range of wt% Cr typical of superaustenitic stainless steels (19–28%) is highlighted in both diagrams (redrawn from [19])

solidification interval to the left of 30% Cr, where there is coexistence of $\delta + \gamma$, narrows with an average lowering of the liquidus line. In light of the two boundary situations just described, it is reasonable to think that at Ni percentages intermediate between the two reported, the situation of the two mentioned fields would also be intermediate between those represented. Referring again to the two SASSs mentioned above, and relying only on the ternary system, it cannot be ruled out that $\delta$ formation occurs during solidification. This phenomenon could be traced back to the Cr segregation in the interdendritic liquid during solidification. This condition could then lead to a situation whereby $\delta$ would be stable, at least at high temperature. However, what has just been stated cannot be considered totally reliable because the discussion is based on a great simplification of the system referred to, reducing it to only three components. In fact, in order to be able to draw more effective conclusions it would also be necessary to consider a third alloying element, the presence of which defines the very existence of SASSs: Mo. As found in the literature, and as will be explained below, Mo has a high tendency to segregate in the liquid during solidification (see Fig. 2.12). Therefore, it is not surprising that the redistribution of Mo during solidification correlates with the development of the correct microstructure and performance required for this family of alloys. Moreover, it is not possible to overlook the effects that the addition of Mo has on the mechanisms involved in the solidification process, and thus from being ternary, the system analyzed must necessarily be considered as multicomponent (i.e., at least quaternary). Help in this regard comes from the countless publications produced to understand the formation of phases other than austenite during solidification of SASSs where, however, the focus has often been on the formation of another deleterious phase: $\sigma$. These studies are largely carried out to understand the post-welding situation of joints obtained by fusion welding processes applied on superaustenitic stainless steels. Although the solidification process is analyzed with deliberate attention on the formation of this deleterious phase, the detail of each step from the beginning of solidification is usually reported.

### *2.4.2 Analysis of 254 SMO Solidification Through QDS Process*

Several approaches have been employed to understand the phenomenon of $\sigma$ phase precipitation in SASSs during solidification, including quenching during directional solidification (QDS) experiments performed on members of this family, including 254 SMO [20]. Directional solidification (DS) is a widely used technological process to produce superalloy components for high-temperature applications that have greater creep resistance than their polycrystalline counterparts. The directionality of the microstructure is based on the fact that grain growth occurs along the heat extraction direction. The method that first achieved directionality in heat removal, and consequently in microstructure development, was the Bridgman method [5]. The purpose behind the application of QDS is to replicate the conditions occurring in a finite volume of material that solidifies after casting or welding process once the solid–liquid interface has formed. With reference to traditional solidification processes such as ingot casting, once the liquid has been poured into the mold, a thin equiaxed-grained layer forms in the contact zone between the liquid and the mold itself. After that, grain growth tends to proceed in a columnar fashion toward the center of the ingot with direction parallel to that of maximum heat removal. Then, depending on the chemical composition of the alloy, the columnar structure may reach to the center or an additional central equiaxed-type zone could form (see Fig. 2.9). The DS process could be employed experimentally to reproduce the columnar solidification process to scale and thus recreate, with properly calculated and determined process parameters, the conditions that could exist at the industrial level. This could make it possible to perform the reproduction of an in-situ study of the solidification process and better understand the segregative issues of materials exhibiting this tendency. In case there is the possibility of cooling fast enough to freeze the situation at a specific instant, this would allow to have a picture of the situation at that instant and make the study of the segregative phenomena and resulting microstructure possible. A study published by Marin et al. [21] applies this latter experimental procedure using an equipment with a Bridgman-type set-up and a sample of 254 SMO produced via hot rolling and makes the comparison with ThermoCalc© simulations. Summarizing the experimental procedure, the QDS process starts essentially like any directional solidification process having defined the extraction velocity in the withdrawal direction and the thermal gradient (controlled by the temperature difference between the furnace and the cooling box) [5, 21]. The major difference with the complete DS process is that after a certain portion of material with desired length has solidified, it is quenched in order to literally freeze the microstructure at that time, as well as the eventual microsegregation. In addition, the study also included the employment of thermal gradients in the liquid and in the mushy zone that would replicate solidification conditions in the zone near the center of a hypothetical industrially-produced ingot (see Fig. 2.9 for the general structure to take as reference) [21].

The QDS process was aided by SEM observations and computational thermodynamics simulations. Calculations were performed to simulate both equilibrium and non-equilibrium solidification. Furthermore, the computational thermodynamic approach was used for the simulation of alloy solidification and, consequently, for the prediction of related phenomena, including microsegregation. Returning for a moment to what is the main purpose of many studies, as it can already be deduced from the references made in this chapter, the focus is often the understanding σ phase formation during heat treatment of different types of stainless steels (i.e., ferritic, austenitic and duplex). This has led to the identification of two main mechanisms for the formation of this deleterious phase:

- Nucleation in austenite.
- Nucleation from eutectoid decomposition process of δ ferrite (for those steels capable of forming it because of their chemical composition).

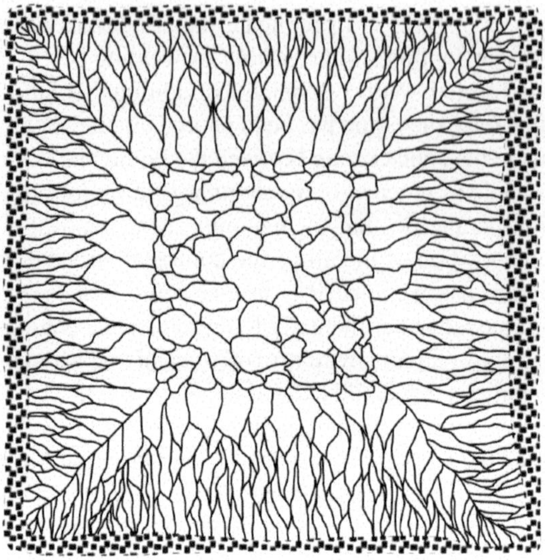

**Fig. 2.9** Schematic of the general solidification structure of an ingot resulting from casting processes. Please note that the outer part is characterized by a fine-grained equiaxial microstructure that, moving toward the center, gives way to a columnar structure with growth direction parallel to that of heat extraction. Finally, in the center an equiaxed structure is there once again, this time with coarser grains. Note also that this is the generic structure that can be obtained at the end of solidification; process parameters and chemical composition can have different influences on it leading, for example, to the absence of the central equiaxed zone

In contrast, less publications have focused on the mechanism of σ phase formation during solidification. This process is of interest for the formation of secondary phases both when dealing with casting processes, and as a consequence of welding processes. In this case, also considering the fact that Mo tends to segregate in the liquid during solidification, two main hypotheses are found:

- Formation of σ in the last stages of solidification through eutectic reaction due to the enrichment of the liquid of σ-forming elements during solidification itself.
- Formation of σ in the IDRs as a result of solid-state phase transformation (decomposition of δ ferrite dendrites) [20].

Thanks to the experimental setup and the use of QDS, it was possible to observe the evolution of the microstructure during solidification and make appropriate correlations with related temperatures. Phase discrimination and crystallography determination was performed thanks to electron backscattered diffraction (EBSD). Several areas were evaluated, such as those that could be defined as the mushy zone and solidified zone before freezing [21]. With reference to Fig. 2.10, the analysis of the QDS samples highlighted that, for 254 SMO, solidification begins with the formation of austenitic dendrites at temperatures around 1430 °C (the exact solidification temperature is dependent on both the cooling rate employed ad on the specific chemical composition of the steel). It is worth remembering that for real processes the reference should be made to heterogeneous solidification and that dendrites, once formed, grow along a direction close to the thermal gradient. At temperatures about 30 °C lower, a second phase forms in the interdendritic region that should be identified as δ ferrite, especially in light of the chemical composition of this steel in terms of Cr, Mo and Ni contents [20, 21]. With reference to Fig. 2.10, this result corresponds to the $L + \gamma + \delta$ region in the figure and, moreover, would be consistent with what is reported with reference to Fig. 2.8. At complete solidification (around 1345 °C) there was no σ phase, only austenite with a certain δ ferrite amount in the interdendritic spaces. As a consequence of this evidence, it would appear that for the solidification of this steel grade under the specific experimental conditions, no σ phase formation occurs directly from the liquid. It follows that the second hypothesis of formation as a result of solid-state reaction could be the confirmed one for the specific case [20, 21].

At temperatures near 1200 °C the interdendritic phase undergoes a solid-state transformation to give rise to two additional phases, the morphology of which is lamellar. From the point of view of the microstructural characterization, the lamellae appear to consist of two phases, one appearing generally light gray and the other dark grey. These two phases seem to nucleate within the interdendritic δ ferrite region. SEM–EDS analyses showed that the light grey phase appeared to be rich in Mo and Cr, while it was instead depleted in Ni. The other phase presented opposite situation, with Ni enrichment and depletion of the other two elements. Literature gives numerous evidences of similar phenomena for other stainless steel types (i.e., duplex and ferritic). Hence, in agreement with the results of SEM–EDS analysis, it was possible to identify the first phase as σ, while the second appeared to be eutectoid austenite (often referred to as $\gamma_2$). The formation mechanism of the σ- and $\gamma_2$-containing regions is the eutectoid reaction, which occurs in the former interdendritic regions leading to the situation of transition from δ to $\sigma + \gamma_2$ surrounded by γ (decomposition of δ to $\sigma + \gamma_2$ occurs from a single grain of δ). This created an ensemble of crystalline phase structures involving the *fcc* γ, the *bcc* of δ and the tetragonal σ (see also Appendix B). δ ferrite grains were also found to have an orientation relationship toward austenite dendrites with a slight deviation of 5° between growth directions. With reference to Fig. 2.10, the final structure, upon completion of solidification and subsequent solid-state transformation, appeared to consist of: primary γ and interdendritic phase composed of σ and $\gamma_2$ [20, 21]. A summary of the sequence just described can be found in the schematic given in Fig. 2.11. In addition to QDS experimental results, those obtained by computational thermodynamics predictions under equilibrium and non-equilibrium conditions were also considered. There are mainly two aspects that were important to be analyzed and compared between experimental and computational approaches: the solidification interval and the prediction related to the σ phase formation. As might be expected, the amplitude of the solidification interval relative to QDS is larger than that predicted by the calculations under equilibrium conditions. The presented specific values were 32 °C for the computational part, and 85 °C for what concerns the QDS sample [20, 21]. This difference is not an unusual event and is already known when studying phase

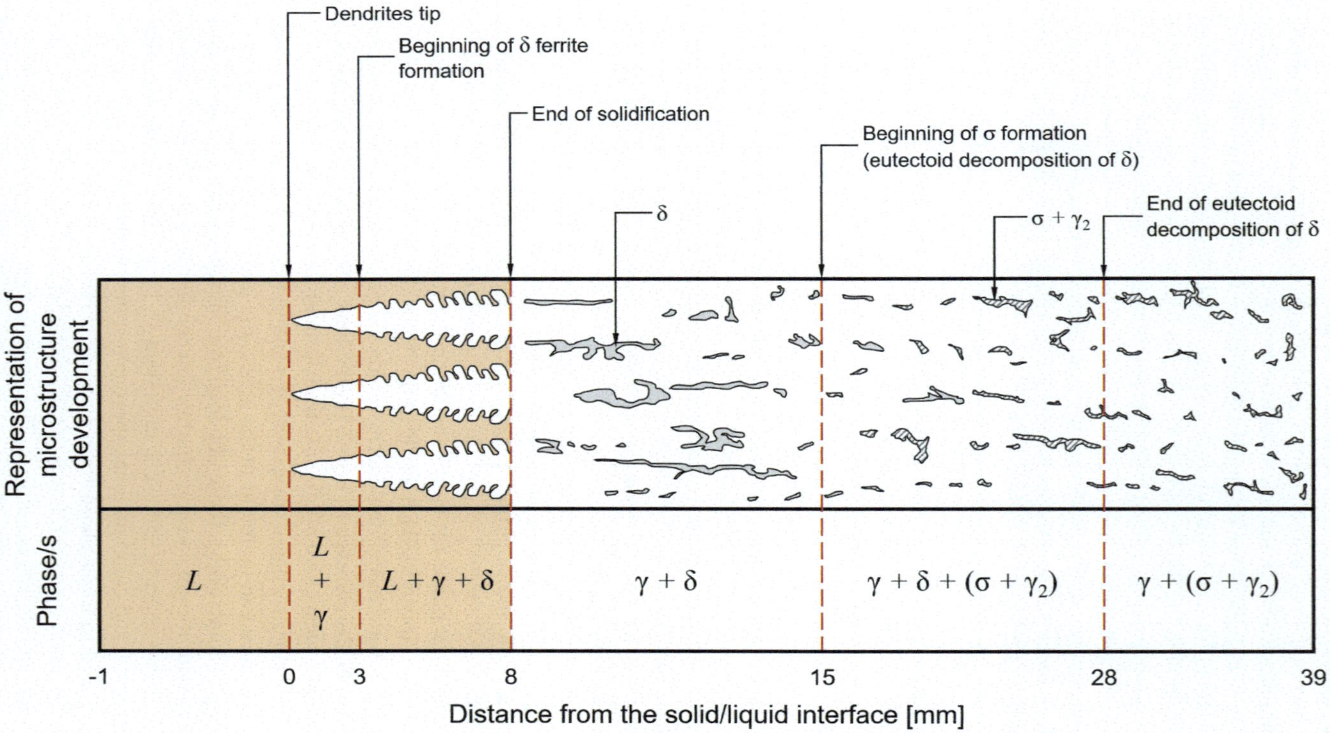


**Fig. 2.10** Schematic representation of the microstructural evolution obtained from QDS samples. The analysis of the figure from left to right not only gives information about the situation as a function of the distance from the solidification front, but also gives an idea of the temporal progress of the process. In fact, the leftmost part can be considered as the starting condition in which the alloy is fully liquid. Movement to the right resembles time passing by and successive instants of the process are seen until complete solidification is achieved (redrawn from [21])

diagrams and the thermodynamics of phase transformations. In fact, with increasing cooling or heating rates that move away the system from equilibrium conditions, the transformation lines in the diagrams tend to shift and the intervals for the transformation themselves tend to widen. The comparison between QDS and non-equilibrium simulation also revealed discrepancies, however, in this case the calculated range is greater (124 °C). This difference can be explained by the assumptions related to the model used for the non-equilibrium calculation. More in details, the non-equilibrium simulation considered no diffusion in the solid after its formation, no back-diffusion and infinite diffusion in the liquid. Under real conditions, the diffusion of solute atoms in the solid is different from zero, while diffusion in the liquid is finite [20, 21].

The analysis of the σ phase formation is also affected by differences between experimental results obtained from QDS sample and calculations. In particular, calculations at equilibrium showed σ phase precipitation from austenite at temperatures of about 100 °C lower than in the QDS sample, where formation of this phase was detected at 1210 °C and starting from δ. This discrepancy highlights the importance of microsegregation during solidification (see also Fig. 2.12) with an increase in σ phase formation temperature, despite the back-diffusion in the solid, that in real processes is different from zero. In fact, the calculation at equilibrium does not take microsegregation into account, which, on the other hand, turns out to be a non-negligible phenomenon as will also be seen later in this section (see Fig. 2.12). Finally, the

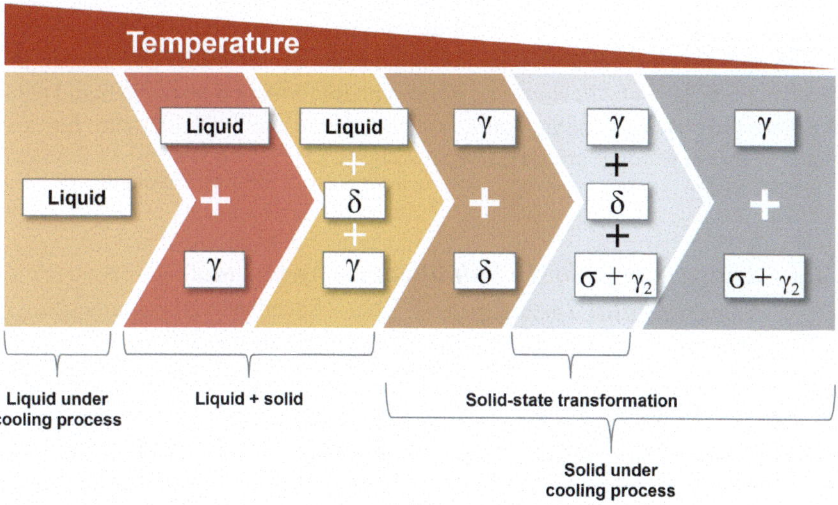


**Fig. 2.11** Proposed solidification sequence for 254 SMO according to the results obtained from the experimental analyses. Reading direction is from left to right. The initial situation involves the alloy being totally liquid. The first step implies that the first solid to form is γ. Subsequently, δ solidifies from the liquid in the enriched interdendritic regions. Once solidification is complete, a solid-state transformation occurs at lower temperatures with formation of σ and $\gamma_2$ from eutectoid decomposition of δ. The final microstructure is made of γ, σ and $\gamma_2$ [20, 21]

solidification path can be said to be in good agreement with the non-equilibrium calculation. In fact, the calculation at equilibrium would predict a fully austenitic structure at the end of solidification, a condition that did not occur for both the non-equilibrium calculation and the QDS sample. In addition to this, it was found that for the process conditions analyzed, no σ phase formation occurred during solidification [20, 21]. It should be pointed out, however, that the analysis of the QDS sample also showed the presence of σ phase in the quenched liquid and in the mushy zone, and this may be at odds with the statement just made. With reference to Fig. 2.10, these two regions correspond to L and $L + \gamma + \delta$ region, respectively. These zones differ in microstructure when compared with those formed during the directional solidification process as a result of the δ phase decomposition. In both cases the morphology of σ phase-containing islands is lamellar with alternating σ and γ, but in the case of what was found in the quenched liquid and in the mushy zone the interlamellar spacing was 300 nm. In case of the σ phase formed during directional solidification, on the other hand, the microstructure was coarser with an interlamellar spacing > 1 μm (in agreement with process conditions associated with a lower cooling rate and more time available for the growth). However, the formation of σ phase in the quenched liquid and in the mushy zone could be traced back to a peculiarity related to the experimental conditions. σ phase was only able to form because of the high cooling rate associated with the quenching process, which, locally and only in some very limited areas, made it the most stable phase. This phenomenon was, therefore, considered as a negligible side-effect for the purpose of analyzing the results when studying a solidification process aimed at simulating industrial conditions. Taking this latter aspect into consideration, however, it can be concluded that the solidification path for conditions simulating an industrial process can be similar to that shown in Figs. 2.8 and 2.9. However, direct σ phase formation from the liquid could take place for very high cooling rates, as reported here and in the literature for cases related to fusion welding processes of SASSs [20, 21].

### *2.4.3 Analysis of 25-6Mo Solidification by Various Experimental Techniques*

From the previous analysis, it is evident that cooling rate and microsegregative tendency play an absolutely important role in the study of solidification processes of SASSs. As seen for the QDS process, some production processes typical of industrial manufacturing can be used to reproduce small-scale solidification processes that allow the study of the phenomena that have the greatest influence on them. Given their importance, it was shown that a further support in understanding the influence of cooling rate, macro- and microsegregative behavior of SASSs could come from another process employed in industrial production: TRSC (twin-roll strip casting). TRSC technology was introduced to produce near-net-shape strips also in stainless steel directly from the melt, despite the fact that it is not a technology created specifically for this type of steels (further details will be presented in Sect. 4.2.4). With this technology, intermediate processes between ingot casting and hot rolling can be avoided. Specifically, for highly-alloyed steels such as SASSs characterized by known segregative issues and deleterious secondary phase formation, high-T solution treatment can be avoided, thus saving time, energy, and cost [22]. Moreover, usually an increased cooling rate is an effective method for limiting and reducing segregation and precipitates formation, especially of the deleterious type. Therefore, a sub-rapid solidification technology such as TRSC would prove to be of definite interest since it has shown the potential of strongly reducing micro-segregation (obviously, what is being discussed here is in general terms, deliberately avoiding special cases such as the one seen above regarding the solidification of 254 SMO). Furthermore, there are not a few difficulties faced in transforming an ingot into a strip and, finally, into a thin sheet. TRSC therefore comes to the aid of this as well. As shown by several previous studies and researches, during solidification of SASSs Mo tends to segregate in the liquid by concentrating in the interdendritic regions (IDRs). Studies conducted on this family of stainless steels led to the conclusion that microsegregation of Mo is unavoidable and results in the formation of eutectics in the as-cast structure [22]. In fact, it should not be forgotten that the high Mo content typical of SASSs can promote the formation of Mo-rich intermetallics according to eutectic or eutectoid transformations, the most known of which certainly are $\sigma$ and $\chi$ (see also Appendix B and Sect. 2.4.6). This results in negative consequences on processability, mechanical properties and corrosion resistance due to compositional inhomogeneity caused by their presence. It is not surprising that some studies show how the presence of these phases results in hot cracking during hot plastic deformation processes [15, 22]. TRSC technology was used by Hao et al. [22] to study the solidification behavior of a steel with chemical composition corresponding to 25-6Mo (see Table 2.2). The experimental activity implied the comparison between results obtained by ingot casting (IC), confocal scanning laser microscopy (CSLM) and TRSC. The aim was both to understand the microsegregative behavior and the influence of cooling rate on solidification structure. Regarding the IC process, the liquid metal was poured into the mold from a temperature of 1580 °C and allowed to solidify naturally, hence with a low cooling rate. The final result was a 40 kg ingot. The cross-section orthogonal to the main axis of the ingot, analyzed after polishing and acid attack, showed that the solidification structure obtained was consistent with that depicted in Fig. 2.9. Analyses on the cross-section were performed at the edge, ½ and center of the ingot also considering morphology and size of primary and secondary dendrites (PD and SD, respectively). Furthermore, the analysis of the ingot microstructure revealed a massive presence of eutectics (about 5% in volumetric fraction) in the interdendritc regions (IDRs) and at the grain boundary (GB) (the low cooling rate gave enough time for the solute elements to segregate in the IDRs). This, combined with the fact that, in any case, solidification begins at the surface of the mold and proceeds toward the center of the ingot, resulted in a distribution of eutectics that is increasing moving toward to the inner part. Examination of the chemical composition of the eutectics by means of electron probe microanalysis (EPMA) pointed out high concentrations of Cr and even higher of Mo. The results of the analysis showed that the average composition of the eutectics differed from that of the alloy (see Table 2.2) and was characterized by around 40% Fe, 16% Ni, 25% Cr and 16% Mo. Furthermore, the morphology of eutectics showed γ and intermetallic compound arranged in a lamellar fashion. These results went in the direction of what has been shown in the literature thanks to another work on a different alloy, always belonging to the 6-Mo family, that will be presented in this section: AL-6XN® (see Sect. 2.4.4). In that case it was shown that, at the end of the solidification, the as-cast structure was characterized by the presence of eutectics made of γ and σ [23, 24]. Further studies have shown similar results with the obvious observation that the content of eutectics may vary from alloy to alloy and is therefore, and obviously, chemistry-dependent. In parallel with the microstructural analysis just mentioned, the examination of the macrosegregation of the preponderant elements (i.e., Fe, Ni, Cr and Mo) of the ingot from the periphery to the center was performed. The analysis still considered a cross-section orthogonal to the main axis and pointed out that the percentages of Fe and Ni gradually decreased. The opposite behavior was exhibited by Cr and Mo, with the latter characterized by more pronounced segregation than Cr. The Mo content in the center of the ingot was shown to be higher than 10%, almost double the typical content for this alloy (see Table 2.2). All of this, of course, results in a serious problem of central macrosegregation of the ingot. This was in agreement with the findings just mentioned above and contributes to the explanation regarding the chemical composition of the eutectics found [15, 22].

It is known that a change in the cooling rate results in a variation in secondary dendrite arm spacing (SDAS) leading to its reduction as the cooling rate increases. In this context, key support came from the use of CSLM, which makes it possible to melt and resolidify a finite volume of the alloy. Furthermore, the liquid alloy was allowed to solidify at different cooling rates, from 90 °C/s down to 0.1 °C/s. Results obtained by confocal scanning laser microscopy (CSLM) at low cooling rates, in the order of 0.1 °C/s, showed continuous eutectics in interdendritic regions (IDRs). As the cooling rate increases, the eutectics become discontinuous and their average size tends to decrease. This mainly occurs as a result of the fact that with higher cooling rate the time available for segregation is reduced, which contributes to the inhibition of eutectics formation. This is also confirmed by changes in the value of the partition coefficient, often used as an indicator for the evaluation of elementary microsegregation during solidification processes. The partition coefficient $k$ (see Eq. 2.6) expresses the ratio between the concentration of a specific element in the liquid under solidification and the concentration of the same element in the solid newly formed [22].

$$k = \frac{C_{iL}}{C_{iS}} \tag{2.6}$$

where

- $k$ is the partition coefficient.

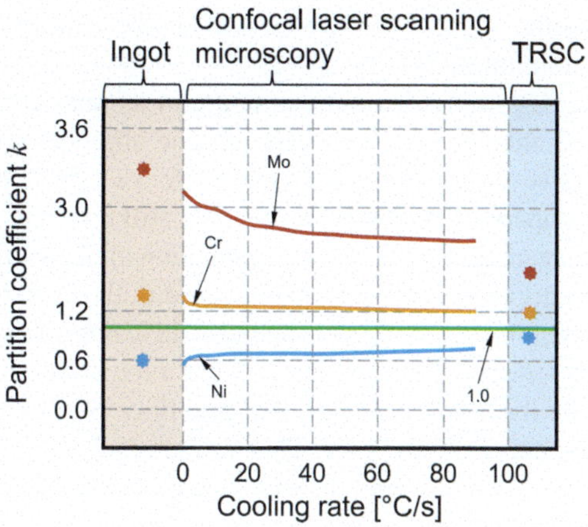


**Fig. 2.12** Solute partition coefficient of Ni, Cr and Mo as a function of the cooling rate during solidification with CLSM. Results of the analysis performed on the ingot and on the material produced via TRSC are also reported for comparison (redrawn from [22])

- $C_{iL}$ is the concentration of the *i* element in the liquid under solidification.
- $C_{iS}$ is the concentration of the *i* element in the solid newly formed.

A value of *k* greater than 1 indicates that the element tends to segregate in the liquid, while a value less than 1 indicates exactly the opposite: segregation in the solid. Thanks to the use of EPMA, the average values of $C_{iL}$ and $C_{iS}$ were measured for the material under investigation in the three conditions: IC, CSLM and TRSC. For the CSLM condition, as mentioned above, different cooling rate values were considered, from 0.1 to 90 °C/s.

The observations led to the results summarized in Fig. 2.12, where it can be seen that the two processes at the left and right boundary are IC and TRSC, for very slow and very fast cooling (i.e., subcritical), respectively. There is good agreement in the literature on the fact that during the solidification processes of SASSs, for low cooling rates, Ni is characterized by a value of $k > 0$, but less than 1. This confirms how Ni has segregative tendency in the solid. For the same alloy system in the same conditions, Cr has a *k* value slightly above 1, indicating a slight segregative tendency in the liquid. Mo, on the other hand, has a *k* value higher than that of Cr for low cooling rates. This confirms the strong segregative tendency in the liquid when compared with Cr. An increased cooling rate leads to a change in the partition coefficient for both Mo, Cr and Ni, specifically, the higher the cooling rate, the more the value of *k* tends to 1. Figure 2.12 shows that the partition coefficient of Ni tends to increase with increasing cooling rate and thus its segregative tendency in the solid decreases as well, but not reversing its behavior since $k < 1$. Hence, Ni segregation in the solid will be less pronounced, but still present. As for Cr and Mo, the trend is the same, and *k* tends to approach 1 by increasing the cooling rate. In this case, however, both elements tend to segregate less in the liquid, particularly Mo, which reduces the value of *k* more markedly. This latter fact is clearly positive with regard to the possible suppression or reduction of eutectics formation in IDRs since compositional constraints that could lead to their presence can be removed. Furthermore, with specific reference to the results of the TRSC process, it can be seen that the Mo partition coefficient is at a minimum [22].

Regarding the microstructure of the material obtained by TRSC, in this case the presence of very fine dendrites and absence of eutectic compounds were observed. However, the predicted volumetric fraction of eutectics should have been very low compared with the IC results, but still different from zero. Various hypotheses have been made regarding the explanation of the results, including errors in data measurements and fitting. Among the reasons, the most interesting one from a manufacturing point of view concerns the intrinsic features of the TRSC process. In fact, in view of the effects of cooling rate on the microstructure, it is expected that the microstructure itself will gradually become finer as cooling rate increases as demonstrated by the experimental part. This would result in a reduction of the IDRs to a very low value, which would make segregation and thus eutectics formation more difficult. In agreement with these statements, the predicted estimation for the volume fraction of eutectics gave values close to 1%. The reason for the absence of eutectics could be attributed to a fundamental difference of the TRSC process from the others involved in the study: the application of a certain rolling force to the material under solidification. The rolling force acts as breaker of the solidification structure and significantly reduces the width of the IDRs by physically preventing segregation through repulsion of the liquid into the molten bath (as if the liquid is being squeezed out). In order to confirm the hypothesis just stated, a test was carried out on the same steel employing the TRSC process with a greatly reduced rolling force. The result was a different microstructure with the presence of a small volume fraction of eutectics detected in the IDRs. Essentially two reasons were hypothesized to explain the result:

- The reduction in rolling force made it insufficient to reduce the size of the IDRs and to be able to redistribute the liquid in the molten bath.
- The reduction in rolling force led to a worsening in the contact conditions and actually reduced the heat transfer efficiency between the liquid-cooled rolls and the material under solidification. This caused the cooling rate to be lower than the standard process and clearly promoted the formation of eutectics.

The two hypotheses should not to be considered independent from each other, but rather related to each other in terms of the result obtained. In conclusion, it was possible to see another pragmatic aspect of how cooling rate has an influence on the solidification structure and microsegregative tendency, and how this can be used to explain the solidification of SASSs. In addition, the TRSC process highlighted the interesting correlation between sub-critical cooling rate and application of mechanical action on the solidification result. In particular, the synergistic action of increasing the cooling rate and applying a certain rolling force proved to be effective in suppressing segregation and eutectics formation [22].

### 2.4.4 *Analysis of AL-6XN® Solidification by GTAW and DTA Measurements*

As already mentioned, in addition to understanding solidification processes for obtaining primary solidification products (e.g., ingots), the importance of studying this mechanism is considerable for understanding the phases that can form at specific temperatures, their formation mechanism and sequence, as well as knowing the consequences of post-weld solidification. It is known, and demonstrated, that conventional welding processes involve the redistribution of alloying elements in the fusion zone with possible deleterious consequences on mechanical resistance and, especially, on the local corrosion resistance. In particular, the segregative tendency of Mo in the interdendritic regions has been extensively mentioned in previous sections and demonstrated in the literature, as well as the greater tendency for preferential corrosion processes in the dendrite cores depleted of this important alloying element (the virtues of Mo have been exposed in Sect. 1.2.1 and also further exploited in Appendix A). In order to be able to find the relationships necessary for a better understanding of the solidification process, studies have been performed also on AL-6XN®. In this context, a useful comparative study aimed at understanding the solidification of AL-6XN®, with specific focus on this process after fusion welding processes, was carried out by Perricone and Dupont [24]. Based on the composition of this alloy (see Table 2.2), in order to better understand the individual contribution of the alloying elements, additional three experimental compositions were produced using it as reference, but varying the amount of specific alloying elements (see Table 2.4 for comparison with Table 2.2).

Since the specific research was primarily aimed at understanding solidification behavior, but with focus on welding processes, two of the three additional compositions were developed to replicate the behavior of Mo-rich Ni-based filler metal typically used for welding SASSs (reference is made to the materials labeled *filler 1* and *filler 2* in Table 2.4). The remaining alloy, on the other hand, had similar composition as AL-6XN®, but with twice as much Mo and was named *experimental alloy 1*. For the study of specific alloy solidification, the fact that this additional composition is based on the quaternary Fe–Ni–Cr–Mo system is of fundamental importance for the observation of the influence of Mo, but also Fe, on the overall solidification

**Table 2.4** Chemical compositions for alloys used for the experimental campaign aimed at understanding the solidification behavior of AL-6XN®

| Alloy | Fe | C | Cr | Ni | Mo | Mn | N | Cu | Si | O |
|---|---|---|---|---|---|---|---|---|---|---|
| AL-6XN® | 46.8 | 0.02 | 21.6 | 24.2 | 6.1 | 0.4 | 0.24 | 0.3 | 0.5 | 0.005 |
| Experimental alloy 1 | 43.6 | 0.03 | 20.1 | 23.9 | 12.4 | 0.01 | 0.017 | – | – | 0.014 |
| Filler 1 | 0.1 | 0.02 | 20.6 | 66.9 | 12.1 | – | 0.006 | – | 0.1 | 0.011 |
| Filler 2 | 0.1 | 0.02 | 20.2 | 54.5 | 24.3 | – | 0.008 | – | 0.3 | 0.015 |

Values are expressed in wt% [24]

process. It is known that both the Ni–Cr–Mo ternary system and the Fe–Ni–Cr–Mo quaternary system, on which it is also convenient to rely for the study of solidification of SASSs, contain different intermetallics. These compounds have, in turn, different crystal structures and chemical compositions. As a consequence of this, it should be specified that all the alloys analyzed were selected in order to show primary γ solidification with the possibility of subsequent formation of Mo-bearing eutectics in the interdendritic regions. The amount of these latter would, of course, have been dependent on the specific composition of the alloy [24].

The experimental approach included an initial step in which 23 kg ingots were produced for each of the four alloys listed in Table 2.4. Each ingot was then hot-rolled at a process temperature close to 1200 °C until an 8 mm thick sheet was obtained. After the rolling process, the produced samples were heat-treated at the homogenization temperature of the respective alloys for the time needed to minimize any segregative phenomena and to develop a uniform microstructure. A weld seam was made on each specimen thanks to GTAW technique. In addition to this, samples for each alloy were prepared in order to be able to perform differential thermal analysis (DTA) of the materials from solidification down to room temperature [24]. It is important to emphasize that the GTAW process and the DTA are strongly different approaches. However, the combination of the results obtained from them can help in understanding the solidification process. In particular, the welding process can give information about the re-solidification of a small volume of material. The related cooling rate is essentially free and a function of the ambient temperature and welding pool volume, but in this case still greater than that employed in the DTA for the specific case (slow cooling, as it will be shown in the following). The DTA analysis, on the other hand, can provide information regarding the nature of the transformations that occur upon cooling (i.e., exothermic or endothermic). In addition to this, the possibility of determining the cooling rate allows for obtaining a specific microstructure that can be traced back and correlated to this important process parameter. The experimental procedure related to DTA involved each one of the four alloys being heated up to 1100 °C with a heating rate of 50 °C/min. Once this temperature was reached, the heating rate was lowered to 10 °C/min until a temperature 20 °C higher than the theoretical calculated liquidus temperature of each alloy was reached. The same rates were used for the cooling process associated to the solidification process: 10 °C/min down to 1100 °C and 50 °C/min from 1100 °C down to room temperature. The only exception in terms of experimental parameters was for AL-6XN® in the temperature field above 1100 °C. In fact, with this material the heating and cooling rates were 5 °C/min instead of 10 °C/min. The reason for the difference laid in the intention to better maximize and isolate signals related to possible phase transformations [24].

Results of DTA analysis for AL-6XN® and *experimental alloy 1* are shown in form of plots in Fig. 2.13. For what concerns AL-6XN®, solidification upon cooling begins at 1403 °C with formation of primary γ. As cooling proceeds, at a temperature of 1354 °C, an exothermic peak is detected, corresponding to a eutectic-type reaction happening during solidification (Fig. 2.13a). Once this latter reaction is completed,

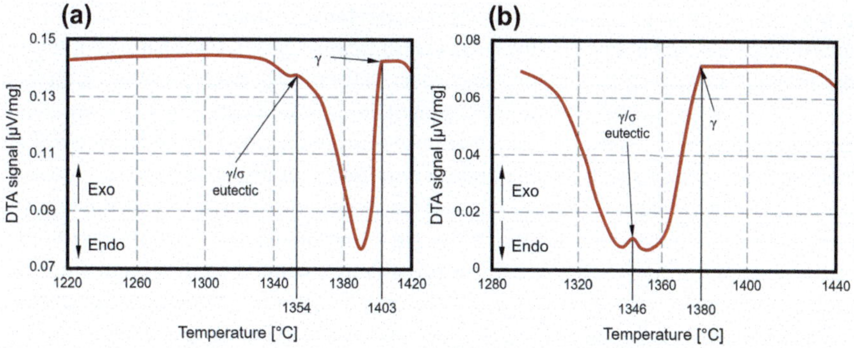


**Fig. 2.13** DTA results on cooling for: **a** AL-6XN® and **b** experimental alloy 1. For heating and cooling rate data please refer to main text. *Exo* and *endo* in the figure refer to exothermic and endothermic phenomena, respectively (redrawn from [24])

cooling continues down to low temperature with no further peaks detected. For *experimental alloy 1*, on the other hand, solidification begins at a lower temperature corresponding to 1380 °C. Furthermore, *experimental alloy 1* also exhibits an exothermic peak after the beginning of solidification at a temperature of 1346 °C (Fig. 2.13b). In this case, however, in comparison with Fig. 2.13a, both amplitude and intensity of the peak were greater. Regarding *filler 1* and *filler 2* alloys (see Table 2.4), the result of differential dilatometric analysis is not shown in Fig. 2.13. In any case, as far as *filler 1* is concerned, no exothermic peaks were detected after the onset of solidification at 1395 °C. The phase formed as the first stable solid was γ, and the absence of exothermic peaks upon cooling was considered as an indication of the absence of eutectic-type reactions. On the other side, *filler 2* (still a Ni-based alloy, but with twice as Mo as filler 1) started the solidification at 1344 °C with formation of primary γ. Then, at 1296 °C, it exhibited a more pronounced exothermic peak than *experimental alloy 1*, both in extension and amplitude [24].

Samples of the four alloys under investigation (see Table 2.4) from both DTA and from the GTAW process were prepared for optical micrographic analysis. Both set of samples were cut, those obtained from GTAW along the transverse direction with respect to the welding direction so as to better observe the fusion zone and the HAZ (heat-affected zone) features. The microstructure of AL-6XN® of the samples from DTA was characterized by coarse dendrites of primary γ. In addition to this, a variable amount of eutectic compound was detected in the interdendritic regions. *Experimental alloy 1*, which main feature was to have practically twice as much Mo as AL-6XN®, exhibited a significantly higher amount of eutectic compound than the superaustenitic stainless steel. This latter phenomenon, in addition to the mere percentage of Mo, can also be explained by the segregative behavior of this alloying element. In fact, experimental observation of the redistribution of Fe, Ni, Cr and Mo in the samples after solidification revealed interesting aspects. Regarding *experimental alloy 1* (12.4% Mo), it has been detected that the average dendrite core composition was 8.6% Mo. The difference between the two values showed that during solidification a large part of Mo is rejected into the liquid, thus resulting in an equally large amount of eutectic formed as solidification proceeds as a result of liquid enrichment. The same phenomenon was observed for the AL-6XN alloy, but with less extent for what concerns the volume fraction of eutectics. In fact, in comparison with a nominal value of 6.1% Mo, the dendrite core was characterized by a value of 4.8%. Therefore, the composition of the dendrite core takes on importance because it can be considered as expression of the composition of the first solid formed as the liquid–solid interface is advancing. Furthermore, since the diffusion of Mo into austenite is insignificant during solidification, it is possible to assume that the composition of the dendrite core is constant over time. This latter assumption becomes a useful data for describing the segregative behavior and also for calculating the solute partition coefficient [24].

Optical microscopy performed on the GTAW samples confirmed what was found on those from the DTA analysis for what concerns the phases detected. However, the structure was shown to be more refined on average, and small differences in the volume fractions of eutectics were found. This latter feature happened despite significant differences in the cooling rates. In fact, with the higher cooling rate of GTAW, a lower volume fraction of eutectics would be expected, which was not the case proportionally to the difference in cooling rate. The data obtained thus seem to indicate that the segregative behavior of Mo is practically insensitive to variation in cooling rate for the range of values analyzed. Regarding filler alloys, no eutectic compounds were detected in *filler 1*. Therefore, the material was constituted by a single phase, i.e., γ. The lack of eutectic compound in *filler 1* alloy was also inconsistent with the multicomponent liquidus projections obtained using ThermoCalc® software combined with the Fe-Data database and the Ni-Data database. In fact, solidification under non-equilibrium conditions was expected to lead to formation of eutectic constituent. However, this phenomenon did not occur, and the reason would seem to lie in the fact that the observed experimental phenomenon deviated from the non-equilibrium situation considered in the calculation. In fact, it turned out that, due to its chemical composition, alloy *filler 1* proved to be extremely sensitive to experimental conditions. As a consequence of this, any slight deviation from the *ideal* non-equilibrium conditions (e.g., variability of partition coefficient *k*, dendrite tip undercooling, back-diffusion etc.) may be responsible for the lack of eutectic compound in the microstructure. For any further investigation on this issue, which is not central in the discussion, the reader is referred to the original source in the reference list. In *filler 2*, on the other hand, the presence of eutectic compounds was detected with a morphology that was marginally different from AL-6XN® and *experimental alloy 1*. The different morphology is justified by the different alloy system referred to by *filler 1* and *filler 2* when compared with the other two materials. In particular, the absence of Fe in both filler alloys should be noted. As a result of this, the system to be referred to is the ternary Ni–Cr–Mo one. Determination of the nature of the secondary phases was done thanks to a combination of micrographic, SEM and EPMA techniques. This made it possible to identify the phases in the eutectic constituents, confirming the presence of γ in all alloys, σ (tetragonal) in AL-6XN® and in *experimental alloy 1* and P (orthorhombic) in alloy *filler* 2. The P phase is

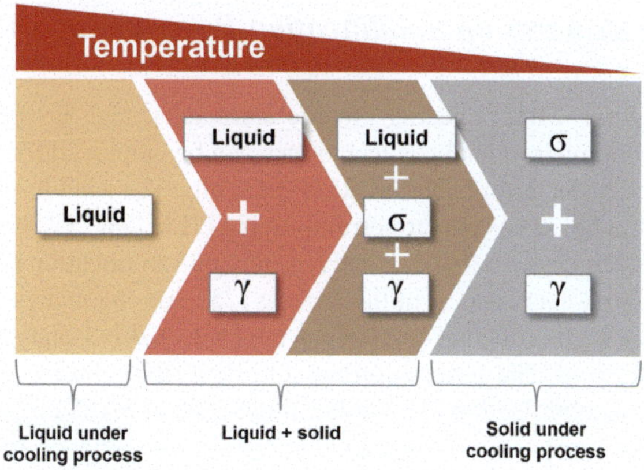


**Fig. 2.14** Proposed solidification sequence for AL-6XN® alloy and alloy 1 according to the results obtained from the extensive materials testing campaign (solidification proceeds temporally from left to right). Reading direction, from left to right [24]

not among those mentioned in Appendix B because it is not typical of steels, but rather of the Ni–Cr–Mo system. The possible presence of μ is also reported in the literature for this system, but it is postulated to be the product of a solid-state transformation of P. As a result of experimental observations and comparison with data resulting from calculations, the proposed solidification sequence for AL-6XN® is shown in Fig. 2.14. The initial situation involves the alloy being totally liquid. The first step implies that the first solid to form is γ. Also referring to Fig. 2.13, the next step involves the formation of σ phase according to a eutectic reaction. The final microstructure implies no liquid, primary γ, and a eutectic solid made of σ and γ [24].

In addition to the experimental results, it is also worth reporting an interesting aspect related to results of multicomponent liquidus projections obtained using ThermoCalc® software. With specific regard to AL-6XN® alloy, the result of the projections reports that Mo enrichment of the liquid should bring the system into the stability field of the δ phase, consequently modifying the solidification sequence proposed in Fig. 2.14. In fact, the simulation would go in conflict with the first formed intermetallic (σ) detected and also with the presence of ternary eutectics, which would include the δ phase. Inconsistencies can be explained by several factors. The first lays probably in the fact that the intersections of the primary solidification paths incorrectly estimate the eutectic compositions. Second, for alloys containing large amounts of Fe, it is important to consider the fact that the relative content of Fe in the liquid changes during solidification. As a result, the boundary lines of phases and eutectics change in position. In fact, it has been seen experimentally that Fe segregates into the solid during solidification, leaving the liquid depleted of this element. For this reason, the transformation lines of phases and eutectics cannot be considered truly stationary in a real multicomponent system and must necessarily be considered mobile as a result of changes in liquid composition [24].

### 2.4.5 Final Remarks on Solidification Processes of SASSs

As already said for what concerns the literature data, and as seen in previous sections, a great number of analysis techniques can be employed for understanding the behavior of SASSs during solidification. In addition to what has been reported, many other studies have been performed employing DTA, DSC, CLSM, SEM, EBSD and TEM. However, it seems that detailed research on solidification processes is still not enough to clarify every aspect of the process [25]. Among many useful literature contributions, three experimental approaches were selected and analyzed, as reported in the previous sections. The dissertation has considered the main aspects and parameters regarding real solidification of complex alloys such as SASSs. As already stated, the solidification process is not only of fundamental importance for studying the production of primary solidification products such as ingots, billets or castings; but it is also of great importance for understanding important aspects related to welding processes. Three among the most widely spread representatives of the SASSs family were chosen (254 SMO, 25-6Mo and AL-6XN®) and according to what has just been stated, their solidification process was analyzed using different techniques in order to cover the greatest number of variables and aspects involved. The QDS process was used to replicate at scale the features and conditions of a real solidification process, to have an "instantaneous picture" showing the progress of solidification from the front to the innermost part. This experimental approach was used to understand the microsegregative tendency and its consequences on possible transformations from the liquid and/or in the solid state. Furthermore, solidification sequence was also reconstructed based on the experimental data. IC, CSLM and TRSC were used to further investigate the influence of cooling rate, of possible external perturbations/loads applied to the structure undergoing solidification (similarly to electromagnetic stirring in continuous casting machines), and to understand additional aspects and consequences related to the microsegregative tendency of alloying elements under various conditions. DTA and GTAW were also used to reconstruct the solidification sequence and the nature of the transformations involved. Each analysis technique clearly has advantages and limitations and, of course, their detailed description is out of the aim of this textbook. However, with specific reference to CSLM, it is interesting to consider that an intrinsic limitation is related to the fact that this technique can only perform free surface analysis in the case of opaque samples. As a consequence of this, information about the beginning and time of completion of a phase transformation within the bulk material cannot be determined categorically. On the other hand, another technique employed in the studies reported in the previous sections, the DTA, is limited by the fact that the observable phenomena cannot be local in the sample in which they occur. In fact, only the average response of the entire sample is detectable. This implies that in the interpretation of the results, phenomena such as the relative velocity of interphase boundaries cannot be directly observed. As a consequence of this, they must be assessed employing some pre-assumed data [25]. In this latter concern, a common feature of all the approaches presented in the three previous sections was the employment of calculations and simulations using computational thermodynamics software to support, confirm or extend the results obtained from the experimental work.

Considering the evidence exposed in the previous sections, it can be said that there are at least two possible solidification routes for SASSs, as depicted by Fig. 2.15. The two mechanisms identified are called *solidification route 1* and *solidification route 2*. Solidification route 1 involves a three-stage process and is the one resulting from the interpretation of the results coming from the analysis of the solidification process associated with 25-6Mo and AL6-XN®. In this case, solidification begins with the formation of austenite from the liquid within stage 1. During stage 2, a eutectic transformation takes place due to the enrichment of the liquid as solidification proceeds. The result is the formation of a eutectic solid with lamellar microstructure consisting in alternating $\gamma$ and $\sigma$. Solidification ends during stage 3 after the liquid is fully converted into solid. The resulting microstructure is austenitic with a certain volume fraction of eutectic containing $\sigma$, the amount of which depends on alloy's chemical composition and on parameters specific to the solidification process itself. On the other hand, solidification route 2 proved to be the appropriate mechanism to describe the solidification process of 254 SMO according to the structure of the experimental activity employed. In this case, stage 1 during which solidification begins is identical to that of solidification route 1 and involves the formation of $\gamma$ as the first solid. On the other hand, stage 2 is different in that $\delta$ ferrite formation in the interdendritc regions (IDRs) occurs due to liquid enrichment resulting from segregation of Mo and Cr in the liquid itself. Stage 3 involves two phenomena, the first being the completion of solidification as a result of the complete conversion of the liquid to solid at high temperature, while the second is a eutectoid reaction at lower temperature. The eutectoid reaction involves $\gamma$ and $\delta$ and originates a lamellar solid consisting of $\gamma_2$ and $\sigma$. The complete conversion of $\delta$ leads to the final microstructure that characterizes stage 4: austenitic grains with a certain volume fraction of $\sigma$ resulting from the decomposition process of $\delta$.

Although SASSs are a particular family of austenitic stainless steels, it is worth to declare that for austenitic stainless steels the solidification mode is often described by means of practical methods based on the calculation of Ni equivalent ($Ni_{eq}$) and Cr equivalent ($Cr_{eq}$) values and on the resulting $Cr_{eq}/Ni_{eq}$ ratio. This is done in order to reduce a complex multicomponent system down to the relatively simpler ternary Fe–Cr–Ni one. This is usually supported by a diagram with iso-ferrite lines for predicting the amount of $\delta$ ferrite that can form in castings. As a result of this, the solidification modes for austenitic stainless steels are usually the following four types [26, 27]:

- A mode (A stands for austenitic): solidification sequence is $L \rightarrow L + \gamma \rightarrow \gamma$. For promoting this solidification mode, a $Cr_{eq}/Ni_{eq} < 1.25$ is usually needed.
- AF mode (A stands for austenitic, F for ferritic): solidification sequence is $L \rightarrow L + \gamma \rightarrow L + \gamma + \delta \rightarrow \gamma + \delta \rightarrow \gamma$. AF mode takes place when $1.25 < Cr_{eq}/Ni_{eq} < 1.48$.
- FA mode (F stands for ferritic, A for austenitic): solidification sequence is $L \rightarrow L + \delta \rightarrow L + \delta + \gamma \rightarrow \delta + \gamma \rightarrow \gamma$. FA mode takes place when $1.48 < Cr_{eq}/Ni_{eq} < 1.95$.

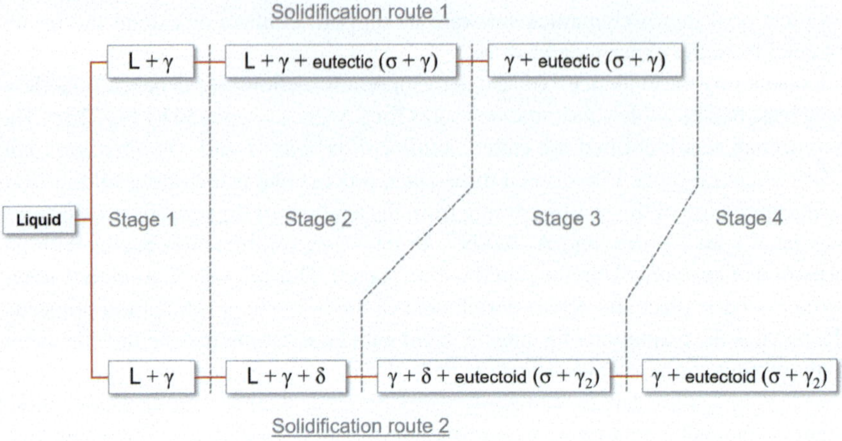


**Fig. 2.15** Possible solidification paths for superaustenitic stainless steels according to the data given in Sects. 2.4.2–2.4.4. Eutectic $\sigma + \gamma$ indicates the lamellar solid resulting from the eutectic transformation that characterizes solidification route 1. Eutectoid $\sigma + \gamma_2$ is the lamellar solid resulting from the eutectoid transformation that characterizes solidification route 2

- F mode (F stands for ferritic): solidification sequence is $L \rightarrow L + \delta \rightarrow \delta \rightarrow \delta + \gamma \rightarrow \gamma$. F mode takes place when $Cr_{eq}/Ni_{eq} > 1.95$.

The reference equations for calculating both Ni and Cr equivalent are a less extensive and, in some details, different version of Eqs. 1.1 and 1.2 [26]:

$$Ni_{eq} = \%Ni + (30 \times \%C) + (30 \times \%N) + (0.5 \times \%Mn) \quad (2.7)$$

$$Cr_{eq} = \%Cr + (1.5 \times \%Si) + \%Mo + (0.5 \times \%Nb) \quad (2.8)$$

Based on the two previous equations, Table 2.5 shows the values of $Cr_{eq}$, $Ni_{eq}$ and the $Cr_{eq}/Ni_{eq}$ ratio obtained for 254 SMO, 25-6Mo and AL-6XN®. The alloying element contents are the average values between minimum and maximum shown in Table 2.2. For comparison, the same values of $Cr_{eq}$, $Ni_{eq}$ and the $Cr_{eq}/Ni_{eq}$ ratio obtained from Eqs. 1.1 and 1.2 are also given in the table.

From the results shown in Table. 2.2, the strongly austenitic character of these alloys is evident, as are the slight discrepancies in the values obtained with the different pairs of equations. This is due to the different coefficients involved in the calculation and, in any case, does not lead to macroscopic differences since 254 SMO still remains the one characterized by the highest $Cr_{eq}/Ni_{eq}$ ratio. In agreement with what was stated above, for all three steels considered, the solidification mode should have been A mode since the $Cr_{eq}/Ni_{eq}$ ratio values are all well below 1.25. The results reported in the three previous sections, however, show that this is not completely the case because for 25-6Mo and AL-6XN® the solidification mode turns out to be a sort of variation of A mode in that a eutectic reaction occurs with consequent formation of

**Table 2.5** Results of the $Cr_{eq}$, $Ni_{eq}$ and the $Cr_{eq}/Ni_{eq}$ ratio calculation for 254 SMO, 25-6Mo and AL-6XN®

| | | | | | | | | Equations 2.7 and 2.8 | | | Equations 1.1 and 1.2 | | |
|---|---|---|---|---|---|---|---|---|---|---|---|---|---|
| Steel | C max | Cr | Ni | Mo | Mn | N | Cu | $Cr_{eq}$ | $Ni_{eq}$ | $Cr_{eq}/Ni_{eq}$ | $Cr_{eq}$ | $Ni_{eq}$ | $Cr_{eq}/Ni_{eq}$ |
| 25-6Mo | 0.02 | 20 | 25 | 6.50 | 1 | 0.200 | 1.00 | 26.50 | 32.10 | 0.83 | 29.75 | 31.40 | 0.95 |
| 254 SMO | 0.02 | 20 | 18 | 6.25 | 1 | 0.200 | 0.75 | 26.25 | 25.10 | 1.05 | 29.38 | 24.33 | 1.21 |
| AL-6XN® | 0.03 | 21 | 24.5 | 6.50 | 0 | 0.215 | 0.75 | 27.50 | 31.85 | 0.86 | 30.75 | 31.00 | 0.99 |

Values for alloying elements are expressed in wt%

σ phase. On the other hand, 254 SMO shows a solidification sequence more similar to AF mode, except for the disappearance of δ ferrite is not due to its conversion to γ, but to a eutectic reaction leading to formation of $\gamma_2$ and σ. The combination between the values given in Table 2.5 and the results given in Sects. 2.4.2–2.4.4, seem to highlight the non-total applicability of the criteria explained above regarding the $Cr_{eq}/Ni_{eq}$ ratio value for the determination of the solidification mode since no σ phase formation was predicted. However, it should be also specified that from what is reported in the literature by Martorano et al. [21] regarding the solidification mode, there is no reference to the formation of intermetallics, but only to the various polymorphs of steel. Hence, based on this latter observation, 25-6Mo and AL-6XN® could be said to be consistent with the solidification sequence assigned to them by the evaluation of the $Cr_{eq}/Ni_{eq}$ ratio. On the other hand, the same cannot be confirmed for 254 SMO. In addition to what stated in Fig. 2.15, data are reported in literature that would point out the AF mode as the one best suited to describe its solidification sequence. Evidence of this is shown within Fig. 2.16, where the predicted solidification mode states that the first phase to form is γ, followed by a small volume fraction of δ (indicated as ferrite) that disappears at high temperature converting to γ. Secondary phases such as σ, χ and $Cr_2N$ are formed at lower temperatures when solidification is already complete as a consequence of solid-state reactions. However, looking at the results of the calculation of the $Cr_{eq}/Ni_{eq}$ ratio given in Table 2.5, even using Eqs. 1.1 and 1. 2 the resulting value remains below the limit beyond which AF solidification would occur, i.e., 1.25. In fact, according to the $Cr_{eq}/Ni_{eq}$ ratio, the solidification mode should be A (fully austenitic). In order to explain the inconsistency just highlighted, various approaches can be employed. For example, to clarify the solidification mode of SASSs, and of 254 SMO in particular, a possibility could be to go even further into details regarding the influence of alloying elements. In particular, it may also be useful to analyze the specific influence of N.

The effect of this interstitial element is already partially accounted for in the equation for the calculation of $Ni_{eq}$ (see Eqs. 1.1 and 2.6). However, a closer look may still be useful for the purposes of the discussion. As specified within Appendix A, one of the key roles of N in steel is to increase the stability range of austenite, i.e., make it thermodynamically more stable. Therefore, regarding high-N alloys such as the case of SASSs, introducing and maintaining high amounts of N in the melt is one of the most critical issues of the fabrication process. This fact is further emphasized by the importance of N in the thermodynamic calculations. In fact, N is a key parameter in simulating the solidification sequence and results of thermodynamic calculations. In fact, despite its qualitative nature, Scheil solidification conditions reported in the literature for 254 SMO show that the N content in the liquid might be a discriminating factor for the type of first phase that solidifies and for the subsequent phase transformations. For instance, it has been shown that with N content lower than 0.183%, the stability of γ as the first phase to form from the liquid is lost and δ is the first solid to form instead. On the other hand, in case the N content lays between 0.183% and 0.258%, γ is formed first and δ forms afterwards [25]. This would be the case for 254 SMO, which should be characterized by this type of solidification with subsequent decomposition of δ into γ as shown in Fig. 2.16, because of its N

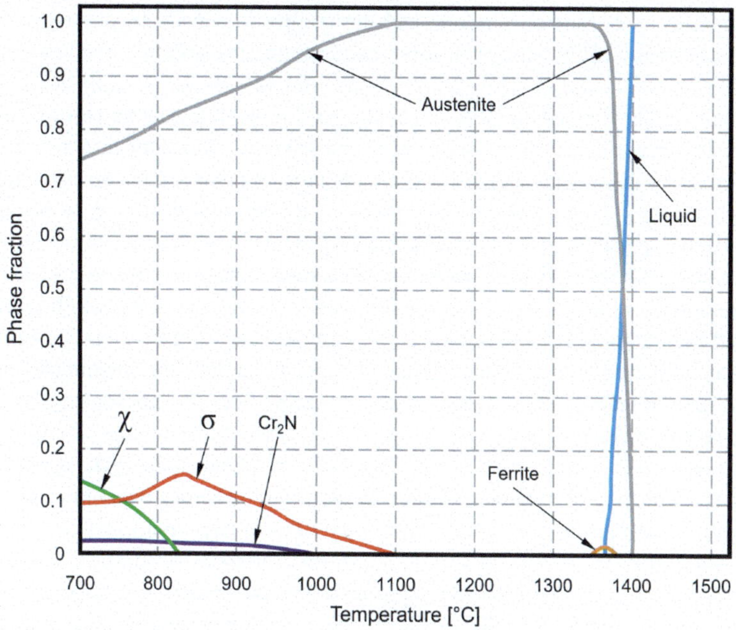


**Fig. 2.16** Calculated phase fractions as a function of temperature for the 254 SMO superaustenitic stainless steel (redrawn from [19])

content, as given in Tables 2.2 and 2.5. Finally, when the N content is above 0.258%, no more δ would form and the solidification is only austenitic, as in the A mode [25].

It is important to underline herewith that the conditions just described regarding the N ranges, however, do not apply to 25-6Mo and AL-6XN® because the values have been specifically calculated for 254 SMO. Furthermore, although they have similar N content, a different $Cr_{eq}/Ni_{eq}$ ratio results from their chemical composition. In conclusion, evaluations made on the basis of N content alone would confirm partly the results given in Fig. 2.15 and totally the information given in Fig. 2.16. The bias related to Fig. 2.15 stems from the fact that the criterion used merely describes which phase will form first and does not consider any subsequent reactions involving for example the liquid, such as those shown in Fig. 2.15 [25]. However, the outcome of the N content evaluation is in contrast to what is associated with the $Cr_{eq}/Ni_{eq}$ ratio value. The discrepancy obtained can be explained by the fact that the $Cr_{eq}/Ni_{eq}$ ratio alone is not totally able to provide an accurate prediction of the solidification mode of austenitic alloys with complex chemical composition such as SASSs, particularly because of the high N content [21]. As highlighted by Martorano et al. [21], accurate prediction of solidification mode and δ ferrite formation in stainless steel castings requires a more in-depth analysis and understanding of individual analytical methods available. Furthermore, the combination of the necessary equations with graphical methods (e.g., Schäffler or Siewert) and the comparison with an extensive experimental campaign allows for even more accurate results and lower relative errors [21].

Further insights from the analysis of Sects. 2.4.2–2.4.4 relate to both the usefulness of simulations under equilibrium and non-equilibrium conditions and the possible inaccuracies or discrepancies these have with experimental results. In fact, equilibrium calculations were not able to fully predict experimental results, as in the case of the exposed experimental activity related to AL-6XN® and 254 SMO. On the other hand, non-equilibrium calculations have been shown to be more effective, albeit with some deviations related to the models employed in the calculation itself. It is clearly expected that deviations from equilibrium conditions impliy discrepancies between what was predicted and what was obtained experimentally. The greater this deviation, the more markedly the system will evolve toward non-equilibrium solidification conditions [8]. This deviation is a function of several parameters, many of which have been exposed in the previous sections. The extent of deviation from the equilibrium conditions of each individual parameter is what has to be carefully calibrated in non-equilibrium calculations, along with the choice of the correct parameters to include. As a result of this, in order to have enough information for describing and predicting the solidification process of SASSs, a possible analysis matrix to be applied on the solidification of each SASS will therefore have to take first the effective distance of the analyzed experimental or study conditions from the ideal conditions into account. Then, other possible influencing factors need to be analyzed. As a result of this, four points emerge from the previous sections and must be considered:

- Influence of the cooling rate.
- Influence of the segregative tendency of alloying elements.
- Diffusion in the solid.
- Phases formation from liquid or through solid-state reactions.
- Sensitivity of the system under solidification to external perturbations, in particular those of mechanical nature.

The actual distance of the conditions under study from the ideal conditions turns out to be a concept dependent on what has been explained so far in this section and certainly starts with understanding the consequences of cooling rate variations. Regarding the influence of cooling rate, the analysis of what was reported in Sects. 2.4.2–2.4.4 shows that:

- For what concerns microstructure in terms of austenitic grain size, a higher cooling rate leads to a finer microstructure, as expected and seen for 25-6Mo [15, 22].
- At very fast cooling rates, $\sigma$ phase formation can occur directly from the liquid as seen for 254 SMO. This situation is in agreement with what is found in the literature regarding welding processes (see also Sect. 4.7) [20, 21].
- The increase in cooling rate leads to a reduction in the secondary dendrite arm spacing (SDAS), as expected and as seen for 25-6Mo [15, 22].
- At low cooling rates, the structure of eutectic compounds obtained during solidification is of the continuous type and the size is generally coarse. An increase

in the cooling rate results in a more discontinuous morphology of the eutectics and, in general, in a reduction in their average size as seen for 25-6Mo [15, 22]. This can be considered as a direct result of the lower microsegregation caused by the higher cooling rate. Thus, the inverse proportionality relationship that exists between cooling rate and microsegregation magnitude is confirmed, and this is well represented by Fig. 2.12.

With regard to the issue of segregative tendency of alloying elements, the most important considerations for the determination of post-solidification structure can be summarized as:

- Microsegregation has influence on transformation temperatures and phases present [20, 21, 28].
- Mo and Cr are alloying elements that segregate in the liquid during solidification (see also Fig. 2.12).
- Mo microsegregation can promote δ phase formation in interdendritic regions during solidification [20, 21].
- Theoretically, multicomponent liquidus projections for AL-6XN® had predicted δ phase formation from the liquid, but this did not happen. The condition for having δ formation from the liquid, as seen for 254 SMO, would have been less segregation of Fe in the solid (as the liquid becomes depleted of it during solidification). As seen for 25-6Mo, in the event that the cooling rate is increased, the segregation in the solid would also decrease, as would that of Mo in the liquid (see Fig. 2.12) [24].
- Even exacerbating the cooling rate, it is not possible to reverse the segregative tendency of Mo or even that of Cr. However, it is possible to approach more to unit values of the partition coefficient *k* obtaining, thus, less segregation in the liquid. This result is positive for the reduction of segregation in the interdendritic liquid and leads to a lower volume fraction and size of eutectic compounds that can form. This is achieved also due to a reduction in SDAS and interdendritic regions (IDRs) extension [15, 22].
- In view of the fact that dendrite core composition can be considered as an expression of the composition of the very first solid to form, and since the diffusion of Mo in austenite is insignificant during solidification, it is possible to say that the composition of the dendrite core is constant over time. Consequently, the determination of dendrite core composition can be regarded as a key parameter for understanding the extent of Mo segregation [24].
- In addition to microsegregation, for industrial processes such as ingot casting (IC), macrosegregation is also a relevant and non-negligible phenomenon for SASSs. Cr and Mo tend to macrosegregate in the liquid and thus show large variations in composition even at the macroscopic level in the products of solidification, Mo much more than Cr [15, 22].

Regarding the third issue on the list, the phases formation from the liquid or through solid-state reactions, it is possible to summarize what follows:

- According to what has been reported in Sects. 2.4.2–2.4.4, and considering the $Cr_{eq}/Ni_{eq}$ ratio values given in Tables 2.5 and 2.6, the first phase formed from the liquid for SASSs turns out to be $\gamma$. This is also highlighted in Fig. 2.15.
- For specific compositions (see $Cr_{eq}/Ni_{eq}$ ratio close to 1.25), $\delta$ phase can form directly from the liquid during solidification due to microsegregation [20, 21].
- From the data shown in the previous sections, only 254 SMO was found to be characterized by $\delta$ formation during solidification. Among the steels in Tables 2.5 and 2.6, it is the one with the highest values of $Cr_{eq}/Ni_{eq}$ ratio, 1.05 and 1.21 depending on the equation pair used for the calculation of the value. However, the study of the $Cr_{eq}/Ni_{eq}$ ratio alone has been reported to be not sufficient for confident prediction of the presence of $\delta$ [21].
- On cooling $\delta$ may be involved in a eutectoid reaction to give rise to a lamellar structure that alternates $\gamma_2$ and $\sigma$ intermetallic phase [20, 21].
- $\sigma$ can also form by eutectic reaction between $\gamma$ and the liquid during solidification [22, 24].
- Direct $\sigma$ formation from the liquid appears to be possible when particularly high cooling rates are employed. Evidence of this is seen in the literature for example in the case of welding processes, where solidification of a small volume of liquid results in high cooling rates [20, 21].

Finally, with regard to the fourth point on the list (i.e., sensitivity of the system to external mechanical perturbations), some considerations can be summarized as follows:

- External stresses, if any, can effectively affect the microstructure during solidification process, both in terms of size and morphology and in terms of the volume fraction of the phases present.
- Very positive results can also be achieved with regard to segregation when the applied external load is such that it interferes physically with solidification. This is accomplished by breaking up of the primary solidification structure in a similar way to what happens in continuous casting molds thanks to stirring.
- A physical breaking of the primary solidification structure leads to a reduction in the IDRs extension and, in the case of externally exerted compressive loads, to a squeeze of the interdendritic liquid, which is repelled into the molten bath.
- The beneficial influence of the external application of a load is directly proportional to the intensity of the load itself. Higher stress applied would result in a more noticeable improvement in the quality of the resulting microstructure, i.e., less segregation and limitation of the presence of deleterious intermetallic phases. The direct consequences can be summarized first of all in the possibility of markedly reducing or suppressing interdendritic segregation, and, as a result of this, reducing the volume fraction or eliminating undesired second phases [15, 22].

**Table 2.6** Results of the $Cr_{eq}$, $Ni_{eq}$ and the $Cr_{eq}/Ni_{eq}$ ratio calculation for the remaining superaustenitic grades of Table 2.2

| | | | | | | | | Equations 2.7 and 2.8 | | | Equations 1.1 and 1.2 | | |
|---|---|---|---|---|---|---|---|---|---|---|---|---|---|
| Steel | C max | Cr | Ni | Mo | Mn | N | Cu | $Cr_{eq}$ | $Ni_{eq}$ | $Cr_{eq}/Ni_{eq}$ | $Cr_{eq}$ | $Ni_{eq}$ | $Cr_{eq}/Ni_{eq}$ |
| 20Mo-6 | 0.03 | 24 | 35 | 5.85 | 1 | 0.13 | 3 | 29.85 | 40.30 | 0.74 | 32.78 | 40.55 | 0.81 |
| 1925hMo (UNS N08925) | 0.02 | 20 | 25 | 6.75 | – | 0.15 | 1.15 | 26.75 | 30.10 | 0.89 | 30.13 | 29.70 | 1.01 |
| NAS 254N | 0.03 | 23 | 25 | 5.5 | – | 0.2 | | 28.50 | 31.90 | 0.89 | 31.25 | 30.90 | 1.01 |
| 25-6Mo 1925hMo (UNS N08926) | 0.02 | 20 | 25 | 6.5 | 1 | 0.2 | 1 | 26.50 | 32.10 | 0.83 | 29.75 | 31.40 | 0.95 |
| Nicrofer 3127 hMo | 0.02 | 27 | 31 | 6.5 | – | 0.2 | 1.2 | 33.50 | 37.60 | 0.89 | 36.75 | 36.96 | 0.99 |
| Uranus SB8 | 0.03 | 24 | 22.5 | 6 | 3 | 0.475 | 1.75 | 30.00 | 39.15 | 0.77 | 34.50 | 37.30 | 0.92 |
| 654SMO | 0.02 | 25 | 22 | 7.5 | 3 | 0.5 | 0.45 | 32.50 | 39.10 | 0.83 | 36.25 | 36.74 | 0.99 |
| Nirosta® 4565S | 0.03 | 24.5 | 17.5 | 4.25 | 5 | 0.5 | – | 29.40 | 35.90 | 0.82 | 31.14 | 33.40 | 0.93 |

Values for alloying elements are expressed in wt%. The alloying element contents are the average values between minimum and maximum shown in Table 2.2. Please note that Uranus SB8 contains 2% W on average and that Nirosta® 4565S contains up to 0.15% Nb, both taken into account in the calculation of $Cr_{eq}$, $Ni_{eq}$

### 2.4.6 Phases

Within Sect. 2.4 and its sub-sections, solidification was discussed and it was pointed out that the γ phase is not the only phase that exists in SASSs. In fact, exposure to high temperatures is known to cause decomposition of the γ matrix and the associated formation of various types of intermetallic compounds [29]. It is known that the concept of phase is very important in the engineering field of metallurgy, in particular when dealing with complex alloy systems that may undergo transformations during solidification upon cooling, or during high-temperature exposure, and in this case, reference is made only to solid-state transformations. The issue of phases has already been anticipated in the introductory chapter of this textbook, specifically in Sect. 1.2.8. In that section, the main phases found in stainless steels were mentioned in a generalist manner, with no particular focus on SASSs. However, among the several information reported, one is worth repeating as the concept behind the variety of phases found in SASSs: the number of phases that are able to coexist in an alloy is proportional to the number of alloying elements it contains and their concentration. In order to give evidence of the latter statement, reference can be made to Fig. 2.17 where optical micrographs of three stainless steels heat-treated to purposefully precipitate secondary phases are present (chemical composition is given in Table 2.7). The complexity of the chemical analysis increases from Fig. 2.17a–c, in the same way that precipitation becomes more complex. By simplifying the concept and referring to the thermodynamic meaning of the term, a phase can be defined as a macroscopically homogeneous body of matter [17]. For the present context, the term phase will refer to everything that is not the so-called matrix of SASSs, thus everything that is not austenite. As a consequence of this, compounds such as intermetallics, carbides, nitrides, δ phase etc. will end up under the definition of phase.

Within this section, specific features of the phases that most commonly can form in SASSs, both in service and experimental conditions, will be exposed. Reference will also be made to Appendix B, where a as comprehensive list as possible of all phases documented stainless steels, thus also in SASSs, is given. Hence, continuing along the line just stated, the phases that can be detected in SASSs are dependent on their chemical composition and the conditions they are subjected to. As a result, from literature data and field experience, the documented phases prove to be:

- δ ferrite.
- σ phase.
- χ phase.
- R phase.
- Laves phases.
- Carbides (in particular $M_{23}C_6$, $M_6C$).
- Nitrides (in particular $M_2N$).

Before reviewing the characteristics of the members of the above list, it is worth understanding, at least in a general way, what conditions can lead to the formation

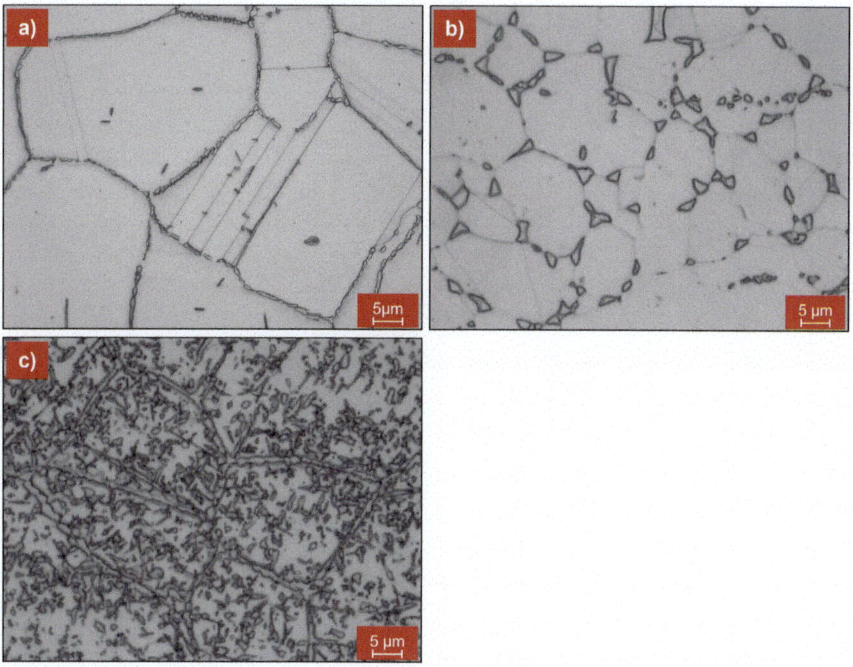


**Fig. 2.17** Optical micrograph of: **a** UNS S31050 (tube) after heating at 800 °C for 40 h and furnace-cooling. Microstructure composed of a matrix of austenitic grains with little precipitation of σ phase and χ phase at the grain boundaries (occasionally in the matrix). Presence of complex chromium carbides and nitrides at grain boundaries; **b** AISI 904L after thermal treatment at 970 °C for 2.5 h and water quenching followed by heating at 970 °C for 5.5 h and furnace-cooling. Microstructure composed of a matrix of austenitic grains with precipitation of plate-like σ phase islands and complex chromium carbides at the grain boundary. Occasional presence of χ phase at the grain boundaries; **c** 654SMO after heating at 800 °C for 40 h and furnace-cooling. Microstructure composed of a matrix of austenitic grains with precipitation of plate-like σ phase islands, χ phase and possible η Laves phase at the grain boundaries and in the matrix. Possible, but difficult to be detected, presence of carbides and nitrides both at grain boundaries and in the matrix. Comparison of the chemical composition of the three alloys is given in Table 2.7 (courtesy of OMECO S.r.l.)

of phases other than γ in these steels. This step takes on fundamental importance not only for a better understanding of the nature and characteristics of the phases themselves, but also for implications related to the prevention of formation and explanation of unexpected or accidental formation. The latter, could be seen as a failure in the production process and therefore what stated above becomes meaningful as part of the related failure analysis. The latter statements are justified because the supply status of semi-finished products (sometimes simply semiproducts) made of SASSs is often that of solution annealed, theoretically free from secondary phases. Not only that, even with regard to finished products, reference is made at most to work-hardened (i.e., cold-worked) products, but still not hardened by means of

**Table 2.7** Chemical composition of steels whose optical micrographs are given in Fig. 2.18 [1–4, 9, 10, 30]

| Most common name | WN | UNS | C max | Cr | Ni | Mo | Mn | N | Others |
|---|---|---|---|---|---|---|---|---|---|
| AISI 310MoLN2 | 1.4466 | S31050 | 0.01 | 25.0 | 22.3 | 2.1 | – | 0.15 | – |
| AISI 904L | 1.4539 | N08904 | 0.02 | 19.0–23.0 | 23.0–28.0 | 4.0–5.0 | 2.0 | – | Cu = 1.0–2.0 |
| 654 SMO | 1.4652 | S32654 | 0.02 | 24.0–26.0 | 21.0–23.0 | 7.0–8.0 | 2.0–4.0 | 0.45–0.55 | Cu = 0.3–0.6 |

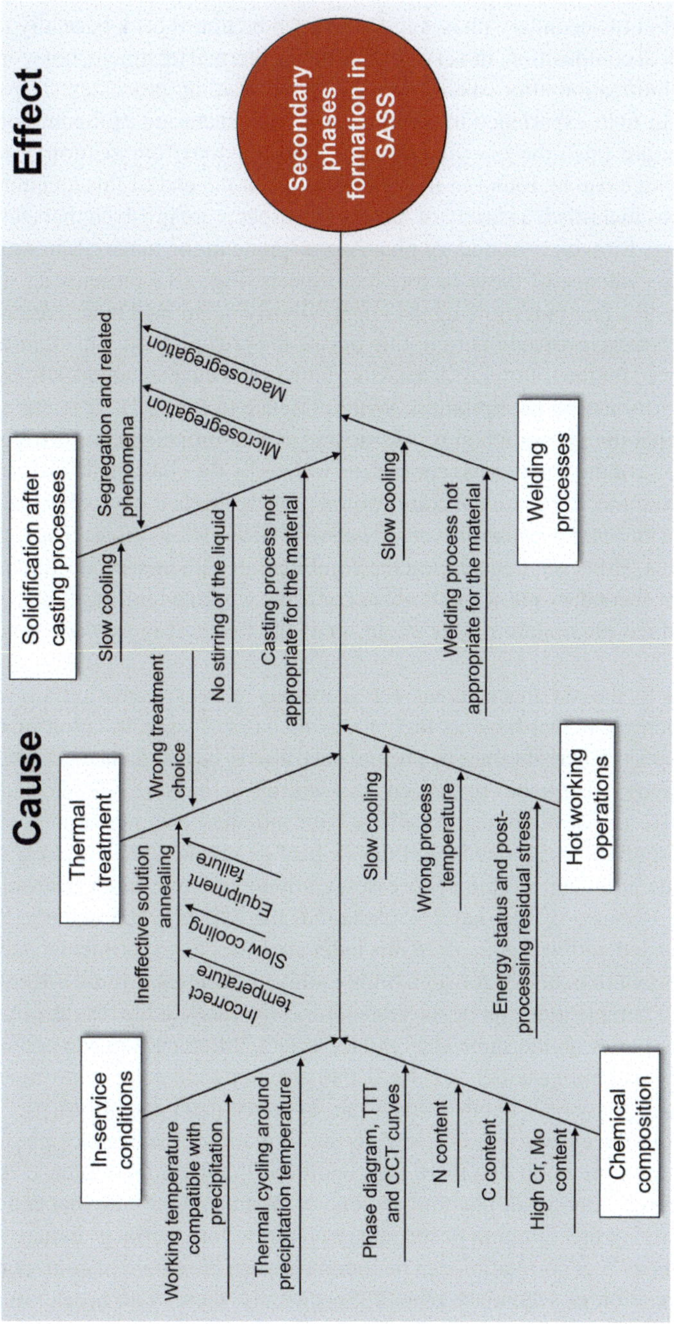


**Fig. 2.18** Ishikawa-like diagram for the root cause analysis of the formation of secondary phases in SASSs

precipitation hardening (i.e., aging treatments) or phase transition (e.g., martensitic transformation).

The formation of secondary phases in SASSs can be traced back basically to six aspects: chemical composition, in-service conditions, thermal treatment, hot working operations, solidification after casting processes and welding processes. These six areas arise from field experience in semiproducts manufacturing of the author and from the literature (also the one mentioned in the previous four sections), many examples of which can be found in the reference list at the end of this chapter. For each of the areas identified, a variety of factors and aspects are involved that can lead to the conditions for which secondary phases precipitate in the material. In order to be able to give evidence of these factors and aspects, Fig. 2.18 presents an image of an Ishikawa-like diagram showing the six previously mentioned macro-areas and the respective details involved in promoting or causing precipitation. This figure will serve as reference for the following discussion, both in this chapter and in subsequent chapters when discussing precipitation. With reference to Fig. 2.18, it is important to emphasize that the phenomenon of secondary phases formation can be triggered either by just one of the conditions reported, as well as by the combination of several of them. For example, the chemical composition of these steels is a possible cause of precipitation in any case because the very existence of these alloys is based on high Cr and Mo contents. However, considering the combination with in-service conditions, precipitation of secondary phases will not occur if the working temperature is lower than that at which precipitation may occur, or if the material is properly solution annealed.

In addition to the six macro-areas for secondary phases formation shown in Fig. 2.18, experiments that have as their focus the study of precipitation or other phenomena/aspects where it might be present have also to be taken into account. The experimental activity has to be considered cross-cutting to the Ishikawa-like diagram and may involve the development of alloys with modified chemical composition, simulation or study of in-service conditions or heat treatments, scale replication or study of solidification and welding processes, study of structures after hot working operation etc. Because of this latter explanation, the experimental activity is not explicitly indicated within Fig. 2.18. In this latter context, i.e., experimental activity, it is very important to note that for such highly alloyed steels even small differences in the chemical composition can be the reason for differences in the results obtained from different studies on the same steel grade. In fact, differences in Mo content in the order of 1% and even smaller in N ($\pm$ 0.1%) are enough to delay or promote the formation of one phase over another under the same treatment conditions.

In addition to the causes that can lead to the formation of secondary phases, it is also important to mention that there are conditions that have an influence on the causes themselves. Among all phenomena, one of the most important that can have an effect on onset of precipitation of secondary phases is cold working. In fact, it has been demonstrated that correlation can be found between the extent of cold working and the precipitation of secondary phases. Specifically, there is an acceleration of precipitation kinetics for the same temperature and aging time when comparing solution annealed and cold worked material. In particular, the greater the degree of deformation undergone by the material, the greater the time anticipation of precipitate nucleation. This suggests how precipitation kinetics accelerates, just as precipitate coarsening occurs earlier for cold-worked steels and for all test temperatures [31]. As a consequence of this, it can be stated that an additional aspect that has to be considered cross-cutting to the diagram given in Fig. 2.18 is the previous history of the material, for instance in the case it has been subjected to cold work before being put into service.

Furthermore, before addressing the upcoming discussion, it is worth to make a specification on the terminology that will be used below. In this section, the term aging will be used in connection with heat treatments or processes below the solution annealing temperature aimed to precipitate secondary phases. It is important to remember that SASSs cannot be aged as is the case with austenitic PH stainless steels, thus the common definition is not consistent in the present context (i.e., sub-critical heat treatment that relies on changes in solid solubility with temperature and whose aim is the precipitation of second phases in order to cause an increase in mechanical properties). Consequently, the precipitates that will be formed do not have positive consequences for mechanical properties. Aging will be a term used to refer to heat treatments required to simulate the conditions necessary for the precipitation of deleterious secondary phases and their study, as in the case of cooling from the solidification process or under high-temperature working conditions and so on.

#### 2.4.6.1 δ Ferrite

δ ferrite is generally considered as an undesirable high-T phase in steels. It is characterized by a *bcc* lattice like α ferrite. Considering a simple Fe–C system, δ will undergo dissolution by the usual diffusional process on first entering the austenitic field [32]. As the degree of alloy complexity increases (i.e., the alloy is richer in alloying elements), the stability range of δ ferrite can expand, and for certain conditions it is possible to find this phase in the material even at room temperature. In fact, the formation of δ ferrite is promoted by the presence of alloying elements that have a ferrite-stabilization effect, such as Cr, Mo, Nb and W (see Appendix A). Hence, the presence of δ ferrite essentially depends on the steel chemical composition, but also on solidification parameters and, possibly, the subsequent thermomechanical processing to which the material can be subjected. The amount and morphology of δ ferrite have influence on several features of the material such as weldability, mechanical properties, corrosion properties and hot formability [32]. In this latter context, the softening mechanism of γ and δ at hot processing temperatures are different. Specifically, γ tends to recrystallize, whereas δ recovers. This difference in behavior might lead to interface fracture [33]. Nevertheless, the presence of δ ferrite can be beneficial or detrimental depending on alloy application, volume fraction, and location (specific regard is made to welding processes with reference to base metal or weld bead). Certain amount of δ ferrite in stainless steels increases yield and tensile strength (see Eq. 3.2), resistance to SCC, reduces hot-tearing in castings and reduces hot-cracking susceptibility in weld joints [34]. For welding applications in particular, it is often desirable to have a small amount of δ ferrite in the weld metal because this phase has a higher solubility for P and S than γ. In fact, a fully γ solidification would result in the rejection of P and S in the liquid under solidification, thus leading to formation of low-melting temperature compounds. These compounds would be trapped in the interdendritic region during γ solidification leading to possible crack formation along GBs (i.e., solidification cracking). The presence of a certain volume fraction of δ ferrite could be therefore desirable because it allows to retain P and S in solid solution and avoid formation of low-melting compounds, with the deleterious consequences just described. However, when a certain volume fraction of δ ferrite is present in the weld beads and the material is exposed to high temperature, δ may transform. Specifically, content and shape may change, and the phase may turn into austenite, σ, or give raise to other solid-state reactions (as already summarized in Sect. 2.4.5) [35]. Furthermore, adverse effects of δ ferrite in some applications might include decrease in impact strength during long-time high-T service through an increase in the rate of σ phase formation. Another consequence of presence of δ ferrite is a higher magnetic permeability of alloys containing it [34]. For this latter reason, it is common practice for suppliers and customers to define a maximum amount of δ ferrite in the material when defining product specifications. Finally, in most cases, δ ferrite can be dissolved by heat treatment in the temperature range between 1050 and 1250 °C [33].

#### 2.4.6.2 σ Phase

σ phase can be probably considered as the most studied and among one of the most undesirable intermetallic phases. As it can be inferred from Fig. 1.6, σ exists in the Fe–Cr system and its presence was theorized even before the large-scale market introduction of stainless steels in the nineteenth century. In fact, in 1907 W. Treitschke and G. Tamman proposed that a compound in the 30–50 wt% Cr range should exist in the Fe–Cr binary system. In 1927 E. C. Bain and E. W. Griffiths found a hard and brittle phase within the Fe–Cr–Ni system ad initially named it B. The letter after which the phase was named was meant to recall the brittle nature of the compound, and sigma was given only later by E. R. Jett and F. Foote in 1936. The crystallographic structure of the compound in the Fe–Cr system was obtained by G. Bergman and D. P. Shoemaker in 1951 [36]. In 1966, thanks to the work of B. Hattersley, W. Hume-Rothery, E. O. Hall and S. H. Algie; σ phase was detected in the interdendritic regions of austenitic stainless ingots in the as-cast condition, thus making solubilization treatment necessary after solidification [20]. As pointed out above and in previous sections, σ can form in different binary, ternary, quaternary and more complex systems. In addition to the Fe–Cr system, some examples are: Fe–Mn, Fe–Mo, Fe–V, Fe–Cr–Mo, Fe–Cr–Ni and Fe–Cr–Ni–Mo. The occurrence of σ has been documented to show specific conditions, specifically for electron/atom ratio between 5.6 and 7 and for atomic radius ratios in binary systems between 0.96 and 1.16 [36]. Furthermore, the elemental composition of σ has been shown to be insensitive to temperature variations, while that of other deleterious phases such as χ phase has been observed to vary with temperature [29]. σ phase belongs to the family of TCP phases, where TCP means topologically close-packed. Given the importance of σ among the TCP phases, it is intended to take advantage of the paragraph devoted to it for a brief discussion related to details characterizing this family of compounds. The feature that gives these phases their name is that the space filling principle is the optimization of the atomic packing factor. This results in coordination polyhedra allowing for CN equal to 12, 13, 14, 15 and 16. This means that the Frank-Kasper polyhedra have high number of nearest neighbor atoms, thus leading to very tight packing. When compared with geometrically packed phases such as *bcc*, *fcc* and *hcp*, TCP phases are characterized by different coordination polyhedra that can be thought of as suitable combinations of octahedral and tetrahedral voids (see Fig. 2.19) [5].

TCP phases can reach relatively high atomic packing, as well as rather complex geometries. This latter feature can be easily seen in Fig. 2.20 with reference to C36-Laves, μ and R phases. One of the key characteristics of TCP phases is the lack of active slip systems, which gives an explanation for their intrinsic brittleness that can lead to generalized decrease in the fracture toughness of the alloy in which they are present if the volume fraction exceeds a certain threshold value. As a result, a hypothetical fracture can nucleate at stress concentration points introduced by the TCP phase and propagate easily through the brittle intermetallic grains (reference is often made to *intersigmatic fracture*).

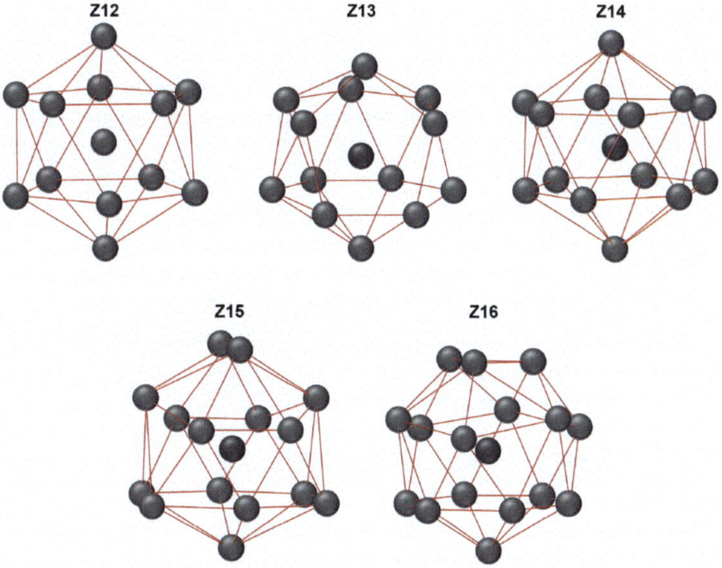


**Fig. 2.19** Frank-Kasper coordination polyhedra that characterize TCP phases (redrawn from [5])

**Fig. 2.20** The complex structures of the TCP phases arising from the combination of the coordination polyhedra represented in Fig. 2.19 and exemplified in the figure by C36-Laves phase, μ and R phase (redrawn from [5, 37])

As already mentioned, in addition to pure mechanical behavior there are other consequences in the formation of TCP phases. The richer composition in substitutional atoms such as Cr and Mo with respect to the matrix is achieved by diffusive flow when the material is exposed to relatively high temperatures. As seen in Fig. 2.18, this condition might occur not only during service, but also as a result of cooling from high-T plastic deformation processes, incorrect cooling after heat treatment, cooling from liquid, or as a consequence of welding processes. In addition to the formation, and subsequent presence of brittle phases in the matrix, one of the consequences of these heat-induced transformations is the matrix depletion near TCP phases of the phase-forming elements themselves. This results in a twofold problem in addition to the increased brittleness of the material. In fact, on one hand a reduction in solid solution hardening given by substitutional atoms subtracted from the matrix and a consequent general decrease in mechanical strength takes place. On the other hand, the lack of important elements such as Cr and Mo causes alterations in corrosion resistance with a worsening of this property [5]. In addition to Cr and Mo, σ phase formation is promoted by the presence of Mn, Nb, Si, Ta, Ti, and V. On the other hand, formation-inhibiting action is exerted by Al, C, Co, N and Ni [36]. Confirmation of this has been obtained by various studies, especially aimed at understanding the solidification behavior after welding processes [23]. For instance, the precipitation behavior of σ phase as a function of changes in Ni and Mn content was studied by Lee et al. [38] on SASS weld metal using an alloy of composition 23% Cr, 17% Ni, 6% Mo and 0.35% N as a base. Specifically, the composition of the samples was varied by increasing Ni by 2% and Mn by 4% for each of the trials. It was found that the formed σ phase precipitated directly from the liquid phase during solidification, which is consistent with the information given in Sect. 2.4.2 with reference to welding processes. As is clear to expect, both Ni and Mn variation influence the precipitation behavior of the σ phase in the weld metal. In particular, the increase in Ni leads to a Mo-richer σ phase, while on the other hand, the increase in Mn leads to a partial substitution of Mo for Mn in the σ phase [38]. Out of the elements just mentioned, both promoters and inhibitors; Al, Co, Si and Ti are not present in typical compositions of SASSs (see Table 2.2), and C is present in small amounts [36].

Beyond the more *traditional* alloying elements, as is often the case, the influence of other more *exotic* elements has also been evaluated for SASSs. For example, the use of rare earth elements (REs) to improve characteristics of steel is a common practice. It has been observed that the addition of small amounts of Ce to 254 SMO brings interesting modifications regarding the solidification sequence. Because of its low solubility in austenite, Ce tends to accumulate in the solute-rich interdendritic liquid during solidification. This has also an important effect in suppressing σ phase precipitation. Analysis of experimental data from observation of the solidification sequence of 254 SMO showed that an addition of 0.016% Ce leads to accelerated δ phase nucleation in the interdendritic liquid, which is solute rich. Not only that, Ce is also capable to refine the solidification microstructure and to reduce Mo and Cr segregation. This latter aspect becomes particularly interesting as it leads to a limitation or suppression of σ phase formation. Moving from σ to δ formation as secondary phase does not solve the problem of the presence of deleterious phases. However, it leads to interesting advantages for the production steps following solidification. In fact, the annealing treatment required to homogenize the structure and free it from δ is shorter and more resolving than that associated with the homogenization of a microstructure characterized by the presence of σ phase [39].

Concerning any external driving forces for phase precipitation that are not related to temperature increase, cold work anticipates the onset of σ phase precipitation, particularly in cases where the material is recrystallized after heat treatment [36]. The fact that σ is perhaps the most characteristic secondary phase of SASSs is demonstrated by the countless experimental investigations conducted on all members of this family where its presence is evidenced. As a result, not only the causes and conditions leading to its formation have been analyzed in the literature, but also the effects on mechanical properties and corrosion resistance. A study conducted by Cardoso et al. [40] on AL-6XN® PLUS involved exposing the material to 600 °C up to a maximum of 2000 h. AL-6XN® PLUS a version of AL-6XN® owing Cr, Mo and N contents at the top of the range given in Table 2.2. Specifically, the Cr content is close to 22%, the Mo content is close to 7%, or even higher depending on the manufacturer. Finally, the N content is close to 0.25%. The objective of the study was to evaluate the microstructural, mechanical strength and corrosion resistance changes of the alloy after the exposure to relatively high temperature for prolonged time. The temperature of 600 °C was chosen because many industrial components work near this value. Given the excellent corrosion resistance of SASSs, and the need to employ them in applications involving aggressive environment and relatively high temperatures, simulation of operating conditions is paramount to understand the material's own reaction to these types of situations. Regarding the microstructural characterization, after 500 h at 600 °C, the presence of coarse precipitates preferentially localized at the triple points of the grain structure was evidenced. These precipitates correspond, by chemical analysis performed through SEM–EDS to σ phase. For longer time at 600 °C, up to 2000 h, the precipitation became heavier with predominantly lamellar morphology. In some regions, a small tendency for the formation of discontinuously organized globular precipitates along the GB and also at triple points was detected. With regard to mechanical properties, measurement of Vickers hardness (HV30) on specimens treated between 0 and 2000 h did not show substantial changes and was characterized by high dispersion of the values, especially for prolonged exposure. From 0 to 500 h there was an increase in the hardness value with a peak around 500 °C, corresponding, however, to an increase of less than 10% of the initial value measured on the solution annealed material. In any case, the heterogeneity of precipitation caused high scattering of the values obtained from the tests, and thus it cannot be said that the results were neither absolute, nor conclusive. On the other hand, with regard to the toughness of the material, the treatment at 600 °C showed a more marked direct correlation than with hardness. Specifically, Charpy's test showed a decrease in toughness as a function of increasing aging time accompanied by a change in the fracture mode of the material. This phenomenon occurred as a result of the intergranular brittleness caused by the presence of the σ phase [40]. Further evidence of this was given by experimental analyses performed by Anburaj et al. [29] on specimens of a 7% Mo cast SASS aged above 900 °C. In fact, specimens were characterized by brittle fracture with the so-called ideal cleavage morphology. The presence of σ phase at the GB is therefore able to significantly reduce the cohesive strength of the material near the boundaries and promoted decohesive rupture. The fracture surface was characterized by facets that are quite distinctive and indicative of intergranular failure according to ideal cleavage. Furthermore, deep secondary cracks along GBs characterized the intergranular decohesion observed. Nevertheless, the intergranular facets were characterized by a certain roughening caused by σ phase in the γ matrix. It is important to note that a fracture is supposed to be decohesive when little or no plastic deformation occurs and it does not occur by splitting, dimple fracture or fatigue. The detected fracture morphology (ideal cleavage together with facets roughening) resulted from the unique microstructure developed after aging and is associated almost exclusively with fracture along GBs. The decohesion mechanism involved weakening of atomic bonds, reduction of the surface energy required for localized deformation, and breakdown of the protective film. All this can occur in the presence of incoherent precipitates such as σ phase. In fact, for samples aged at the typical σ precipitation temperature (850–1050 °C), the toughness value was very low and the failure occurred by decohesion [29].

As already reported, research effort in studying σ precipitation is useful to understand as many aspects of it as possible by not limiting the research to only one member of the SASSs group. In this regard, an interesting study was performed by Koutsoukis et al. [41] on samples of 254 SMO and 654 SMO in the form of hot rolled strips that were aged at temperatures of 650, 750, 850 and 950 °C. For each temperature, the exposure time was 1, 24, 120, 240, 500 and 1000 h. Regarding the

950 °C treatment, for exposure time over 120 h complete oxidation of the sample was detected due to the fact that the experiment was performed in a reheating equipment with oxidizing atmosphere (see also Sect. 3.3.10). As a result of this, the exposure time was corrected to 48 and 96 h for this temperature. The aging process was followed by water quenching in order to freeze the structure obtained after exposure to high temperature. Experimental results showed rapid σ phase precipitation for higher temperatures, i.e., 850 °C after 24 h and 950 °C after only 1 h. Precipitation also occurred at 650 and 750 °C, for longer exposure when considering the most heavily-alloyed 654 SMO and slightly earlier in 254 SMO. For the two most elevated temperatures (i.e., 850 and 950 °C), it was observed that precipitation occurs both at the GB and intragranularly. Consequently, it can be derived that enough driving force exists for nucleation and growth of σ. In contrast, at the two lower temperatures, intergranular precipitation of σ takes place predominantly. Well-dispersed precipitates with size up to ~10 μm were detected in both steels under study. It has been shown that σ phase precipitation occurs early at high temperatures, whereas it is rather delayed at 650 °C. This highlights and confirms faster precipitation kinetics as the treatment temperature increases. In addition to σ phase, a further Mo-rich intermetallic was detected in the samples of both steels. Since no TEM characterization was employed in this study, according to the morphology, this phase was assumed to be χ phase. The largest volume fraction was detected at 850 °C, especially in 254 SMO and with an elongated longitudinal-type morphology and size up to 5 μm [41].

#### 2.4.6.3 χ Phase

The χ phase is also a member of the TCP family and was first identified by K. W. Andrews in a Cr–Mo–Ni steel. The structure that had been assumed to be associated with the compound identified by Andrews was α-Mn type cubic. Subsequently, J. S. Kasper studied in detail the crystal structure of the χ phase after having synthesized it with composition $Fe_{36}Cr_{12}Mo_{10}$. Later, it has been shown that the χ phase can exist with a fairly wide range of stoichiometries, from $Fe_{36}Cr_{12}Mo_{10}$ to $Fe_{36}Cr_{12}Mo_3Ti_7$, also including $Fe_{35}Ni_3Cr_{13}Ti_7$. The compound can therefore change from tertiary to quaternary, thus leading to an increase in the lattice parameter. Generally, the χ phase forms for values of electron/atom ratio ranging between 6.3 and 7.8. The precipitation range and kinetics are different from those of the σ phase and depend greatly on the Mo content of the steel (generally, the precipitation temperature range is 550–900 °C) [36]. As pointed out above, chemical composition resembles somehow that of σ, but unlike the latter, which is unable to do so, the χ phase can dissolve C within itself. In fact, precisely because of this attitude the χ phase was initially classified as an $M_{18}C$-type carbide. Furthermore, due to its ability to dissolve C and its ease of nucleation, the precipitation of χ phase occurs before that of σ and might be dependent/influential on it [36]. In fact, within the literature there are several ambiguities regarding σ and χ phase formation in the austenitic matrix. Furthermore, the formation temperature data are often overlapping and, consequently, also ambiguous. It is usually assumed that at a certain temperature (σ transus, above 800 °C), the χ phase becomes unstable and fully converts into σ due to the diffusion of transition elements such as Cr, Fe and Mo [42].

Specific studies in this regard have been performed on various members of the SASSs family, including 654 SMO. One study performed by Song and Pu [43] involved heat treatment of specimens of this grade obtained by high-T plastic deformation processes (i.e., forging and subsequent rolling). The material was treated in a temperature range between 800 and 1100 °C for 1 h and then water quenched. Among the precipitates detected in the material, in addition to σ, χ was also present. In the specific case at hand, the χ phase had similar size and morphology to σ, but its volume fraction was smaller. Because of these features, discrimination between the two by transmission microscopy (TEM) was difficult. Moreover, the link with σ according to which the two can mutually transform during the aging process was also highlighted. In the early stage of aging, precipitation of σ (Cr-rich) leads to the formation of a Cr-depleted zone near σ itself. This phenomenon would create favorable conditions for the formation of χ phase. Prolonged exposure to aging temperature would lead χ phase to gradually transform into σ, and for this reason χ tends to be regarded as metastable when detected within 654 SMO [43]. The presence of χ phase was also detected in another study performed by Anburaj et al. [42] in 654 SMO exposed to increasing temperatures between 500 and 1000 °C with steps of 100 °C, for exposure time ranging between 1 and 10 h. After 1 h exposure at 600 °C only χ phase was found to nucleate along grain boundaries and to grow with increasing aging time. χ phase becomes continuous along the GBs at this temperature for a soaking time of 10 h. Exposure of the material to 800 °C was found to lead to the formation of χ phase within the γ matrix regardless of exposure time. Simultaneously, σ phase started to form and its precipitation was enhanced as the aging temperature increased. In addition, experimental and literature evidence seem to confirm how the formation of χ phase would be much faster during heating than upon cooling. The reason should lie in the fact that the driving force for nucleation is greater at low temperatures, while diffusion is faster at higher temperatures. At 1000 °C, both phases have been detected after 1 h of exposure with predominance of χ. After 10 h the situation was markedly changed with complete conversion of χ to σ. This seems to confirm that the χ phase can be considered a precursor to σ phase. It would also appear to be confirmed that the rate of nucleation and growth of σ, as well the dissolution of χ, are greater for increasing high-T exposure time. For what concerns the χ phase, however, microstructural evidence suggests that it forms simply at the austenitic GBs. This process, in turn, promotes diffusion of Cr to the χ phase and then the diffusion of Mo into the same phase. As this sequence proceeds, a situation occurs in which the χ phase becomes enriched in both Cr and Mo. At this point, the σ phase begins its formation by nucleating and growing on the Mo-enriched portion of χ. σ phase promotes Mo diffusion and thus σ is more stable in the later stages of this process sequence [42]. Again in 654 SMO, the χ phase can also form for different conditions such as isothermal treatment at 1050 °C for 5 min and in this particular case χ is shown to have size between 200 and 500 nm with faceted shapes and formed at some GBs [44].

Generally, the χ phase forms first at GBs, then at incoherent twin boundaries, coherent twin boundaries and, at the end, within the grains. Similarly to σ, cold working operations performed on the material appear to promote χ phase precipitation with accelerated kinetics [36]. χ phase is characterized by less research effort and less dedicated literature than σ. As a consequence of this, less information is available on it in comparison with the other TCP phase mentioned. Nevertheless, its presence is still regarded as deleterious to the material for what concerns mechanical properties and corrosion resistance [36, 42]. Once again in the field of correlation between precipitation of secondary phases and change in mechanical properties, studies have also been carried out on noncommercial grades that by composition can be categorized in the SASSs family. Experimental analyses were performed by Anburaj et al. [29] on a material compatible with a 7% Mo cast SASS (already cited in the previous section), first cast and then hot forged at 1200 °C. The experimental alloy had composition corresponding to 19% Cr, 20% Ni, 7.5% Mo and 0.37% N. With reference to Table 2.2, the experimental alloy has such a composition which places it in the high-N superaustenitics group. The investigation was carried out on 1250 °C solution annealed material that was subsequently heat-treated for 1 h or for 10 h in a temperature range between 500 and 1000 °C. Finally, a correlation was proposed between the results of Charpy impact test, precipitation of secondary phases and aging treatment. Aged samples showed microstructures consisting of annealing twins within austenitic grains and secondary phases such as σ, χ and carbides. The maximum amount of secondary phases, regardless of their nature, was around 20 vol% and was observed for samples treated at 800 °C for 10 h. Concerning the χ phase, the specific results pointed out that its formation starts at 500 °C. An increase in its volumetric fraction was detected up to 800 °C with a subsequent reduction beyond this temperature. The morphology associated with χ appeared to be rod-like and was extended along the austenitic GB. Regarding the hardness change, all samples aged for 10 h showed a higher value than those aged for 1 h for all the temperature ranges. At relatively low temperatures, an abrupt decrease in hardness was observed due to stress relaxation. In addition, a hardness peak was observed for the 850–900 °C range presumably due to the volume fraction value of secondary phases formed. The hardness value associated to the peak was about 1.8 times that associated with the solution-annealed material. Concerning the toughness of the material, the trend followed was that of a decreasing value as the aging time increased for all temperatures with a minimum value found in the same range where the peak hardness was obtained (i.e., 850–900 °C). In the χ phase formation range, the toughness value did not fall below half the initial value, and the fracture morphology remained ductile. Hence, it seemed that χ phase is able to reduce the impact energy without changing the fracture morphology [29].

#### 2.4.6.4 R Phase

The R phase is a TCP phase that compared to σ, χ and Laves phases occurs less frequently in stainless steels, similarly to μ and G (G, specifically, does not even form in SASSs since it is a silicide). Furthermore, in some grades such as 654 SMO, R phase is considered as a metastable phase. R phase essentially contains Fe, Cr and Mo, is characterized lower Mo and higher Cr content with respect to Laves phases and by a complex hexagonal crystal structure [36, 43, 44]. In SASSs it can be found in the HAZ (heat-affected zone) in regions that have reached temperatures above 800 °C [36]. Evidence of the presence of R phase was found in research published by Heino et al. [44] on a steel of composition that could be ascribed to 654 SMO with isothermal treatments at 1050 °C for 5 min in Ar atmosphere and subsequent water quenching. In this context, it is also important to declare the previous manufacturing history of the material. In fact, the alloy was produced by ESR (electroslag remelting) process in order to limit segregation, then hot-rolled, cold-rolled and annealed at 1200 °C achieving a grain size of 80 $\mu$m. The material in this condition was found to be free from precipitates both at the GB and intragranularly. The experimental conditions served to partly simulate the results obtainable in the HAZ after welding process. In addition to $\sigma$, $\chi$, Laves phases, rounded shape R phase was also found with a size between 50 and 200 nm at GBs [44].

Further evidence in the literature confirms the possible presence of R in the HAZ of 654 SMO. In this regard, characterization of the intermetallic phases in the HAZ of another SASS with composition matching that of 654 SMO carried out by Heino and Karlsson [45] after welding in order to verify crystal structures, compositions and particle morphology also gave confirmation. Furthermore, the characterization was performed together with the correlation of precipitation with the HAZ's own thermal history after the welding process. Results showed two precipitates: $\sigma$ and R phase. $\sigma$ phase was detected to have a composition 30Cr–32Fe–27Mo–8.3Ni–2.2Mn–1.1Si. R phase was detected to have a composition 45Mo–25Fe–6.4Ni–2.6Si–1.4Mn. Their distribution was uniform along the GBs and their shape was plate-like. The average particle size of $\sigma$ has proved to be higher than that of R, specifically 200–900 nm in the former case and 30–300 nm in the latter. Moreover, near the fusion zone there was coexistence of $\sigma$ and R, while at greater distances from the melting zone only the presence of R was detected indicating the lower temperatures reached (which are, precisely, compatible with the formation of this phase alone). High distortion was detected for small precipitates, both $\sigma$ and R, and this might be ascribed to come from the driving force associated with lattice strain minimization in the early stages of growth. Regarding the R phase, the average particle size of this phase has been shown to be generally smaller than $\sigma$. This is believed to depend on diffusion issues as the higher Mo content of this phase requires more time and a larger adjacent region of the matrix to be depleted in Mo for its formation. In fact, both precipitates are rich in Mo, but the R phase turns out to possess almost twice the Mo content of the $\sigma$ phase. Consequently, the growth rate in a short time such as that associated with weld joint solidification is very limited. In addition to this, the lower formation temperature of R phase, also with respect to $\sigma$, also contributes to limit diffusion and consequently the growth rate [45].

#### 2.4.6.5 Laves Phases

Laves phases were first studied in the 1930s by Fritz Henning Emil Paul Berndt Laves. Laves phases are compounds that should have been characterized by an ideal atomic radius ratio $r_B/r_A$ of 1.225. Nevertheless, values between 1.10 and 1.46 can be detected. Moreover, it is certain that electronic factors, in addition to atomic factors, also have an influence on the formation of Laves phases. In fact, for the Fe-Mo system, the atomic radius ratio is close to 1.10. However, despite this, $Fe_2Mo$ still forms, thus suggesting that the ratio is not the exclusive indicator for the possibility of formation. Laves phases usually form for electron/atom ratios between 6.33 and 8.0. Laves phases are TCPs and three types of them do exist, all characterized by $A_2B$ stoichiometry: C14 ($MgZn_2$ type), C15 ($Cu_2Mg$ type) and C36 ($MgNi_2$ type, see Fig. 2.20). The Laves phases most frequently found in austenitic stainless steels, and thus also in SASSs, have *hcp* structure and are of the $MgZn_2$ type. The most common stoichiometries are $Fe_2Mo$, $Fe_2Nb$ and $Fe_2Ti$, but combinations of the forming elements such as $Fe_2(Mo, Nb, Ti)$ are also possible. As for SASSs, the typical chemistry is the former ($Fe_2Mo$), while the latter containing Ti cannot form because this metal is not an alloying element in these steels (see Table 2.2). Generally, Laves phases are stable below 815 °C, but this is highly dependent on the amount of the forming alloying elements in the chemistry of the specific alloy [36]. The Laves phase in an alloy such as 654 SMO has $Fe_2Mo$ stoichiometry and is characterized by hexagonal structure. Similar to the $\chi$ phase, the Laves phase is also considered metastable in 654 SMO as it can convert to $\sigma$ or dissolve in the austenite matrix. However, it seems that the dissolution process is the most likely one because the interface energy between the Laves phase and $\sigma$ is higher than between Laves phase and $\gamma$, and this would make the dissolution easier than the conversion process [43].

Although Laves phases are related to Mo in that they contain it, literature data reported that high W content in the steel can promote the formation of Laves phases too. In fact, regarding the high-Mo SASSs, the synergistic effect W together with Cr in increasing the solid-state solubility of N has been exploited for the development of the high-N UNS S31266. The effect of the addition of W to high-N SASSs was experimentally studied by Heino [46] on an alloy with composition resembling that of 654 SMO from which two variants were derived with incremental W content, 2% and 5%, respectively. More in details, aging treatment was carried out at a temperature of 900 °C with exposure time of 5 min or 14 h. In particular, the longer aging time (i.e., 14 h) was employed in order to allow the distinction between transient or metastable secondary phases and more stable ones. Regarding short exposure time, the first compounds formed were intermetallics located at the GBs. As these compounds grew at the GBs, the surrounding $\gamma$ matrix became depleted in Mo and W, since these intermetallics are rich in the two metals. C14 Laves phase was detected in all the three steels, but it was observed that in the W-free alloy the Laves phase resulted to be transient in that for longer exposure it disappeared. On the other hand, for the increasing W compositions, the Laves phase appeared to be stable regardless of time for the conditions investigated. An increasing W to Mo ratio promotes the formation of Laves phases characterized by high Mo + W content. Consequently, the effect of local depletion of important alloying elements seems to be more pronounced in the case where W to Mo ratio was higher. This latter fact underlines the importance derived from knowing the specific conditions of formation of these phases since the limiting factor in corrosion resistance performance of these alloys is almost always the formation of secondary phases. From this factor comes the interest in evaluating the consequences of the presence and possible variation in concentration of W within high-performance alloys [46].

Experimental results indicate that 654 SMO, and possibly UNS S31266, are not the only members of the family of SASSs where Laves phases can be detected under certain conditions. In this regard, Laves phases have been shown to be detectable both in AL-6XN® and 254 SMO when experimentally treated at 700 °C, 800 °C and 900 °C with exposure time between 1 min and 2040 h. Concerning 254 SMO, precipitation of intermetallics was detected already at 700 °C after 1 min. However, the highest volume fraction for Laves phases was detected at 900 °C after 500 h. The Laves phase identified within 254 SMO was rich in Mo. In fact, SEM–EDS analysis showed a composition featuring about 50% Mo. From Fig. 2.21a it can be seen that the volume fraction of Laves phase tends to decrease for both 700 °C, 800 °C and 900 °C treated material after 100 h, 250 h and 500 h, respectively. It is interesting to note that the decrease in the Laves phase volume fraction occurs almost in conjunction with the increase in the volume fraction of the σ phase, which is also Mo-rich with a percentage of this alloying element for the steel analyzed around 12%. Furthermore, it was seen that the Laves phases precipitate tend to assume a more spherical shape as exposure time increased [47].

Differently from 254 SMO, the sample of AL-6XN® has been shown to have a significant macrosegregative level corresponding to the original dendritic cast structure (the macrosegregation of Mo observed in AL-6XN® could have been avoided if the as-cast material had been subjected to solution annealing temperature). This fact strongly affected the accuracy of the obtained data leading to a strong dispersion

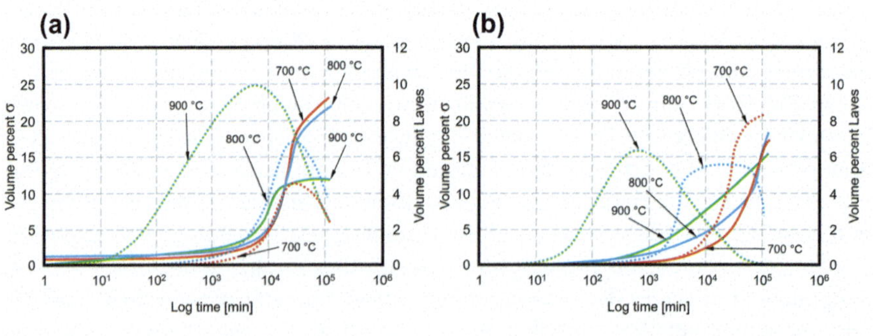


**Fig. 2.21** **a** Experimental Laves volume percentage in 254 SMO (dotted line) and superimposition with that of σ (continuous line) as a function of the treatment temperature represented as best-fit curves; **b** experimental Laves volume percentage in AL-6XN® (dotted line) and superimposition with that of σ (continuous line) as a function of the treatment temperature represented as best-fit curves (redrawn from [47])

which, as a result, influenced the trend obtained from the best-fit curves. Metallographic characterization showed that the interdendritic region (higher Mo content) was observed to be the most favorable for Laves phase nucleation, where the phase actually nucleated. On the other side, the intradendritic region with lower Mo content was the one chosen for σ nucleation. Regarding the trend of the volume fraction of Laves phase, at both 800 and 900 °C there was a decrease after a time that was shorter as the temperature became higher (see Fig. 2.21b for reference). The only exception was for the temperature of 700 °C where there no decreasing trend was observed, probably because the temperature was too low and did not allow enough energy to accumulate to reach dissolution within the time window available [47].

#### 2.4.6.6 Other Less Common Intermetallics

One of the less common deleterious intermetallic phases that can form in SASSs is the μ phase. Often, in the literature, the μ phase is classified as, or grouped into, the Laves phases. This happens because it is considered to belong to the group of the Laves-related phases [46]. The μ phase is a TCP that exists in ternary systems like Fe–Cr–Mo and Fe–Cr–W. In the Fe-Mo system the stoichiometry of μ is $Fe_7Mo_6$, while in ternary systems is $(Cr, Fe)_7(Mo, W)_2(Fe, Mo, W)_4$ [36]. Its existence has been reported for prolonged high-T exposure of alloys that by their nature are close relatives of SASSs: Ni-based alloys. Specifically, this phase has been detected in this family of alloys when their chemical composition is characterized by a significant amount of elements such as Mo, Nb, Re, Ru and Nb [43]. For what concerns SASSs, evidence of the presence of the μ phase was obtained in W-containing SASSs (e.g., UNS S31266) for prolonged exposure at 900 °C [46]. Likewise, in the case of 654 SMO exposed to similar temperature (i.e., 850 °C) for 1 h, the presence of coarse μ phase with irregular morphology at GB's triple points, as well as continuously distributed carbides along the GB, was detected. In light of the chemical composition of 654 SMO (see Table 2.2), the presence of the μ phase can be justified by the high Mo content in simultaneous presence with the other mandatory forming elements like Cr, Fe and Ni. Furthermore, the formation of this phase is closely related to the formation of another compound: $M_6C$ carbide. This carbide can act as a preferential nucleation site for the μ phase because it markedly reduces the energy required for the formation of the new phase. In case of similarities between the existing and the new-forming phase (e.g., chemical composition or crystallographic features), the associated formation energy can be further reduced leading to greater ease of formation. The fact that the μ phase is quite uncommon under non-experimental conditions is justified in the literature by the fact that this phase, along with others such as the R phase, is classified as metastable [43].

#### 2.4.6.7 Carbides

For exposure temperatures ranging from 700 to 1100 °C the first compounds to precipitate are carbides, and for longer time substantial carbide replacement by intermetallics takes place [47]. As reported in Sect. 1.2.8.1, carbides are compounds of C and at least one metal, and their presence is intrinsically linked to the presence of C in steels. Out of the five types of carbides that, as reported in Chap. 1, can form in stainless steels, literature and experimental data seem to suggest that only two are usually detected in SASSs: $M_{23}C_6$ and $M_6C$ [36]. Nevertheless, it seems that for specific temperature and exposure conditions like a temperature of 850 °C for 1 h, the formation continuously distributed orthorhombic $M_3C$ carbide along GBs in 654 SMO is also possible. The precipitation of this carbide in 654 SMO might be caused by the segregation of C and as reported by Padilha and Rios [36]. $M_3C$ is usually not found in austenitic stainless steels, therefore its presence is said to be rare in these alloys [43]. $M_{23}C_6$ carbide is typical of virtually all stainless steels and is usually the first phase formed under the conditions described in Fig. 2.18. Depending on the chemical composition, non-negligible amounts of carbides can form as early as after a short time at temperatures between 650 and 750 °C; however, the presence of $\chi$ phase and $M_{23}C_6$ carbide was detected along GBs in the form of irregular blocks and approximate disks also in 654 SMO aged at 900 °C [36, 43]. In the past, $M_{23}C_6$ was believed to have stoichiometry $M_4C$, where M is Cr. In 1933 A. Westgren demonstrated that $M_{23}C_6$ was the correct chemistry of the compound he was studying. $M_{23}C_6$ carbide has a *fcc*-type structure and typically Cr as the metallic element. Fe and Mo can replace Cr, while B and N can partially replace C. The lattice parameter of this carbide is three times that of austenite, which produces a characteristic diffraction pattern during TEM analysis. The lattice parameter increases in case Mo is present in the compound, while decreases in case Fe is present. The thermal history of the material in which the carbide is contained appears to have strong influence in the composition of $M_{23}C_6$. In fact, it has been seen that in experimentally aged Mo-containing stainless steels, the lattice parameter of the detected carbides increases with treatment time, likely indicating the increase of Mo in the carbide itself. $M_{23}C_6$ carbide nucleation can occur at different sites: GBs, incoherent twin boundaries, coherent twin boundaries and dislocations within the grains. The most favorable nucleation site is the GB, followed by all others in the order in which they were mentioned [36].

High-T solution annealing treatments performed on the material can accelerate the beginning of precipitation. Cold working operations performed after solution annealing can accelerate precipitation in case the material is subsequently exposed to temperatures compatible with the phenomenon. $M_{23}C_6$ carbide is generally undesirable in steels as it is associated with decreased ductility and toughness, as well as reduced corrosion resistance due to the promotion of sensitization phenomena. However, it is also reported in the literature that $M_{23}C_6$ carbide presence may bring beneficial effects. This would be due to the possibility for the carbide to limit GB sliding, with a consequent positive effect on creep resistance properties [36].

On the other hand, $M_6C$ carbide has a diamond-type *fcc* structure, it is also known as η carbide and is typical of Mo-, Nb-, and W-containing stainless steels; particularly Mo-rich ones. Metal M can be Cr, Fe, Mo, Nb, W or V. As for SASSs, the group is narrowed by including only Cr, Fe and Mo and possibly Nb and W for grades containing them (see Table 2.2). In addition to this, it has been shown that $M_6C$ carbide inevitably contains at least two metallic elements, implying the need for the presence of at least three chemical spices, one of which must obviously be C. This leads the carbide to be represented by the formulas $A_3B_3C$ or $A_4B_2C$. High N content in the steel is supposed to promote the formation of $M_6C$ carbide at the expense of $M_{23}C_6$ because the former would appear to be able to dissolve larger amounts of the interstitial element than the latter. In addition, N would replace some of the C in the lattice making $M_6C$ more stable. Generally, there is little literature related to $M_6C$ carbide because the attention on this compound has never been high due to the fact that in the "traditional" stainless steels it is not much detected. The advent of SASSs has brought more attention and thus also increased investigation of the carbide itself. In fact, one of the pieces of evidence found in the literature and related to $M_6C$ is that it can be a suitable nucleation site for the μ phase due some similarities between the two [36, 43].

#### 2.4.6.8 Nitrides

As explained in Sect. 1.2.8.3, nitrides are N compounds where the element is found in in the form of the $N^{3-}$ ion. They feature high hardness, chemical inertia and several similarities with borides and carbides. Except for the classification in four types already given in Chap. 1, nitrides that usually form in austenitic stainless steels are of two types: primary and secondary. The first group is of the MN type, where M can be Nb, Ti, V or Zr. Primary nitrides are usually found in stabilized grades containing residual amounts of N, typically less than 0.1%. MN nitrides are characterized by same crystalline structure as MC carbides, but they feature a higher stability that leads to impossibility of their dissolution during solution annealing treatments [36]. According to Table 2.2, the only SASS that could form primary MN nitrides due to its chemical composition would be Nirosta® 4565S, as it contains high amount of N as well as Nb up to 0.15 wt%. However, there appears to be no evidence of primary nitride formation in SASSs in the literature to date. Secondary nitrides, on the other hand, are usually of the $M_2N$ type with M being either Cr or Fe. They can precipitate in grades containing high amounts of N, such as SASSs [36].

The presence of nitrides should be explained in conjunction with the description of N behavior in alloys containing high amount of it. First of all, high N content could delay carbide formation because the diffusion coefficient of N in austenite is higher than that of C. In addition, N segregation occurs preferentially at defects or discontinuities such as GBs during aging. An additional action of high N contents is the reduction in the diffusion coefficient of Cr, thus hindering the formation of carbides based on this element. For the above reasons, the nucleation of nitrides is favored in comparison to that of carbides, and this in any case in addition to the difficulty of carbides formation resulting from the very low %C featured by SASSs. It is worth noting, however, that an increase in nitride precipitation reduces the level of N saturation in the matrix, gradually making the conditions described above lack and increasing the driving force for carbide formation [43]. As for the secondary nitride $Cr_2N$, its formation in SASSs tends to be related to prolonged exposure time at relatively high temperature. For example, evidence of its presence has been found in both 254 SMO and 654 SMO with exposure time up to 3000 h at temperatures in the 650–950 °C. It is interesting to note that there is a peculiarity of formation behavior. In fact, this nitride always tends to nucleate adjacent to the Laves phases in the austenitic matrix since nitride nucleation is promoted by Mo depletion as a consequence of the formation of Laves phases [48].

The π phase is another nitride with primitive cubic structure that can be detected in SASSs. It is stable in the ternary Cr–Ni–N and Cr–Mo–N system, but also in the quaternary Cr–Fe–Ni–N. Its typical chemistry, as stated in Appendix B, is $Mo_{13}Fe_7N_4$. π phase is also detected in duplex stainless steels [49]. Evidence of the presence of the π phase has been experimentally reported for a steel with composition similar to that of 654 SMO, from which two additional alloys with varied Mo and W content were obtained. An experimental alloy called 6Mo-2W and 5Mo-5W, respectively. These batch steels were developed in order to resemble steels such as 654 SMO and W-containing SASSs such as UNS S31266 (see Table 2.2 for the chemical composition of 654 SMO and UNS S31266). The experimental alloys were subjected to 900 °C exposure for increasing time. Specifically, for exposure time between a minimum of 5 min and a maximum of 14 h, the π phase was detected for exposure of 14 h. The precipitation has been shown to take place in areas next to previously-formed Mo-rich secondary phases (σ, χ, Laves-related phases). This phenomenon would find justification in the depleted austenite that has lost Mo due to the formation of precipitates rich in this element. As exposure time increases, a decrease in the Mo and W content in all secondary phases that are formed is detected. This can be likely explained by the decreased availability of both elements in the surrounding γ matrix as precipitates grow. This sequence of events seems to be the basis for justifying the possibility for Cr-rich phases to form. In particular, $Cr_2N$ and π phase are allowed to become stable under the conditions just mentioned [46]. In addition to relatively high temperatures such as 900 °C, the π phase has also been detected at 700 °C in steels such as 254 SMO and 654 SMO for exposure time of more than 6000 h [50].

## References


1. Kovach, C.W.: High-Performance Stainless Steels. The Nickel Institute (2000)
2. Schweitzer, P.A.: Fundamentals of Metallic Corrosion—Atmospheric and Media Corrosion of Metals. CRC Press (2006)
3. Burt, V.: Corrosion in the Petrochemical Industry. ASM International, Materials Park, Ohio (2015)

4. Farrar, J.C.M.: The Alloy Tree: A Guide to Low-Alloy Steels, Stainless Steels and Nickel-Base Alloys. Woodhead Publishing Limited (2004)
5. Gialanella, S., Malandruccolo, A.: Aerospace Alloys. Springer (2020)
6. Davidson, R.M., Redmond, J.D.: Practical guide to using 6 Mo austenitic stainless steels. Materials Performance December, 39–43 (1988)
7. Kutz, M.: Handbook of Materials Selection. John Wiley & Sons, Inc (2002)
8. Haidemenopoulos, G.N.: Physical Metallurgy—Principles and Design. CRC Press (2018)
9. Capelli, F.: L'Acciaio INOX. Centro Inox Servizi S.r.l. (2022)
10. ThyssenKrupp Nirosta: NIROSTA® 4565 (UNS S 34565) Nitrogen-alloyed austenitic stainless steel with maximum corrosion resistance and high strength. ThyssenKrupp Nirosta GmbH, Krefeld, Germany (2002)
11. Heubner, U., et al.: Alloy 31, A new high-alloyed nickel-chromium-molybdenum steel for the refinery industry and related applications. Paper presented at the National Association of Corrosion Engineers (NACE) international annual conference and corrosion show, Orlando (FL), 26–31 March 1991
12. Jasner, M., Heubner, U.: Alloy 31, a new 6 moly stainless steel with improved corrosion resistance in seawater. Paper presented at the National Association of Corrosion Engineers (NACE) international annual conference and corrosion show, Orlando (FL), 26–31 March 1995
13. Redmond, J.D., Davison, R.M.: The Evolution of High Performance Stainless Steels. International Molybdenum Association (IMOA), London, UK (1993)
14. Wallén, B., et al.: AVESTA 654 SMO™—a new nitrogen-enhanced superaustenitic stainless steel. Mater. Corros. **44**(3), 83–88 (1993)
15. Hao, Y., et al.: Solidification structures of Fe–Cr–Ni–Mo–N super-austenitic stainless steel processed by twin-roll strip casting and ingot casting and their segregation evolution behaviors. ISIJ Int. **58**(10), 1801–1810 (2018)
16. Hosford, W.F.: Physical Metallurgy, 2nd edn. CRC Press (2010)
17. Abbaschian, R., et al.: Physical Metallurgy Principles, 4th ed. Cengage Learning (2008)
18. Cahn, R.W., Haasen, P.: Physical Metallurgy Volume I, 4th ed. North Holland (1996)
19. Outokumpu: Handbook of Stainless Steels. Outokumpu Oyj, Espoo, Finland (2013)
20. Marin, R., et al.: Solidification path and phase transformation in super-austenitic stainless steel UNS S31254. IOP Conf. Ser.: Mater. Sci. Eng. **529**(1), 012008 (2019)
21. Marin, R., et al.: $\sigma$-phase formation in super austenitic stainless steel during directional solidification and subsequent phase transformations. Metall. Mater. Trans. A **51**(7), 3526–3534 (2020)
22. Hao, Y., et al.: Influences of cooling rates on solidification and segregation characteristics of Fe-Cr-Ni-Mo-N super austenitic stainless steel. J. Mater. Process. Technol. **275**, 116326 (2020)
23. Perricone, M.J., et al.: Effect of composition on the formation of sigma during single-pass welding of Mo-bearing stainless steel. Metall. Mater. Trans. A **38A**(9), 1976–1990 (2007)
24. Perricone, M.J., Dupont, J.: Effect of composition on the solidification behavior of several Ni-Cr-Mo and Fe-Ni-Cr-Mo alloys. Metall. Mater. Trans. A **37**(4), 1267–1280 (2006)
25. Wang, T., et al.: Detailed analysis of nucleation and subsequent $\delta$ to $\gamma$ phase transformation of a high N super austenitic stainless steel. Mater. Charact. **195**, 112485 (2023)
26. Fu, W.J., et al.: Formation of a two-phase microstructure in Fe-Cr-Ni alloy during directional solidification. J. Cryst. Growth **311**, 132–136 (2008)
27. Martorano, M., et al.: Predicting delta ferrite content in stainless steel castings. ISIJ Int. **52**(6), 1054–1065 (2012)
28. Weiping, G., Zhanpeng, J.: Phase equilibria in the Mo-Ni-Co and Mo-Ni-Fe systems at 1000 °C. J. Less Common Metals **160**(1), 29–33 (1990)
29. Anburaj, J., et al.: Ageing of forged superaustenitic stainless steel: precipitate phases and mechanical properties. Mater. Sci. Eng., A **535**, 99–107 (2012)
30. Outokumpu: Ultra range datasheet. Outokumpu. https://otke-cdn.outokumpu.com/-/media/files/products/ultra/outokumpu_ultra_range_datasheet.pdf?revision=bd68fc24-324f-445c-b36a-51f82c60647e&modified=20221018121314. Accessed 11 November 2023

31. Koutsoukis, T., et al.: Precipitation sequence in cold deformed superaustenitic stainless steels. Mater. Sci. Technol. **26**(9), 1041–1048 (2010)
32. Totten, G. E.: Steel Heat Treatment—Metallurgy and Technologies, 2nd ed. CRC Press (2007)
33. Plaut, L.P., et al.: A short review on wrought austenitic stainless steels at high temperatures: processing, microstructure: properties and performance. Mater. Res. **10**(4), 453–460 (2007)
34. Hull, F.C.: Delta ferrite and martensite formation in stainless steels. Weld. J. **52**, 193s–203s (1973)
35. Kokawa, H., et al.: Crystallographic characteristics of delta-ferrite transformations in a 304L weld metal at elevated temperatures. Weld. J. **68**(1–6), 92–101 (1989)
36. Padilha, A.F., Rios, P.R.: Decomposition of austenite in austenitic stainless steels. ISIJ Int. **42**(4), 325–337 (2002)
37. Barrett, C.S., Massalski, T.: Structure of Metals, 3rd ed. Pergamon Press (1980)
38. Lee, C., et al.: Precipitation behavior of the sigma phase with Ni and Mn content variations in superaustenitic stainless steel weld metal. Mater. Charact. **144**, 148–154 (2018)
39. Wang, Q., et al.: The influence of Ce Micro-alloying on the precipitation of intermetallic sigma phase during solidification of superaustenitic stainless steels. J. Alloy. Compd. **815**, 152418 (2020)
40. Cardoso, A.S.M., et al.: Thermal aging effects on mechanical and intergranular corrosion resistance of super-austenitic stainless steel exposed at 600 °C. Mater. Res. **25**(suppl.), e20220079 (2022)
41. Koutsoukis, T., et al.: Comparative study of precipitation effects during aging in superaustenitic and superferritic stainless steels. Mater. Sci. Technol. **27**(5), 943–950 (2011)
42. Anburaj, J., et al.: Quantitative analysis of intermetallic precipitates in high Mo superaustenitic stainless steel. Appl. Mech. Mater. **592–594**, 711–715 (2014)
43. Song, Z., Pu, E.: Precipitated phases of superaustenitic stainless steel 654SMO. Journal of Iron and Steel Research **24**, 743–749 (2017)
44. Heino, S., et al.: Precipitation in a high nitrogen superaustenitic stainless steel. Mater. Sci. Forum **318–320**, 143–150 (1999)
45. Heino, S., Karlsson, B.: Precipitation behavior in heat affected zone of welded superaustenitic stainless steel. Mater. Sci. Technol. **15**(1), 101–108 (1999)
46. Heino, S.: Role of Mo and W during sensitization of superaustenitic stainless steel—crystallography and composition of precipitates. Metall. Mater. Trans. A **31**(8), 1893–1905 (2000)
47. Phillips, N.S.L., et al.: Phase transformations in cast superaustenitic stainless steels. J. Mater. Eng. Perform. **18**(9), 1285–1293 (2009)
48. Koutsoukis, T., et al.: Phase transformations and mechanical properties in heat treated superaustenitic stainless steels. Mater. Sci. Eng., A **561**, 477–485 (2013)
49. Knyazeva, M., Pohl, M.: Duplex steels. Part II: Carbides and nitrides. Metall. Microstruct. Anal. **2**(5), 343–351 (2013)
50. Svoboda, M., et al.: Phase changes in superaustenitic steels after long-term annealing. Z. Met. **95**(11), 1025–1030 (2004)

## *Further Readings*

ASM International: ASM Specialty Handbook: Stainless Steels. ASM International, Materials Park, Ohio (1994)

Campbell, F.: Elements of Metallurgy and Engineering Alloys. ASM International, Materials Park, Ohio (2008)

Cefalu, S., Krane, M.: Comparison of predictions of microsegregation in the Ni-Cr-Mo system to experimental measurements. Mater. Sci. Eng., A **454–455**, 371–378 (2007)

Mitchell, B.S.: An Introduction to Materials Engineering and Science for Chemical and Materials Engineers. Wiley Interscience (2004)

# Chapter 3
# Properties of Superaustenitic Stainless Steels

**Abstract** Materials are characterized by various properties, including physical, mechanical, and corrosion resistance-related ones. In-depth knowledge of superaustenitic stainless steels also comes through analysis of such characteristics. Material properties represent a key information for designers because in the design process, the product features, its use, and its expected useful lifetime are affected by material attitudes. The characteristics of the material certainly influence its usability, but the reverse is also true in that constraints related to service conditions may force the choice to a specific material because others do not possess the appropriate combination of properties for effective use. Once the chemical nature of a material is clear, knowledge of its physical properties becomes important for understanding how it reacts to external stimuli such as a change in temperature or the flow of an electron current through it. The application of external stresses under various conditions helps to understand how the material reacts from a mechanical point of view. Finally, the interaction of the material with the external environment and the way the environment modifies the material provide insight into the behavior from the perspective of corrosion resistance.

**Keywords** Physical properties · Mechanical properties · Corrosion resistance · Corrosion environments · Corrosion types

## 3.1 Introduction

Within the previous chapters, information was first provided on stainless steels in general, and then specifically on the family of SASSs. In fact, Chap. 2 was structured to adequately describe superaustenitic stainless steels by reporting the grades belonging to this family, as well as their chemical composition. The chemical composition of SASSs was used as the basis for the discussion of the solidification process and possible phases that can be observed in these steels. However, in addition to what is stated in Chap. 2, it is known that chemical composition has other fundamental consequences besides defining the specific alloy to which it refers. In particular,

 
A. Malandruccolo, *Superaustenitic Stainless Steels*, Engineering Materials,
https://doi.org/10.1007/978-3-031-68744-0_3

as mentioned in previous chapters, chemical composition can be used as starting point for the definition of additional material characteristics that are fundamental in the selection process for specific uses, particularly for high-demanding applications where requirements are very restrictive. Although the driving force for the development of SASSs has always been the quest to improve corrosion resistance performance, the change in mechanical properties resulting from a chemical composition optimized for corrosion resistance cannot be neglected, and this is why it is necessary to explain it. Within this chapter, there will be an initial link to what has been presented in Table 2.2 in order to define how SASSs respond to some specific conditions. Specifically, in a first part the way these materials behave in relation to external physical factors (e.g., gravity or heat flow) will be exposed and the most relevant physical properties will then be defined. Immediately after, and almost concurrently, the attitudes of SASSs to withstand applied loads of various nature will be exposed. This will allow the mechanical properties of these materials to be defined. Finally, an additional feature resulting from the chemical composition will be analyzed and exposed more in detail: resistance to corrosive phenomena. Corrosion resistance, however, cannot be defined solely through chemical composition, but also depends on other factors related to the material (e.g., surface roughness) and also the environment in which the material itself is placed. In this regard, general concepts related to corrosion and the main types of this phenomenon will be exposed first. Subsequently, a part will be devoted to indicators that allow materials to be classified according to corrosion resistance and to provide a preventive estimation of the related performance. Finally, the behavior under specific environmental conditions will be exposed with emphasis on the comparison with other stainless steels families for uses where corrosion is a non-negligible driver. Analysis of the properties of SASSs in terms of mechanical and corrosion response turns out to be of fundamental importance in explaining their behavior under certain experimental and operating conditions. This is necessary, for example, in the process of designing components, products and systems for all kinds of applications and particularly for those where the requirements for high mechanical properties go hand to hand with the need to ensure outstanding resistance to corrosion. In the design process, the correct material's choice for the specific application makes the difference between being able to guarantee the expected service life (and in some cases exceeding it), and premature failure with all the negative consequences this occurrence brings. Of course, there can be no perfect material able to satisfy the requirements of any application. However, the selection of the one best suited for the specific use is a process that cannot be disregarded without a preliminary, and often experimental, assessment of the characteristics of interest. Consequently, the correct selection process must necessarily go through the characteristic data of the material in order to be able to best meet all the requirements along with an additional issue of no small importance: the evaluation of the associated cost. In fact, in this latter context, it has been declared several times in previous chapters that SASSs aim to be able to be an alternative to the more expansive Ni-based alloys. Furthermore, for many applications they are in direct competition with high-performance dual-phase stainless steels (i.e., duplex, super- and hyperduplex grades, see Sects. 1.2.6 and 1.2.7). Hence, this chapter stands as a collection of all this information in order to provide the reader with enough tools to get as complete picture as possible regarding the issues just stated. Not only that, the information that will be given below will also be propaedeutic for a better and deeper understanding of the final chapter related to processing and specific uses of SASSs.

## 3.2 Physical and Mechanical Properties

Physical properties of materials can be defined as the characteristics resulting from the response to physical stimuli (e.g., temperature variations, gravity, electron flow). Hence, the physical behavior of materials is related to electrical, magnetic, optical and thermal properties. Knowledge of the physical properties of materials is of primary importance for both manufacturing and for the context related to design of products. For example, the knowledge of the thermal properties of a material is important in machining operations because it helps determine the cutting temperature, which in turn influences the tool life before the tool becomes unable of cutting. Designers and engineers are daily faced with the issue of material selection for their projects. In addition to cost considerations, the selection of the material with the right physical properties for the specific application also plays an important role [1, 2]. For a more in-depth discussion on the topic of materials selection for design, the reader is referred to the texts at the end of this chapter in the further reading section. Regarding SASSs and with the grades whose composition is given in Table 2.2, the following physical properties will be presented:

- Density.
- Electrical resistivity.
- Specific heat.
- Mean coefficient of thermal expansion.
- Thermal conductivity.

In order to provide a more representative picture of these characteristics of SASSs, the physical properties of two standard grades will also be reported for comparison: the ferritic AISI 430 and the austenitic AISI 304. Unfortunately, it is not always possible to find reliable literature data for all members of the SASSs family (see Table 2.2), mainly due to the fact that experimental research is often focused on the grades most widely available on the market. For this reason, it has been decided to report information only for those steels that had known and documentable values for each of the physical properties given in the list above. The first physical property reported is density. It is known that in the engineering field density is defined by the mass of a material per unit volume. The term specific gravity, on the other hand, is used to express the density of a material with respect to that of water. For this reason, specific gravity is dimensionless (i.e., it is a ratio with no units). Density plays an important role in the materials selection process for a given application, but often it is not considered directly as single property. In fact, for structural applications, density is used as property to obtain other specific properties, such as specific strength (also known as strength-to-weight ratio) resulting from the tensile strength of a material divided by its density. The density of an element is determined through parameters such as atomic number and radius, but also considering atomic packing. The density of an alloy is influenced by the type and quantity of alloying elements that constitute it [1]. Density data given in Table 3.1 show that SASSs turn out to have a higher density with respect to steel defined as simple Fe–C alloy (conventionally, the value is assumed to be 7.86 kg/dm$^3$). The values of SASSs are consequently higher than the ferritic grade reported for comparison. In fact, for what concerns AISI 430, its lower density than standard steel can be easily explained through its chemical composition (see Table 1.3). AISI 430 steel is essentially a Fe–Cr alloy, with small amounts of other alloying elements. The density of Cr is 7.14 kg/dm$^3$, and this, combined with the high Cr content in AISI 430, leads to a lower value than that of C steel. On the other hand, the chemical composition of SASSs is strongly different and features high Ni and Mo content. Not only that, Cu is also an alloying element in these steels and in some cases even W is present. The density of Ni is 8.908 kg/dm$^3$, that of Mo is 10.28 kg/dm$^3$; the value associated to Cu is 8.92 kg/dm$^3$ and for W the value reaches 19.25 kg/dm$^3$. Without going too much into detail, the density of an alloy is not simply the result of the weighted average of the densities of constituent elements and the non-applicability of the weighted average is mainly related to the affinity, or lack thereof, between the different atoms that combine to form the alloy [3].

**Table 3.1** Density and electrical resistivity of selected SASSs together with AISI 430, AISI 304 and AISI 904L [4–6]

| Group | Most common name | UNS | Density (kg/dm$^3$) | Electrical resistivity (m$\Omega \cdot$ m) |
|---|---|---|---|---|
| Ferritics | AISI 430 | S43000 | 7.70 | 0.60 |
| Austenitics | AISI 304 | S30400 | 7.90 | 0.73 |
| High-performance austenitic stainless steels | AISI 904L | N08904 | 7.95 | 0.95 |
| 6Mo superaustenitics | 1925 hMo | N08925 | 8.15 | 1.00 |
| | NAS 254N | S32053 | 8.06 | 0.95 |
| | 25-6Mo | N08926 | 8.15 | 1.00 |
| | 1925 hMo | | 8.15 | 1.00 |
| | 254 SMO | S31254 | 7.95 | 0.85 |
| | Nicrofer 3127 hMo | N08031 | 8.03 | 1.00 |
| | AL-6XN® | N08367 | 8.06 | 0.89 |
| High-N superaustenitics | Uranus B66 | S31266 | 8.20 | 1.00 |
| | 654 SMO | S32654 | 8.00 | 0.78 |
| | Nirosta® 4565S | S34565 | 8.00 | 0.92 |

In any case, with proper adjustments it is possible to say that the addition of an element with higher density than that of the initial alloy matrix will inevitably lead to an increase in density. As a result of this, the high Ni content of austenitics, and especially of SASSs, leads these steels to have higher densities than ferritics. SASSs, in particular, also due to their Mo content, rank all above the value of AISI 304. Of course, density is not a constant property, but varies with temperature. Specifically, the value of density tends to decrease with temperature and the magnitude of this reduction is a function of the specific chemical composition of the material [1, 7].

The other property shown in Table 3.1 is electrical resistivity. In a simple, and probably reductive way, resistivity can be defined as the physical property of the material that expresses its ability to resist a current flow. As with other material properties, resistivity varies with temperature, and for metallic materials an increase in temperature is followed by an increase in resistivity; at low temperature resistivity decreases. For metals and alloys, electrical properties are of great importance, particularly for applications implying heating, power and signal transmission, temperature measurement, thin film devices and semiconducting, precision resistances and switching devices or simply specification of purity [1, 7]. Concerning SASSs, it can be seen that the resistivity value is higher than both the ferritic and the austenitic grades taken as reference for comparison. This can be easily explained through the more complex chemical composition. In fact, conduction in real (i.e., inhomogeneous) solids is influenced by microstructure and chemical composition, and different atoms in the alloy lead to different behavior in terms of scattering of conduction electrons in the solid being passed through. The more heterogeneous the composition becomes; the greater amplification can be expected for this phenomenon [7].

Thermal properties turn out to be of particular interest in materials science and engineering, manufacturing and design because heat generation, heat transfer and heat use are extremely common phenomena in these contexts. Not only that, an absorption or transfer of heat by the material has dimensional consequences because of the associated volumetric expansion/contraction [1]. Table 3.2 shows the main thermal properties of SASSs starting from specific heat. Synthetically, specific heat is defined as the amount of heat required to raise a unit mass of material by one degree [1]. Regarding the values associated with SASSs, all the alloys in Table 3.2 are characterized by lower specific heat than AISI 304, except for 654 SMO and S34565. For sure this has consequences in usual manufacturing situation such as a hypothetical reheating process prior to hot working. In fact, comparing two materials with different specific heat values, the one with the lower value will need less energy to raise its temperature. This latter statement can only be considered correct if the materials being compared have the same ability to conduct heat. Conduction is one of the three mechanisms of heat transfer along with convection and radiation. When talking about thermal conduction for a material, specific reference is made to the ability of the material to conduct heat through itself by purely thermal motions, without mass transfer occurring [1]. From the data given in Table 3.2, it can be seen that the thermal conductivity of SASSs is significantly lower than that of AISI 430 and also than that of AISI 304. Among all family members, 654 SMO is found to have the lowest thermal conductivity value. A comparatively lower thermal conductivity leads to increased difficulty for heat to flow within the material and this, in some contexts, can be disadvantageous. Again, with reference to the example related to reheating the material before hot working operations, a material with low thermal conductivity will need more time to achieve thermal homogenization in the furnace. This becomes relevant for processes such as hot rolling or forging, where in the past the setting of soaking time in furnace was mainly based on experience and specific cases of production sequence. Currently, even more manufacturers are using dedicated software to calculate the temperature of the material in the furnace in order to be able to best calibrate the heating cycle and start processing only when the temperature is actually within the ideal processing range for the material. These software are customized for each plant as they have to consider parameters such as:

- Furnace type.
- Furnace size and geometry.
- Type, power, arrangement of burners in the furnace.
- Internal furnace fluid dynamics.

**Table 3.2** Specific heat, thermal conductivity and mean coefficient of thermal expansion of selected SASSs together with AISI 430, AISI 304 and AISI 904L

| Group | Most common name | UNS | Specific heat (J/(kg · K)) | Thermal conductivity at 20 °C (W/(m · K)) | Mean coefficient of thermal expansion ($10^{-6} \cdot K^{-1}$) | |
|---|---|---|---|---|---|---|
| | | | | | T = 20–200 °C | T = 20–400 °C |
| Ferritics | AISI 430 | S43000 | 460 | 25.0 | 10.0 | 10.5 |
| Austenitics | AISI 304 | S30400 | 500 | 15.0 | 16.5 | 17.5 |
| High-performance austenitic stainless steels | AISI 904L | N08904 | 461 | 12.0 | 16.1 | 16.9 |
| 6Mo superaustenitics | 1925 hMo | N08925 | 461 | 12.0 | 16.1 | 16.9 |
| | NAS 254N* | S32053 | 457 | 11.9 | 15.2 | 15.8 |
| | 25-6Mo | N08926 | 461 | 12.0 | 16.1 | 16.9 |
| | 1925 hMo | | 461 | 12.0 | 16.1 | 16.9 |
| | 254 SMO | S31254 | 498 | 14.0 | 17.0 | 18.0 |
| | Nicrofer 3127 hMo | N08031 | 440 | 12.0 | 14.7 | 15.5 |
| | AL-6XN® ** | N08367 | 461 | 13.7 | 15.3 | 16.0 |
| High-N superaustenitics | Uranus B66 *** | S31266 | 450 | 12.0 | 15.0 | 16.0 |
| | 654 SMO | S32654 | 510 | 8.6 | 15.4 | 16.2 |
| | Nirosta® 4565S | S34565 | 510 | 14.5 | 15.5 | 16.8 |

*mean coefficient of thermal expansion related to T = 30–200 °C and 30–400 °C. **mean coefficient of thermal expansion related to T = 20–100 °C and 20–500 °C. *** mean coefficient of thermal expansion related to T = 20–100 °C and 20–300 °C [4–6]

- Balance between conduction through furnace components, convection through hot gases and radiation from burners.
- Position of the material in the furnace.
- Type of material and related thermal conductivity.

Using the parameters listed above, the software allows for the process engineer and operators to calibrate the production sequence and heating cycle in order to maximize productivity, without sacrificing the final product quality. In fact, some of the quality problems deriving from incorrect heating include: formation of surface defects due to lack in plasticity caused by too low temperature; internal defects caused by incipient melting resulting from an incorrect starting temperature to which the rise caused by the heat produced by plastic deformation process is added. This latter situation, in case of incorrect process conditions in terms of strain rate leads to the adiabatic heating predicted by the deformation maps of the material (see Sect. 4.3.1). On the other hand, in a use-related context, low thermal conductivity could determine the choice of a material with a higher value in specific applications. For this reason, design engineers might choose a ferritic grade with comparable corrosion resistance for heat exchangers applications [5].

The last property shown in Table 3.2 is the mean thermal expansion coefficient of the material. As already declared, the density of a material decreases with increasing temperature. Since the calculation of density implies the mass to be constant over time, an increase in volume results in a decrease in density. The increase in volume per unit mass with increasing temperature is the easy way to explain a phenomenon called thermal expansion. Thermal expansion is usually expressed by means of a parameter that indicates the change in length per degree of temperature called coefficient of thermal expansion. The coefficient of thermal expansion is also affected by temperature, and the dependence is not always linear. For stainless steels, increasing temperature leads to an increase in the coefficient of thermal expansion. For this reason, two temperature ranges for its value are given in Table 3.2, as well as the average value of the coefficient itself in the specific temperature range [1, 5]. From the data given in Table 3.2, it can be seen that the family of austenitics, of which SASSs are members, is characterized by higher coefficients than ferritics. There is also substantial homogeneity among the austenitics and altogether little variation among the various grades. Thermal expansion is a parameter to be considered in various situations, for example when heat treatment or welding processes are performed on the material. In fact, thermal stresses can develop during the cooling phase and this can affect the integrity of the material leading to dimensional changes and, in some cases, crack formation. On the other hand, thermal expansion can be positively exploited for assemblies involving shrink fit and expansion fit. In such cases, usually one part is cooled so that it contracts and is easily inserted into another one, which is instead heated to accommodate it more easily. Temperature equalization leads to expansion of the former and contraction of the latter, and this results in an interference coupling [1].

When reference is made to mechanical properties, the term is used for those characteristics of materials that show the response to the application of external stresses. In fact, by definition, mechanical behavior is something that has to do with the reaction of a body to the application of a force or load. Materials have two principal reactions when subjected to external forces or loads: deformation and fracture. Deformation can be elastic (reversible), plastic (irreversible) or time-dependent. In the latter case reference is made to viscoelastic deformation when dealing with time-dependent elastic deformation and creep in case the time-dependent deformation is plastic. On the other hand, fracture may occur after repeated application of the load (i.e., fatigue) or suddenly once the material exceeded its maximum resistance to the applied load. In a generalized and simplified way, both the deformation and fracture modes of materials are influenced by temperature, the rate of load application and also the presence of defects. As with physical properties, mechanical properties of materials have important influence in both design and manufacturing, with often conflicting implications between the two areas [1, 2, 8]. In fact, from a design perspective, the material selection process is often seeking a material that can guarantee mechanical properties that can prevent permanent deformation or make it predictable and useful for a purpose (e.g., energy dissipation in case of impact). Thus, the main objective is to select a material that is capable of resisting deformation when loaded or, otherwise, that does not undergo major geometric alterations and this is dependent from the elastic characteristics. On the other hand, in manufacturing, the ultimate purpose is often exactly the opposite: altering the shape of a material to obtain a finished product or a product to be finished in later stages. It is therefore obvious that in the case of metal forming processes the goal is to apply loads that exceed the elastic limit and allow permanent deformation. At this point, the existence of a dilemma again becomes evident: mechanical properties that are useful to the designer (e.g., high yield strength) usually lead to challenges for manufacturing the material, making the process more difficult [1]. Hence the need to know the material properties is highlighted, both for what concerns the material selection process for design, and for what concerns the proper settings of manufacturing processes (see Chap. 4).

Table 3.3 gives the main mechanical properties of SASSs and, as in the previous tables, also the comparison with a ferritic grade, a standard austenitic, and AISI 904L. Although the development of SASSs has had as its main driving force the search for superior corrosion resistance over standard austenitics, mechanical properties also turn out to be higher than these latter grades. Specifically, it can be seen that in all cases the minimum tensile strength values are higher than those of AISI 430 and AISI 304. SASSs turn out to be better performing from the point of view of the mechanical properties by virtue of their chemical composition, which allows these alloys to benefit from considerable solid solution hardening (SASSs cannot be aged like PHSSs since the precipitates that would form would be deleterious and not reinforcing, see Sect. 2.4.6 for further details). The effect of alloying elements in steels is given in Appendix 1, however, as for the solid solution hardening effect of the main alloying elements, a graphical representation is given in Fig. 3.1. Within Fig. 3.1, the solid solution hardening effect is represented as an increase in the yield strength (which is one of the most easily detected effects, along with the increase in the hardness value). In agreement with Table 2.2, SASSs are alloys mainly rich in Cr, Ni and Mo. In general, most of the alloying elements employed for improving corrosion resistance or for phase balance control act also as solid solution strengtheners [5]. Excluding Cr, which is not reported, it can be seen that the presence of Ni would tend to decrease the yield strength. However, the reinforcing action performed by other elements such as Mo compensates this phenomenon.

By analyzing what is presented in Fig. 3.1, it can be seen that the alloying element with the greatest solid solution strengthening effect is N. Furthermore, as mentioned in Chap. 2, N is also an important alloying element in improving corrosion resistance and suppressing/delaying the precipitation of deleterious phases in SASSs. Obviously, the effect of N cannot be solely positive for SASSs. In fact, while strength properties increase, a higher N content results also in a decrease in the percent elongation at break, which in turn is reflected in lower deformability than standard austenitics such as AISI 304. Further detail of the strengthening effect operated by N in austenitic stainless steels is given in Fig. 3.2. It becomes even more evident at this point how an increase in N content results in substantial difference with lower-performance austenitics, but also within the same family of SASSs. In fact, 25-6Mo turns out to have lower strength characteristics than 654 SMO, and this can also be easily explained by the average N content of the alloy (i.e., less than half that of 654 SMO). Keeping the focus on the yield strength of these alloys, in addition to any findings from laboratory tests, Eq. 3.1 can be used for evaluating the yield strength in advance [9].

$$\sigma_Y = 67.8 + 20\%\text{Si} + 3.7\%\text{Cr} + 14.5\%\text{Mo} + 354\%\text{C} + 494\%\text{N} + \frac{7}{\sqrt{d}} + 2.45\%\delta \tag{3.1}$$

where

- $\sigma_Y$ is the calculated yield strength in MPa.
- $d$ is the grain size usually expressed in μm.
- $\%\delta$ is the δ ferrite percentage of the considered steel.

Equation 3.1 is experimental in nature and easily shows how the limited action of Ni is not taken into account. Moreover, the influence of grain size also appears to be limited. Both these latter issues can be traced back to the *fcc* structure of these steels implying a large number of slip systems. As it is well known, this latter fact allows for an easy movement of dislocations. Interestingly, the possible presence of δ ferrite leads to an increase in yield strength [9]. Equation 3.1 thus provides justification for what has been reported in Sect. 2.4.6.1.

One of the things that certainly stands out the most in comparison with the ferritic AISI 430 is that the minimum percent elongation is practically always double. This fact is mainly explained through the *fcc* cell that austenitics have (see Sect. 2.2), which allows them to deform significantly before fracture. This also results in the ability to lend themselves better to even intense plastic deformation processes (e.g., deep drawing). However, in comparison with a standard austenitic grade, the minimum

**Table 3.3** Mechanical properties at room temperature of selected SASSs together with AISI 430, AISI 304 and AISI 904L

| Most common name | UNS | Minimum tensile strength $\sigma_{UTSmin}$ (MPa) | Minimum yield strength $\sigma_{Ymin}$ (MPa) | Minimum elongation (%) | Maximum hardness | Minimum Charpy impact resistance KV (J) | |
|---|---|---|---|---|---|---|---|
| | | | | | | Transverse | Longitudinal |
| AISI 430 | S43000 | 430 | 260 | 18 | 183 HB | – | – |
| AISI 304 | S30400 | 520 | 200 | 45 | 201 HB | 100 | 60 |
| AISI 904L | N08904 | 530 | 230 | 35 | 230 HB | 100 | 60 |
| 1925 hMo | N08925 | 650 | 295 | 35 | – | – | – |
| 25-6Mo | N08926 | 650 | 295 | 35 | 250 HB | 100 | 60 |
| 1925 hMo | | 650 | 295 | 35 | 250 HB | 100 | 60 |
| Uranus SB8 | N08932 | 550 | 250 | 35 | – | – | – |
| 254 SMO | S31254 | 650 | 300 | 35 | 223 HB | 100 | 60 |
| Nicrofer 3127 hMo | N08031 | 650 | 276 | 35 | – | – | – |
| AL-6XN® | N08367 | 690 | 310 | 30 | 240 HB | – | – |
| Uranus B66 | S31266 | 750 | 420 | – | – | – | – |
| 654 SMO | S32654 | 750 | 430 | 35 | 250 HB | 100 | 60 |
| Nirosta® 4565S | S34565 | 800 | 420 | 35 | – | – | – |

For all the austenitic grades, the supply status is intended to be the annealed condition [4, 5].

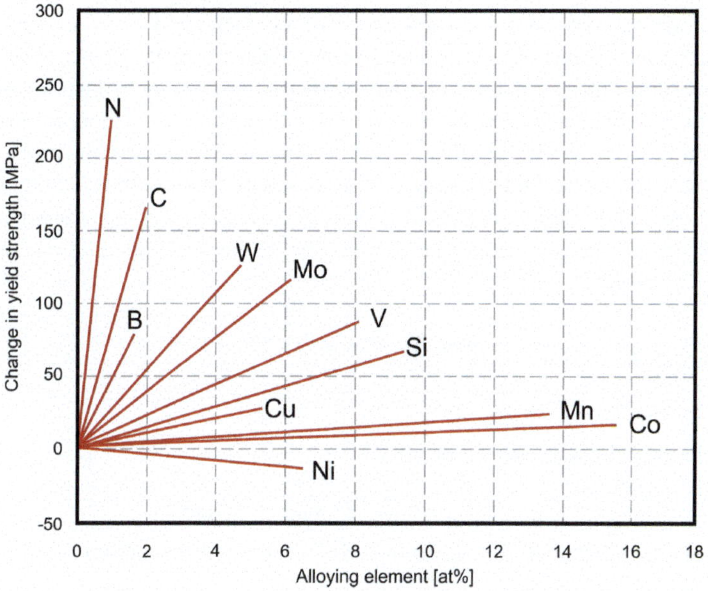


**Fig. 3.1** Solid solution strengthening effect of selected alloying elements in austenitic stainless steels (redrawn from [5])

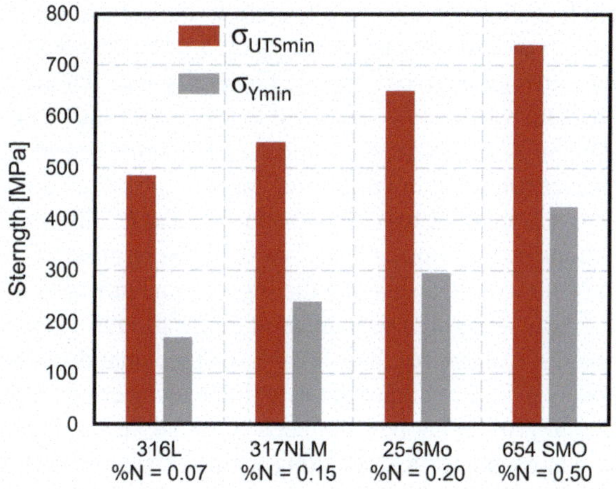


**Fig. 3.2** Variation of the ultimate tensile strength and yield strength as a function of increasing average N content [5]

elongation is lower, and this is mainly explained through the chemical composition, as already mentioned. Nevertheless, these alloys retain sufficient ductility to be subjected to the most common cold forming operations (see Sect. 4.3.2).

Austenitic stainless steels are known to be characterized by strain hardening, a phenomenon that can be summarized as the increase of the stress level necessary to continue plastic deformation (the terms strain hardening and work hardening are generally used interchangeably). Austenitic stainless steels are not the only materials to show strain hardening; this is also the case for Al and Cu alloys. However, among the stainless steel families, they are the ones that exhibit the phenomenon the most. For austenitic stainless steels, it is possible to account for the strain hardening with reference to the true stress-true strain curve by means of the relationship given in Eq. 3.2 [9, 10].

$$\sigma = K \cdot \varepsilon^{n} + e^{(K_L + n_L \cdot \varepsilon)} \tag{3.2}$$

where

- $\sigma$ is the true stress expressed in MPa.
- $\varepsilon$ is the true strain.
- $K$ is the strength coefficient of the material given in MPa.
- $K_L$ and $n_L$ are two materials constants of the Ludwigson equation [9].

The strain hardening tendency of austenitics increases with the addition of specific alloying elements, particularly N. Consequently, it is not surprising that SASSs are characterized by high strain hardening, which is higher than the other members of the austenitic family. For example, a consequence of this is that their machinability is also affected. In fact, during machining operations a localized work hardening due to the material's response to the tool action takes place. This leads in turn to less machinability than other grades (see Chap. 4). Generally, the high strain rate tendency of high-N stainless steels tends to be ascribed to the promotion of dislocation accumulation caused by high N content (see Table 2.2). However, recent studies based on the analysis of dislocation accumulation behavior through the modified Williamson-Hall/Warren–Averbach method have shown no noticeable difference between the increase in dislocation density during deformation between austenitic stainless steels with low or high N content. It seems that the high strain hardening might be related to the fact that the cross-slipping mechanism of dislocation is suppressed in high-N steels, and moving dislocations result back-stressed by planar dislocation array. This phenomenon appears to induce higher deformation resistance and higher work hardening in the material [10]. What has just been reported, again underlines how one of the key alloying elements for SASSs along with Mo is confirmed to be N. This is highlighted by its weight in Eq. 3.1 and by the data shown in Fig. 3.2; especially for what concerns the mechanical behavior. Regarding the fracture toughness of SASSs, the values are high and essentially insensitive to N increase for the typical content range of SASSs as shown in Fig. 3.3 [5].

As is the case with *fcc* metals and alloys, SASSs do not exhibit DBTT (ductile–brittle transition temperature) and therefore their resistance to Charpy test decreases progressively and not suddenly with decreasing temperature, while still remaining at high levels as shown by Fig. 3.4 for AL-6XN® (it should be pointed out that

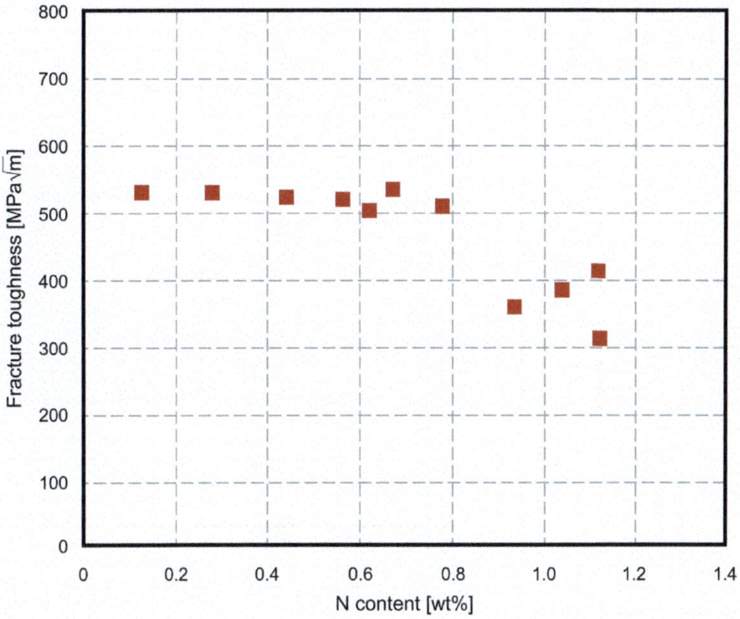


**Fig. 3.3** Variation of the fracture toughness as a function of increasing average N content (redrawn from [5])

the values shown in Table 3.3 are the minimum values guaranteed by the material according to the cited references).

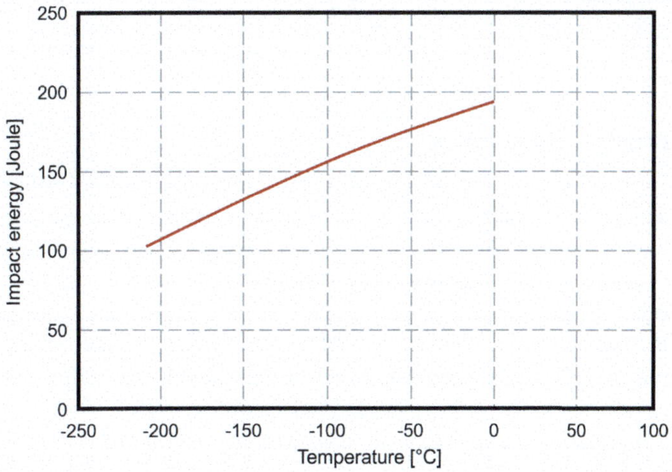


**Fig. 3.4** Variation of the Charpy impact resistance KV of AL-6XN® with temperature (redrawn from [5])

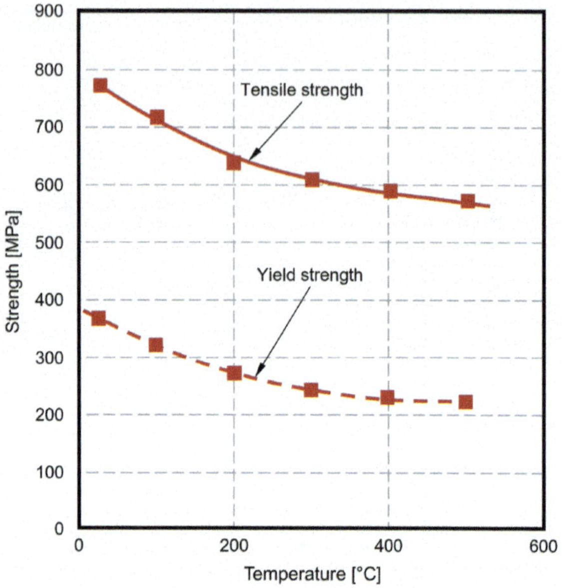


**Fig. 3.5** Variation of tensile and yield strength of AL-6XN® with temperature (redrawn from [5])

On the other hand, in terms of mechanical properties at high temperature, SASSs can retain their strength up to significantly higher service temperatures than ferritic and duplex grades (see Fig. 3.5 for AL-6XN® behavior). One of the reasons why for this is that SASSs do not show the so-called 475 °C embrittlement (see also Sect. 1.2.2) [5].

Obviously, with regard to the service temperature of SASSs, it will be necessary to avoid temperature ranges that could result in the formation of secondary phases deleterious to mechanical and corrosion resistance properties (see Sect. 2.4.6).

## 3.3 Corrosion Resistance

Although advances in materials science have rebalanced the proportion among the major families of materials (i.e., metals, ceramics, polymers, and composites) leading to a better balance of use, people of industrialized nations still live in a metal-based society. This becomes especially evident when walking down the streets and observing the surroundings: steels and metal alloys are used in automobiles, to build trains, as reinforcement in concrete, to build bridges, and for an infinity of components such as screws, nuts, washers and so on. Regardless of their use and except for few cases, all metals and alloys tend to react with the environment they are placed in. This obviously occurs to different extents and at different rates according to the material-environmental parameters combination (e.g., temperature, contact with liquids etc.) [11]. Interaction with the environment can lead to positive phenomena for the material (e.g., passivity, see Sect. 1.2.1), or to its deterioration. Talking in general terms and with specific reference to metals and their alloys, corrosion can be defined as a process of deterioration of materials when they are exposed to specific environmental conditions. This damage process involves chemical reactions that can significantly alter some of the original characteristics of the material itself. Severity and extent of the corrosive phenomenon depend on the relationship between the material and environmental conditions [12]. Nowadays, the concept of corrosion as material deterioration process is also applied beyond metal-based materials and is used to describe the degradation of nonmetallic materials too, as is the case of polymers and ceramics [12, 13]. Excluding noble metals, the other members of the metal family of elements are usually found in nature in the form of compounds such as carbonates, oxides, silicates, sulfates and sulfides; the main components of a class of substances called ores. These same compounds are often present as products of corrosion reactions. Corrosion can be considered as the consequence of the natural tendency of materials, or their constituent elements, to move or return to a thermodynamically more stable state. This is the reason why, in some respects, the corrosive process can be though as the reverse of the metallurgical process of separating metals from the other atoms that form the ores [12–14].

Corrosion has impact on virtually every technological field, from complex ones such as aerospace, energy production, chemical and petrochemical, to those closer to the common understanding such as the food industry, construction and transportation. A key step in product development is to consider the environment in which it will operate and the required corrosion resistance that allows it to provide the expected service life. However, it is not always easy to accurately predict and anticipate all service conditions with their associated corrosion protection strategies. This can obviously lead to failure in the moment when the component has lost its bearing characteristics [12]. The study and understanding of corrosion are fundamental processes because of the effects on human life and safety, the costs associated with issues arising from corrosion and its prevention, and the preservation of materials over time. Premature and/or unexpected failures of artifacts used by humans can result in damage to health or life loss, costs due to consequences and interruption of services, and ultimately the need for product replacement. Finally, an additional motivation lies in the fact that corrosive phenomena can be very complex and complicated to explain, so their study is itself interesting and also a challenge [11, 14]. Corrosion is a thermodynamically unsolvable problem and therefore cannot be avoided. Obviously, magnitude and time of development may be so insignificant that the corrosive process results to be irrelevant to the materials that are subjected to it. However, it is clear that the opposite situation is equally possible: corrosive phenomenon intense and rapid enough to quickly damage and make the material fail. Corrosion is a relevant issue for virtually all classes of alloys, and specifically, for stainless steels as well. Indeed, for these materials it is of fundamental importance because as explained in Chap. 1, they were developed precisely to deal with corrosion by resisting it more than other types of steels [6]. In fact, one of the strategies to tackle corrosion is to use materials that exhibit low corrosion rates for the specific environment considered. With regard to metal alloys, having materials with high corrosion resistance is the result of a development and alloy design process that necessarily begins with the chemical composition. As mentioned in Chap. 1, the development of the so-called *super* versions of stainless steels has had precisely the need for more corrosion-resistant materials as its main driving force, but less expansive when compared to Ni alloys [12, 13].

In terms of corrosion mechanisms, two main types can be identified: electrochemical corrosion and high-temperature corrosion. In the former case, the mechanism is also referred to as wet corrosion or aqueous corrosion because the electrochemical process needs an electrolyte in the liquid state, which is in most cases $H_2O$ or $H_2O$-based. On the other hand, high-temperature corrosion is intended to be the condition in which the material is in contact with an environment above 400 °C in which liquid $H_2O$ is absent. The difference between the two types is therefore evident. On one hand, wet corrosion is an electrochemical process, while high-temperature corrosion is a process governed by chemical process kinetics in the gas phase [12, 13].

Regarding the analysis related to corrosion resistance and behavior characteristics of SASSs, in view of the properties of these steels and possible uses (see also Chap. 4), the discussion will be focused more on the part related to the electrochemical process. In this regard, the current section is organized in such a way as to briefly recall basic concepts that will be useful to the reader in order to have a complete picture and the correct references for understanding the behavior of SASSs in the various environmental conditions that will be presented hereafter. High-temperature corrosion basics will also be summarized since, by virtue of their nature, SASSs can also be employed in few high-T applications under specific conditions. This, of course, should be done considering that their strength does not derive from precipitation hardening such in the case of PHSSs or some Ni-based superalloys, but from solid solution hardening and work hardening. Work hardening, in particular, is known to loose its effectiveness at high-T because it is recovered by the temperature increase itself (basic concept behind stress relief annealing). Furthermore, as explained in Sect. 2.4.6, SASSs are subjected to the precipitation of deleterious secondary phases at high temperature and this limits their use in applications employing such condition. Afterwards, concepts related to indicators suitable for describing the performance of SASSs in order to classify their corrosion resistance will be exposed. Finally, the behavior under specific environmental conditions will be exposed and analyzed. This information will serve to complete the picture on the corrosion resistance of SASSs by adding to what has already been anticipated within the Sect. 2.3.

### 3.3.1 Reactions and Environments

In agreement with the preliminary definition given in the previous section, it can be said that electrochemical corrosion is a fairly common process for metallic materials exposed to the environment. As its name suggests, electrochemical corrosion involves the transfer of electrons from and to the metals involved in the phenomenon. In fact, in the basic explanation of how the electrochemical process works, the reference for understanding the mechanism is very often the Cu–Zn cell developed by John Frederic Daniell. The so-called Daniell battery is a very useful setup that reproduces an electrochemical corrosion process of a Zn anode electrically connected with a Cu cathode, both immersed in solutions of their own salts (i.e., sulfates). To complete the electrochemical circuit, a salt bridge connects the two solutions [11–13]. From a conceptual point of view, Daniell's cell is useful in understanding the features common to all electrochemical corrosion processes:

- The overall corrosion reaction can be split into two *semi-reactions* or *half reactions*, the anodic and the cathodic one.
- According to the local electrochemical potential, the two half reactions occur either at the same region at the interface between the material and the environment, or in separate spots.
- There is only one corrosion rate common to the two half reactions, this is because the two are sustained by electron release and trapping.
- A liquid electrolyte is necessary for the corrosive process to take place. It can be $H_2O$ (most common case), or an equivalent substance. In case of aqueous solutions, it is also important to consider $H_2O$ electrolysis reactions as they play an active role in the corrosive process [12].

With reference to the above list, and based on what is depicted in Fig. 3.6, the anodic half reaction (i.e., oxidation) to which the metal is subjected turns out to be:

$$M \rightarrow M^{z+} + ze^- \tag{3.3}$$

Equation 3.3 describes the oxidation reaction, where $z$ indicates the number of electrons generated and also the oxidation number, i.e., the valence of the metal involved in the reaction that has influence on the density of charge carriers. On the other hand, the reduction reactions involving water are:

$$\frac{1}{2}\mathrm{O_2} + \mathrm{H_2O} + 2e^- \rightarrow 2\mathrm{OH}^- \tag{3.4}$$

$$2\mathrm{H}^+ + 2e^- \rightarrow \mathrm{H_2} \tag{3.5}$$

Equations 3.4 and 3.5 describe the reduction reactions taking place in a basic and acid environment, respectively (see also Fig. 3.6).

From the reactions just described, it is evident that electrochemical corrosion is a redox process in which electron transfer from the anode to the cathode occurs. The spontaneity of the redox process is generally expressed on the basis of the equilibrium potentials of the half reactions that constitute the process. The electrical potential of each half reaction, $E^0$, is generally measured with respect to a standard electrode, to which a zero value is associated. Resulting potential values are usually placed in a series, called the standard potential series. The series is obtained considering each chemical species in a 1 mol/L concentration solution of the relevant ion. Since the series is obtained under specific and limiting conditions, it cannot take into account

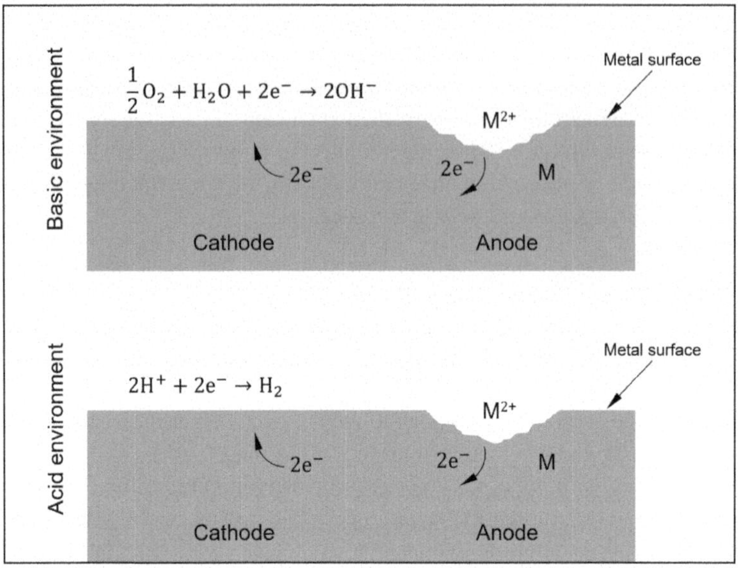


**Fig. 3.6** Corrosion reactions in basic and acid aqueous solutions involving a metal M. The number $z$ of electrons generated according to Eq. 3.3 is equal to 2 (redrawn from [12])

more complex aspects of corrosive processes, such as kinetic ones that are also related to passivation phenomena. Furthermore, the standard potential series has additional restrictions such as unit activity, the fact that it is obtained from solutions containing cations of the only metal of interest and, last but not least, the fact that it relates only to pure metals and not alloys [11–13].

The galvanic series (in seawater, see Fig. 3.7) is an ordered list of experimentally measured potentials in natural seawater for both pure metals and alloys and can provide information about the behavior of the material, for instance in relation to the contact with another one. Metals and alloys characterized by more negative potential are called active, while those characterized by more positive potential are referred to as noble. The utility of such series can be summarized in the fact that in a galvanic coupling, the metal or alloy associated with the more negative potential will be the anode and thus subjected to dissolution. On the other hand, the metal or alloy with the more positive potential will be the cathode and thus subjected to the reduction reaction [11–13]. The galvanic series in seawater can be used to predict corrosion tendencies in seawater or similar solutions. The use of this instrument for different solutions, such as those containing HF or $H_2SO_4$, must be considered more carefully and with specific considerations [11]. Despite the usefulness of galvanic series, an intrinsic limitation to them is that they are able to provide information of thermodynamic type, i.e., related to the driving force that determines the spontaneity (i.e., tendency) of the reaction. Information regarding the kinetics of the corrosive process, and therefore its rate, cannot be derived and must be determined separately by tests or experiments. Simplifying and summarizing the related discussion, the kinetics of the corrosive process depends on the electron flow associated to the reactions and can be described by a current value ($I_{corr}$), which is directly proportional to the mass of metal or alloy corroded. Therefore, the higher the $I_{corr}$, the more severe the degenerative phenomenon will be. Since the corrosion process can be seen as consisting of two half reactions occurring simultaneously, $I_{corr}$ will be equal to the current associated with the anodic half reaction ($I_A$), which in turn will be equal to the current associated to the cathodic half reaction ($I_C$). With reference to Fig. 3.6, in addition to anodic and cathodic reactions, corrosion implies two additional processes that have to be considered. The first is the electron flow within the metal from the anodic area where they are produced to the cathodic area, where electrons are consumed (in this representation the electron flow direction is opposite to the conventional current direction). A current value is therefore associated also to this ($I_m$). The second process is that of ion transportation in the electrolyte, to which a current value $I_{el}$ is also associated. The four processes take place at the same rate and thus, according to Eq. 3.6, the current values are equal to each other [11, 13].

$$I_{corr} = I_A = I_C = I_m = I_{el} \tag{3.6}$$

A direct consequence of Eq. 3.6 and the considerations so far is that to slow down and limit the corrosion process it is sufficient to limit at least one of the four current terms by acting on the related process. The way materials undergo corrosion is thus a more complex function of the associated potential and the environment where the process takes place as the latter influences the kinetics in various ways (e.g., type of electrolyte, temperature, radiation, presence of microorganisms etc.). In addition to this, the presence of stress fields also influences the corrosion behavior, both in case of stresses internal to the material (e.g., cold working) or imposed from outside due to the service conditions (e.g., tensile stresses) [12, 13]. The combination of factors just mentioned allows some main types of corrosion to be defined. The most important of them are extensively discussed in the textbooks of Pedeferri [13] and McCafferty [11] in the reference list of this chapter. However, in order to provide a complete context for corrosive processes and behavior for SASSs, the main aspects of the following types will be briefly recalled:

- General corrosion.
- Pitting corrosion.
- Crevice corrosion.
- Selective corrosion.
- Galvanic corrosion.
- Mechanochemical corrosion.
- Hydrogen induced damage.
- Microbiologically induced corrosion.
- High-temperature corrosion.

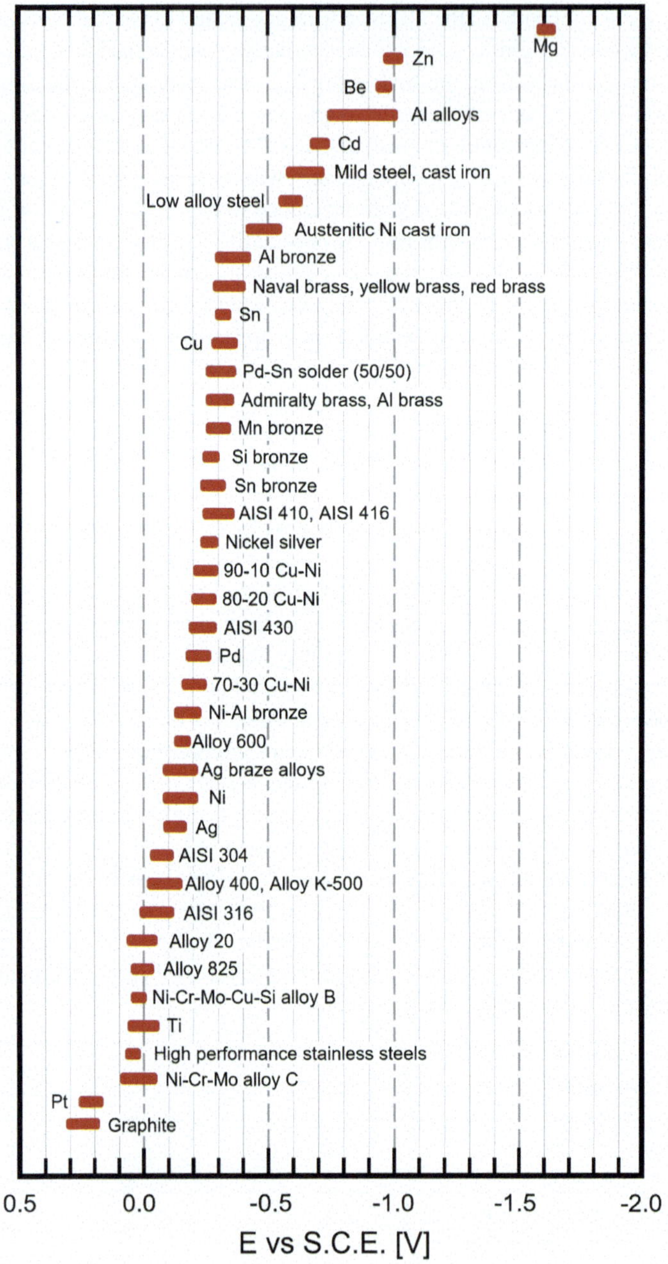


**Fig. 3.7** Galvanic series in seawater. S.C.E. means standard calomelan electrode (redrawn from [11, 15])

### 3.3.2 General Corrosion

General corrosion implies homogeneous corrosion of the metal or alloy subjected to it. The corrosive attack affects the entire surface of thc component as shown in Fig. 3.8, or at least a very wide part of it. The formation of corrosion products results in uniform weight change (increase or decrease) over the entire area affected by the corrosion phenomenon (anodic and cathodic reactions change dynamically over time). The possible weight loss associated with generalized corrosion phenomena is caused by the non-adherent nature of corrosion products that may spall off after some time under the conditions in which the material operates [12, 13].

General corrosion may occur uniformly over the surface, or non-uniformly. Uniform general corrosion is usually tolerated when the degenerative process takes place at a sufficiently low rate so as to guarantee the expected or adequate lifetime to the component. Nonuniform general corrosion is characterized by an uneven spatial distribution of the corrosive attack and related weight change. When general corrosion persists over time it can lead to a weakening of the base metal or alloy with local redistribution of the stresses to which the component is subjected. This can lead to a local concentration of stresses that can result in deformation or failure of the component. However, it is not always true that general corrosion is a deleterious phenomenon. In fact, it can lead to positive effects in some cases, and therefore can be voluntarily induced in some application cases [12, 13]:

- Surface roughness increase in case of specific applications, e.g., osteogenesis improvement for bio-implants.
- Pickling of the surface.

In the event that corrosion-affected areas are accessible for inspection, once the phenomenon has been detected, protection and remediation actions can be put in place. On the other hand, in case the corrosive attack involves internal or difficult-to-access parts, the damage can become serious and jeopardize the material characteristics. For this reason, inspection methods and practices are always recommended in order to early detect the problem and be able to put appropriate maintenance and protective actions in place [12, 13].

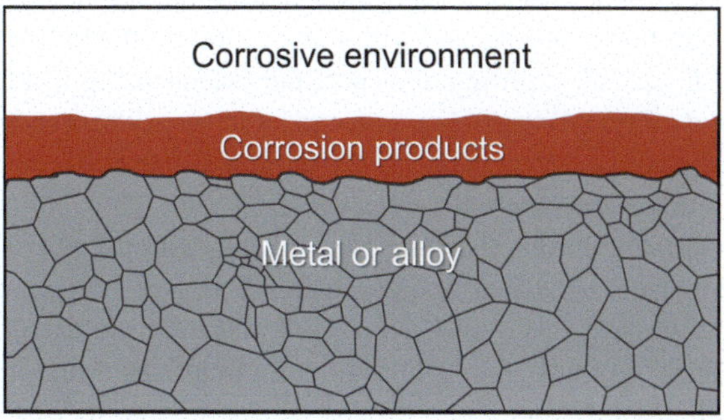


**Fig. 3.8** Schematic of general corrosion

### 3.3.3 Pitting Corrosion

Pitting is a localized corrosion attack that can affect self-passivating metals and alloys such as Al alloys, stainless steels and Ti alloys, as well as reactive metals and alloys such as those based on Mg in the moment when their surface is subjected to local wetting. According to the ASTM G46 standard, the pit can be defined as a defect that can be extensive shallow or narrow deep or even elliptical, transverse, sub-skin, vertical or horizontal. Pitting is a severe localized corrosion that, for metals and alloys with passive behavior, occurs when the passive film breaks locally (Fig. 3.9). The presence of halide ions, particularly $Cl^-$, promotes the pitting occurrence. Corrosion starts at the damaged spot and leads to the formation of a macrocell with the intact surroundings of the passive material as shown in Fig. 3.9. The mechanism involved implies that the circulation of macrocell current promotes a series of reactions and chemical modifications in the affected region. In particular, concentration of oxygen in the surface cavity is comparatively low and this leads to a local inhibition of the passive film formation. In addition to this, two reactions take place inside the defect: reduction of hydrogen with formation of $H^+$ ions and consequent decrease in the pH value and an anodic reaction involving the metal. Furthermore, reduction of H is promoted by the increasing concentration of metallic ions resulting from the anodic process. Metal ions diffuse out of the defect, while a counter-flux of chloride ions is drawn back into the cavity thanks to electrophoretic transport driven by the electric field associated with the ionic charges [12, 13].

The situation just described enhances metal dissolution and limits the possibility of passive film reformation. The anodic process keeps releasing metal ions into solution, thus promoting localized corrosion and pit extension [12, 13]. Pitting corrosion occurs according to two successive stages: pit initiation and pit propagation. The first stage is essentially the time required for local damage and subsequent breakdown of the passive film by the action of specific chemical spices such as chloride ions. Other halides such as $Br^-$, $F^-$ and $I^-$ are reported to have similar effect, but with lesser extent. The necessary electrochemical condition for what has just been described to occur is that the cathodic process must be more noble than a parameter defined as pitting potential ($E_{pit}$). Above $E_{pit}$ the protective film is damaged by perforation and this allows the corrosive process to happen. So, if the free corrosion potential ($E_{corr}$) is greater than $E_{pit}$, pitting corrosion can start. This condition occurs for highly oxidizing solutions and with the presence of the halide ions listed earlier. Otherwise, if $E_{corr} < E_{pit}$, pitting does not start (see also Fig. 1.4 for passive behavior of metals and alloys). Pitting initiation can have very different durations, up to several months depending on the material considered and operating conditions. Once the stable pit initiates, the macrocell current starts to circulate and the corrosive process proceeds at a rate expressed by the value of the current itself. The inherent danger of pitting corrosion is twofold: on one hand, there is a stochastic nature of the location where the passive film will be damaged and the time it will take for the initiation itself. This, of course, makes prediction complicated and certainly probabilistic. On the other hand, once the corrosive attack begins, its penetration rate is so high that it

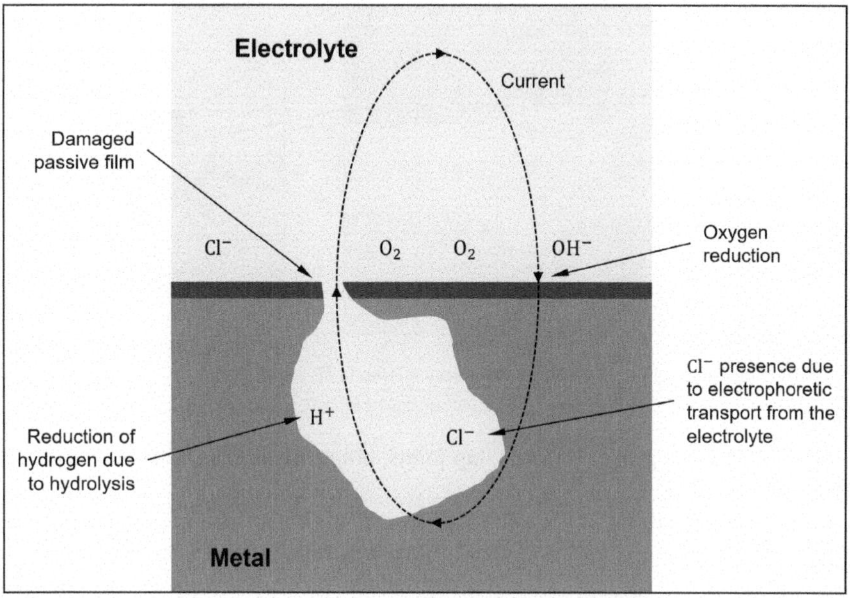


**Fig. 3.9** Schematic of pitting corrosion in environment containing chloride ions (redrawn from [12])

can affect the entire thickness of the involved object in a relatively short time. This, of course, can lead to sudden and unpredicted failure if the mistake of associating the damage with what is visible from the surface has been made. With regard to stainless steels, pitting corrosion in chloride-containing solutions and environments is one of the most dangerous forms of localized corrosion to which the material can be subjected. Therefore, prevention and development of increasingly resistant alloys such as SASSs became of paramount importance. In this regard, in addition to electrochemical conditions promoting pitting or protecting from it, useful indices have been developed to quantify the resistance to pitting and the boundary conditions that can promote it. Thus, reference is made to pitting resistance equivalent number, critical pitting temperature (CPT) and critical pitting chloride concentration (CPCC). A more detailed explanation of these parameters will be offered within Sect. 3.3.11 [13].

### *3.3.4 Crevice Corrosion*

Crevice corrosion belongs to the group of localized corrosive phenomena. Its occurrence is related to the presence of small (typically sub-millimetric) interstices on the surface of the metal or alloy. These interstices can be surface defects or, typically, small gaps between two metal surfaces as in the case of stagnant aqueous solutions

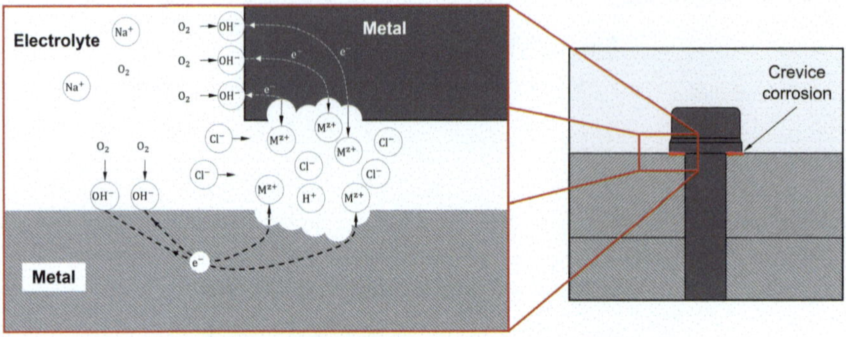


**Fig. 3.10** Schematic of crevice corrosion (redrawn from [12])

trapped in between gasket surfaces, lap joints, holes, rivets or bolts (Fig. 3.10) [12, 13].

Any metal or alloy can be subjected to crevice corrosion, irrespective of the composition of the electrolyte. The way in which the corrosive process occurs is somewhat similar to pitting. In fact, the degradation process is stimulated by changes in electrolyte composition between the inner and outer parts of the interstice. These variations may be caused by corrosion reactions, or by obstacles in the ion transport phenomena that would otherwise homogenize the electrolyte composition, thus reducing the incidence of crevice corrosion. In this regard, Fig. 3.10 represents a schematic of a crevice corrosion process that occurs in the gap between the head of a screw and the surface of the component where it is mounted. At the beginning of the process, the oxidation of the metal surface with the formation of $M^{z+}$ ions takes place. The resulting electron flow allows the formation of $OH^-$ at the interface between the metal and the electrolyte. After an initial stage of the process, limitations to mass transport from and to the interstice lead oxygen to be consumed. As a result, the composition of the electrolyte inside and outside the gap becomes progressively different. In particular, inside the crevice, due to the very low concentration of oxygen, cathodic semi-reaction can no longer occur. As a result, anodic semi-reaction is localized inside the crevice, while the cathodic process continues to occur outside. The production of $M^{z+}$ and $H^+$ is balanced by $OH^-$ and in this way the corrosive process is sustained. As a consequence of this phenomenology, the anodic area is far way smaller than the anodic one, and the corrosive process is therefore very severe and localized between the two metal parts. Because of the way it occurs, crevice corrosion is also called *differential aeration corrosion*. However, unlike pitting, crevice corrosion can be effectively prevented by careful design of the components to be joined. This process aims to minimize the presence of interstices through, for example, the use of rubber gaskets [12, 13].

### 3.3.5 Selective Corrosion

Selective corrosion is a degradation phenomenon involving the preferential dissolution of alloying elements, alloy phases or alloy parts, characterized by a lower electrochemical potential. An interesting example of selective corrosion is intergranular corrosion, a process that occurs at the grain boundary of the alloy or metal involved. The GB is a region of the material with different characteristics when compared with the inner part of the grain, particularly because of its distorted crystallographic structure and possible segregation of impurities and secondary phases (see also Sect. 2.4.6). These characteristics make the grain boundary a region of non-equilibrium and result in a higher reactivity associated with it. This leads the GB to show weakness in terms of corrosion resistance, and therefore intergranular attack can occur when the material gets in contact with the corrosive environment. Selective corrosion of the intergranular type is particularly dangerous as it is severe with the possibility of grain detachment. Although material consumption as a result of corrosion is limited, grain detachment leads to a condition of reduced mechanical strength. In some environments, and in presence of tensile stresses, this condition can promote the onset of stress corrosion cracking [12, 13].

A selective corrosion phenomenon that affects austenitic stainless steels, and therefore also SASSs, is sensitization. This family of steels is usually supplied in the solution annealed condition, which implies that C and other alloying elements capable of forming secondary phases are dissolved in the metal matrix. As seen in Sect. 2.4.6, when exposed to particular temperature ranges, these steels can precipitate Cr-rich secondary phases at the GB, including carbides such as $Cr_{23}C_6$. One of the situations in which this occurs the most frequently is welding. The precipitation of Cr-rich phases leads to a depletion of the surrounding regions, which find themselves unable to ensure the development of passivity. The Cr-depleted region thus assumes active behavior regarding corrosion phenomena and allows the anodic half reaction to concentrate there [12, 13]. There are several strategies to avoid or prevent the occurrence of sensitization, the most common of which are:

- Rapid cooling from high temperature in order to avoid exposure to temperature ranges where Cr-rich secondary phases can form.
- Development and use of alloys with alloying elements with higher affinity to C than Cr so as to avoid the phenomenon of Cr depletion resulting from the formation of Cr-rich phases. These alloys are called stabilized grades with elements used for this purpose being for example Nb and Ti.
- Low C content for reducing the possibility for Cr to combine with it.
- In case of welding, use of techniques employing rapid cooling and with a heat source characterized by high power density so as to reduce the extension of the heat-affected zone (HAZ) [12, 13].

### 3.3.6 Galvanic Corrosion

From a macroscopic point of view, galvanic corrosion occurs when metals and alloys with different practical nobility (see Fig. 3.7) are in contact in a corrosive environment and the contact allows them to be electrically connected. For this reason, galvanic corrosion is also called *bimetallic corrosion* or *two-metal corrosion.* However, galvanic corrosion does not necessarily require contact between different metals or alloys to occur. In fact, regions of the same material with different free corrosion potentials can also give rise to galvanic corrosion. These regions may originate from different residual stresses, mechanical working or presence of precipitates with different nobility with respect to the matrix. Thus, the requirements for galvanic corrosion to happen are a suitable electrolyte and the difference in electrochemical potential. The less noble metal, alloy or region acts as the anode of the electrochemical cell and corrodes, while the more noble metal, alloy or region acts as cathode and is protected in the long run. This type of situation conceptually mimics that typical of the Daniell battery, where the Cu cathode results protected, while the Zn anode dissolves. Possible mitigation or protection strategies involve, for example, designing couplings where components do not have excessive differences in practical nobility. Other possibilities are insulating flanges/gaskets, proper coatings avoiding electrical coupling or cathodic protection [12, 13].

### 3.3.7 Mechanochemical Corrosion

Mechanochemical corrosion can be defined as the degradation process that takes place when mechanical stresses are combined with a corrosive environment. For this reason, mechanochemical corrosion is also known as mechanically-assisted corrosion. The severity of the corrosive phenomenon is dependent on the balance between the characteristics of the corrosive environment and the mechanical stresses [11]. Because of the conditions that allow mechanochemical corrosion to take place, it is not unique in nature and can occur in different forms. One of the most insidious is the so-called stress corrosion cracking (SCC). Stress corrosion cracking leads to failure due to a slow, environmentally driven crack propagation as a consequence of the synergistic interaction between mechanical stress and corrosion reaction. SCC involves three main stages: slow crack initiation/incubation usually caused by pitting, propagation and sudden failure. SCC is particularly dangerous because its evolution is difficult to be adequately predicted. Not only that, one of the unique features of this corrosion type is that it takes place in environments that on their own would not be particularly aggressive for the material involved. The same applies to stress conditions. In fact, those involved in SCC inherently are not particularly critical as they are below UTS and yield values. However, the synergistic combination of a non-severe stress condition with a non-particularly corrosive environment leads to a condition that goes beyond the sum of the two and is amplified. As for austenitic stainless steels, SCC can be a problem for components made of these materials in service conditions implying tensile stresses in chloride-containing environments [12, 13].

### *3.3.8 Hydrogen Induced Damage*

Hydrogen is the smallest element and as a result of this it is able to dissolve and diffuse in metals and alloys. The extent of these two phenomena depends on concentration gradients, chemical potential and, of course, the structure of the crystal lattice involved. Usually, hydrogen induced damage results in catastrophic failure and is therefore considered as a very dangerous phenomenon. Furthermore, hydrogen induced damage is the term used to refer to a family of degenerative processes involving H among which the best known is hydrogen embrittlement (HE). Although HE has been thoroughly analyzed in the literature, the nature of the atomistic mechanism behind it is still quite controversial and various models have been proposed for its explanation. What is evident from the phenomenology associated with HE is that it implies a radical change in the mechanical behavior of the material undergoing it with a marked decrease in its ductility. HE leads the material to undergo a sort of ductile-to-brittle transition caused by the interaction of H with the metal or alloy, for instance with the behavior of the most important line defects in metals: dislocations [12, 13]. Since detailed description of hydrogen induced damage would need much more space and, as a result, would greatly burden the discussion by skewing it toward one of the degenerative mechanisms; for further details the reader is invited to refer to the texts used as reference in this section.

Usually, HE is a phenomenon that especially affects high-strength steels whose lattice is not *fcc* like that of SASSs. The *fcc* unit cell has a solubility for H that is about one order of magnitude higher than that of the *bcc* cell, and this allows the lattice to avoid distortion and better accommodate H without excess energy required [13]. As a consequence of this, SASSs are characterized by good resistance in environments having high H partial pressure and therefore are often used in handling H over a wide range of pressures and temperatures [5]. However, this does not exclude *fcc* materials from being subjected to HE when H concentration in the alloy becomes very high. Since H is present in the environment of various processes, e.g., melting and welding processes with no atmosphere control, hydrocarbon processing plants etc., metals and alloys have several possibilities to come into contact with H and be subjected to its inward diffusion [12, 13]. For example, in offshore applications where SASSs such as 254 SMO are employed for pipelines components, this family of alloys can be subjected to hydrogen embrittlement because of the presence of $H^+$ ions mainly produced by $H_2S$ present during operations [16].

### 3.3.9 Microbiologically-Induced Corrosion

When the corrosion process is caused by microorganism, it is called microbiologically-induced corrosion (MIC) [12]. This type of corrosive process is also known as microbial corrosion, bacterial corrosion, bio-corrosion. MIC can occur as localized corrosion such as pitting, or as generalized corrosion. The occurrence of one form or another is influenced by the metal substrate and the environment in which the process takes place [17]. The microorganisms involved are usually bacteria, and materials operating in a wide variety of industrial-type plants and processes can be subjected to this type of corrosion. Some examples are: water cooling circuits; fire-fighting systems; production, transport and storage of hydrocarbons; sewage treatment plants and marine and buried structures; in clayey and swampy soils or beneath fouling [12]. MIC takes place thanks to the very wide availability of bacteria in such environments, together with the presence of adequate nutrients and corrosive products [17].

Simplifying the subject for the actual context, bacteria can be divided into two families: aerobic and anaerobic. The former act by lowering the pH, thus promoting acid-related corrosive attacks. The latter, on the other hand, are for instance sulfate-reducing bacteria (SRB), which catalyze the reduction of sulphates to sulphides in environments that are free from oxygen and with a sulphate content exceeding 100 ppm [12]. In some cases, such as that of *Pseudomonas Aeruginosa*, the nature may be aerobic or facultative anaerobic. Currently, research has focused more on MIC phenomena caused by anaerobic bacteria such as SRB or iron-oxidizing bacteria (IOB). However, SRBs are not the dominant spices in the marine environment where high-performance stainless steels may be employed [18].

When MIC occurs, the first step in the corrosive process is the formation of the so-called *biofilm* [12]. IUPAC has given a definition of biofilm that considers it an aggregate of microorganisms in which cells that are often embedded within a self-generated matrix of extracellular polymeric substance and adhere to each other and/or to a surface [17]. In other words, the *biofilm* consists of colonies of microorganisms stuck on the material surface by means of a self-produced gel, which locally changes the pH and oxygen availability in order to allow the bacteria to thrive. The *biofilm* is usually tens of μm thick and characterized by two layers, where the inner one, adherent to the surface, is almost oxygen-free. Stainless steels may be subjected to MIC in seawater or in freshwater. In both cases, MIC occurs due to contamination by aerobic bacteria. The resulting corrosive process is usually pitting corrosion, a process that requires oxygen for propagation. In seawater (i.e., water containing chlorides) contaminated with bacteria they colonize the surface of stainless steel, forming the *biofilm*. The formation of the *biofilm* raises the potential by about 200–300 mV in comparison with sterilized seawater. This condition leads the potential to be higher than the $E_{pit}$ value and thus allow pitting initiation. On the other hand, in freshwater, the presence of the aerobic bacteria like the *Leptospirillum oxidans* is the cause of the increase in potential. These bacteria are also called manganese oxidizing bacteria (MOB) because they produce a $Mn^{3+}/Mn^{2+}$ couple characterized by a + 1.51 V with reference to SHE (standard hydrogen electrode), which is enough to trigger pitting initiation [12].

### 3.3.10 High-Temperature Corrosion

The term high-temperature corrosion refers to a phenomenon that takes place when a metallic material comes into contact with hot gases, typically at temperatures in excess of 400 °C. Under these conditions, no liquid $H_2O$ is present and therefore reactions typical of wet corrosion cannot take place because a key component of the process is missing: the electrolyte. High-temperature corrosion is therefore a chemical process, ruled by chemical process kinetics in the gas phase. Nonetheless, one common feature with the electrochemical mechanism is present: the layer of corrosion products (i.e., the oxide) formed on the surface is affected by ionic diffusion and electronic conduction in the oxide itself. The corrosive attack produces various consequences such as: corrosion products and metal evaporation, thinning of the material due to the formation of non-protective oxide layers, metal degradation by molten salts, localized attack at the GBs, embrittlement, erosion-corrosion assisted by entrained solid particles [12, 13].

As set forth in previous sections and chapters, SASSs are alloys characterized by outstanding wet corrosion resistance, and it is for the purpose of being able to resist this type of degenerative process that they were developed. However, their possible typical application fields may also be characterized by operating conditions involving high temperatures. For example, the chemical and petrochemical sector, as well as the power-generation one, may show in-service conditions characterized by high temperatures (see Table 3.4). Not only that, although high-temperature corrosion does not involve the presence of liquid $H_2O$, other substances may be present and influence the corrosive process. In fact, hot gases typical of conditions in which this type of corrosion takes place contain oxygen, N, water vapor, and variable concentrations of other potentially harmful gases: CO, $CO_2$, $H_2$, HCl, $H_2S$, $NH_3$, $SO_2$ and $SO_3$. These substances can give rise to specific degenerative processes such as carburization, chlorination, nitriding. In addition to this, particularly when Na, S and V are present, molten salts can form that induce further corrosion processes such as high-temperature corrosion with salt deposit [12, 13].

The complete discussion about high-temperature corrosion is, as stated, outside the scope of this section. However, some fundamental concepts need to be outlined. These certainly include the nature of corrosion products. As with wet corrosion, corrosion products can be protective or non-protective. In order to predict the nature of corrosion products, an index called PBR (Pilling-Bedworth ratio) was developed that expresses the ratio between the volume of the oxide formed and the volume of the oxidized metal or alloy. Briefly, a PBR $< 1$ indicates that the oxide is not protective because its volume is not sufficient to cover the entire surface. In case the PBR value is around 1.5, a compact and adherent oxide is formed that is able to protect the surface by slowing down corrosion. If the PBR value is too high, i.e., greater than

**Table 3.4** Examples of high-temperature corrosion in industrial fields [13]

| Temperature (°C) | Industrial field or application | Types of high-temperature corrosive phenomena |
|---|---|---|
| 450–550 | Boilers<br>Gas combustion | Oxidation |
| 550–650 | Ammonia plants<br>Convertors<br>EDC cracker tubes in vinyl chloride monomer production<br>Reactors for hydrocracking | Halide gas corrosion<br>Nitriding<br>H attack<br>$H_2S/H_2$ corrosion |
| 700–850 | Furnace tubes in carbon disulphide production<br>Heater tubes in hydrodealkylation | Carburization<br>Sulfidation<br>$H_2S/H_2$ corrosion |
| 950 | Reactor and catalyst support grid for $HNO_3$ production<br>Superheater supports in oil-fired refinery boilers | Fuel ash corrosion<br>Oxidation<br>Nitriding |
| 1000 | Reformer catalyst tubes for production of: ammonia, methanol, hydrogen, oxo-alcohol | Carburization<br>Combustion-induced corrosion<br>Oxidation |
| 1100 | Ethylene-pyrolysis furnaces<br>Gas turbines | Carburization<br>Oxidation |

1.8, the oxide formed has an excessive volume and, therefore, thickness. This can lead to internal stresses in the layer with subsequent oxide rupture [11–13].

For the situations just described, the kinetics of the corrosion process can be represented by the curves within Fig. 3.11. For non-protective oxides, the trend is linear in that the longer the exposure time, the greater the volume of metal or alloy converted to oxide, resulting in increased thickness. On the other side, in case of oxides with a protective behavior, the trend may be parabolic or logarithmic. In these cases, an initial increase in the oxide thickness is observed. After this, a reduction in the kinetics due to the protective characteristics of the layer takes place and allows the corrosion to be less than proportional to exposure time [11–13]. Considering what has been explained so far, it would be interesting to understand the response to high-temperature exposure of SASSs starting from considerations related to their chemical composition. From this point of view, as seen in Chap. 1, corrosion protection in stainless steels is achieved thanks to a passivity layer mainly based on $Cr_2O_3$. However, approximately above 950 °C $Cr_2O_3$ transforms into the volatile $CrO_3$. As a consequence of this, the protection of the alloy from the corrosive environment will be jeopardized [12]. As explained within Sect. 1.2, another important alloying element for corrosion resistance is Mo. Mo is important because it improves self-healing properties of the oxide layer and promotes increased resistance to pitting and crevice corrosion in reductive and chloride-based media. Mo is the key alloying element for defining SASSs. However, this does not mean that the high Mo content can also be good for high-temperature corrosion resistance. In fact, steels with high

Mo content tend to exhibit serious oxidation problems when exposed to high temperature for long periods such as in the case of prolonged heating in preheating furnaces prior to hot working operations or during heat treatments. In these situations, in addition to surface degradation due to formation of corrosion products, a change in the behavior during plastic deformation takes place as well, leading to a reduced plasticity and higher defect formation probability. This occurs due to an alteration in surface chemistry as a result of the depletion of specific elements from the surface, resulting in a change in workability characteristics. Although there is still no extensive literature in this regard, it is a fact that steels with Mo content above 5 wt% undergo catastrophic oxidation at high temperature. Evidence of this emerges from data concerning the analysis of the oxidation resistance of one of the best-performing members of the SASSs family: 654 SMO. Results of tests conducted at 900, 1000 and 1200 °C in air demonstrate the unsuitability of this alloy for environmental conditions characterized by such temperature values. Indeed, it was found that for exposure at 900 °C, the oxidation kinetics has parabolic trend for the initial 10 h of exposure. Exceeding 10 h of exposure, the behavior changes from parabolic to linear. This behavior indicates an initial protective action of the corrosion products, which, however, is lost after 10 h of exposure. On the other hand, as for the exposure at 1000 and 1200 °C, the oxidation kinetics is linear from the beginning with no protective action. For what concerns the behavior at 900 °C, the change in trend seems to be related to the consequences of the presence of liquid $MoO_3$. In fact, this oxide has a melting temperature of 795 °C, so at 900 °C it can exist in the liquid state. In the initial stages of oxidation, formation of various oxides on the surface takes place: $Cr_2O_3$, $Fe_2O_3$, $Mn_2O_3$, $MoO_2$, $MoO_3$ and NiO. The action of liquid $MoO_3$ involves the dissolution of some oxides including $Cr_2O_3$ and reaction with $Mn_2O_3$ to form $MnMoO_4$. When the exposure at 900 °C exceeds 10 h, cracking occurs in the surface oxide with also corrosion in the molten $MnMoO_4$–$Mn_2O_3$ layer. This results in a change in the oxidation kinetics and catastrophic oxidation characterized by linear trend. At temperatures of 1000 and 1200 °C, the trend is linear from the beginning and in this case the N in the air plays a role, contributing to corrosion. In fact, N at high temperature forms $Cr_2N$ precipitate and, in this specific case, precipitation takes place in a discontinuous way. The formation of $Cr_2N$ reduces the Cr surface content and thus limits the formation of Cr-based protective oxide. In addition to this, the presence of the molten $MnMoO_4$–$Mn_2O_3$ layer and the volatilization of $MoO_3$ lead to increased porosity and crack formation in the oxide layer, which in turn promotes the diffusion of N into the substrate. It is easy to see how this mechanism leads to catastrophic oxidation and the impossibility for the material to protect itself from the environment [19]. What has just been stated must be combined with the phase precipitation considerations made within Sect. 2.4.6. Hence, it becomes evident that SASSs can hardly be employed at temperatures exceeding 650 °C. For what concerns corrosion resistance at high temperature, limiting factors are issues related to the inability to show oxide formation that can give effective protection from the environment. Second, but not less important, phase precipitation together with possible reduction in mechanical properties are also issues that cannot be overseen.

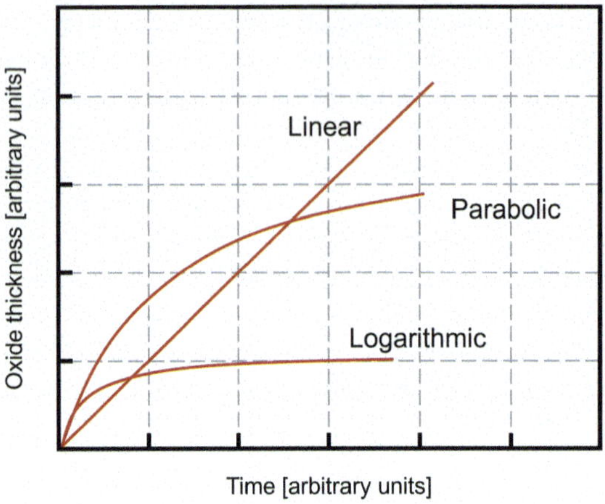


**Fig. 3.11** Linear, parabolic and logarithmic oxidation kinetics (redrawn from [12, 13])

### *3.3.11 Pitting Resistance Equivalent Number and Other Indicators*

It is well known that localized corrosive attack, i.e., pitting, in chloride-containing environments or solutions is one of the most dangerous phenomena for products made of stainless steel. As already stated in previous sections and chapters, the practical and field experience combined with laboratory testing has shown that for stainless steels, but also for Ni alloys with obvious clarifications, the chemical composition determines the susceptibility to pitting corrosion and therefore the resistance of the alloy against this deleterious phenomenon. In addition to this, the characteristics of the corrosive environment in which the material is put into service also play a fundamental role in defining whether or not pitting corrosion will occur. As a consequence of this, the prediction of the behavior of stainless steels with regard to wet corrosion, and specifically pitting, requires the definition of useful indicators that take into account the aforementioned, especially to be able to outline effective prevention strategies. The most known are the free corrosion potential, the pitting potential, the critical pitting temperature (CPT), the critical pitting chloride concentration (CPCC), the repassivation potential and the pitting resistance equivalent. Among these six, the one of wider practical use by manufacturers, suppliers, customers and designers, also because of the ease in its calculation, is surely the pitting resistance equivalent [6, 13].

PRE stands for *pitting resistance equivalent* and is a calculated parameter useful for estimating the expected resistance of a material to localized corrosion (i.e., pitting). Its calculation is made according to an empirical formula considering the content of alloying elements that have beneficial effects on pitting resistance, i.e., those able to increase this property. In order to avoid confusion to the reader who may not fully recognize the nomenclature used below, it is worth to state and anticipate that in many cases the pitting resistance equivalent is also indicated as PREN (pitting resistance equivalent number). In any case, all the variations in the calculation of this parameter will be explained below, as well as the reasons for using one or another. The original formula for the calculation of the pitting resistance equivalent was developed for austenitic stainless steels free from intentional N additions and does not take into consideration nitrogen. This formula is provided by Eq. 3.7 [1, 3].

$$PRE = \%\mathrm{Cr} + (3.3 \times \%\mathrm{Mo}) \tag{3.7}$$

where

- *PRE* is the pitting resistance equivalent.
- %Cr is the chromium weight percentage.
- %Mo is the molybdenum weight percentage.

The development of alloys with more complex chemical analysis than AISI 316, containing higher amounts of nitrogen (e.g., duplex, superduplex, hyperduplex and superaustenitic grades) and with greater corrosion resistance led to the development of an additional formula (Eq. 3.8) that can be referred as PREn [20].

$$PREn = \%\mathrm{Cr} + (3.3 \times \%\mathrm{Mo}) + (16 \times \%\mathrm{N}) \tag{3.8}$$

where

- *PREn* is the pitting resistance equivalent for duplex and superaustenitic grades.
- %Cr is the chromium weight percentage.
- %Mo is the molybdenum weight percentage.
- %N is the nitrogen weight percentage.

Equation 3.8 gives the expression of the most widely used formula for calculating the pitting resistance equivalent referred to duplex, superduplex, hyperduplex and superaustenitic stainless steels. It is also often denoted by the capital letter N, thus changing from PREn to PREN. It is obvious that it is not the fact of writing the letter *n* in upper or lower case that makes the difference in this context, but it is the chemical analysis that determines the exact formula to be used. However, although there is general agreement on the effectiveness of Eq. 3.8 in calculating this useful index, some authors believe that the nitrogen multiplication factor should be increased to 30 for superaustenitics. In this way, the contribution of N would be weighted as in Ni-based alloys. In the event that W is also added to the steel (e.g., some superduplex and superaustenitics), it is necessary to adopt a variation of Eq. 3.8 to account for the presence of this element, thus obtaining Eq. 3.9 [20].

$$PREw = \%\mathrm{Cr} + \{3.3 \times [\%\mathrm{Mo} + (0.5 \times \%\mathrm{W})]\} + (16 \times \%\mathrm{N}) \tag{3.9}$$

where

- *PREw* is the pitting resistance equivalent for W-containing grades.
- %Cr is the chromium weight percentage.

- %Mo is the molybdenum weight percentage.
- %N is the nitrogen weight percentage.
- %W is the tungsten weight percentage.

In addition to what has just been declared above, another possible source of confusion or error regarding the need to write PREn or PREN arises also from the fact that for some sources and authors the use of an indicator for predicting the resistance against pitting corrosion should be extended to another family of alloys that, for chemical analysis, are close relatives of SASSs: corrosion-resistant Ni-based alloys. In fact, a formula has been proposed (see Eq. 3.10) that can be applied to this family of alloys which indicates the parameter as PREN [20].

$$PREN = \%\mathrm{Cr} + [1.5 \times (\%\mathrm{Mo} + \%\mathrm{Nb} + \%\mathrm{W})] + (30 \times \%\mathrm{N}) + (0.5 \times \%\mathrm{Cu}) \tag{3.10}$$

where

- *PREN* is the pitting resistance equivalent for Ni-based alloys.
- %Cr is the chromium weight percentage.
- %Cu is the copper weight percentage.
- %Mo is the molybdenum weight percentage.
- %N is the nitrogen weight percentage.
- %Nb is the copper weight percentage.
- %W is the tungsten weight percentage.

It should be noted, however, that much of the scientific literature, as well as manufacturer's datasheets, makes no distinction between the four names proposed by the four equations for calculating the pitting resistance equivalent and always defines it simply as PREN. What is usually done is simply to apply the correct formula according to the chemical analysis and the type of alloy studied and to declare it in advance. It is therefore emphasized that also in this textbook the pitting resistant equivalent will always be referred to as PREN. The values of pitting resistance equivalent for SASSs are given in Table 3.5 in accordance with the equations just mentioned. The values given in Table 3.5 are calculated using the percentages of alloying elements obtained as the average between the minimum and maximum values given for each element in Table 2.2. The reader may notice that the values provided may differ slightly from those found in the literature, but this is a discrepancy due solely to the fact that the average composition was considered. Furthermore, tables with PREN values in the literature often provide the value and the reference equation, but not the percentages considered. What is important to derive from Table 3.5 is the ranking of the materials according to the resistance parameter. For sake of completeness and comparison, although Eq. 3.7 is not suitable for austenitics with deliberate addition of N, its result was calculated to compare it with Eq. 3.8 and the modified version of this latter. From the given values, it is easy to see that the influence of N leads to higher values and this underlines the importance of N in increasing the steel resistance to pitting. Finally, from the point of view of the value, the grade with the highest pitting resistance equivalent turns out to be 654 SMO. The associated value is greater than 57 in case 16 is used as multiplication factor for N and greater than 64 in case the coefficient is 30. As seen in the equations provided, in addition to Cr and Mo, other elements must also be considered in evaluating the pitting resistance of a specific stainless steel grade. And considering them, it is also necessary to make some comments regarding their influence on corrosion resistance and the way the PREN formula is developed. Tungsten, for example, is used in some grades instead of molybdenum, but it is necessary to remember that its effectiveness is about half that of Mo in improving corrosion resistance. Furthermore, it is worth noting that the PREN calculation does not consider the negative effect on corrosion resistance of certain alloying elements, sulfur being one of them. Finally, the equation cannot consider issues of compositional inhomogeneity, crystallographic structure or allow for effective comparisons between alloys with the same PREN, but different CPT [6, 13, 21].

In this context, in addition to predictive-type parameters such as PREN, laboratory tests for corrosion resistance measurement of alloys are equally important sources of useful parameters for evaluating material performance and for developing improved alloys. These analyses involve measuring appropriate parameters once the material has been placed in aggressive environments, such as corrosive solutions. Of course, these tests are performed according to specific international standards, such as ASTM F 746-04, ASTM G48 and ASTM G150. Most of these tests use temperature as a ranking parameter, or a critical concentration value of specific chemical spices above which the corrosive phenomenon is certain to occur. Among the most commonly used parameters are CPT (*critical pitting temperature*), CCT (*critical crevice temperature*) and CPCC (*critical pitting chloride concentration*); whose values for some SASSs are given in Table 3.5. For fixed potential and environment conditions, CPT is defined in the context of stainless steels as the minimum temperature below which stable pitting cannot occur. In other words, this temperature can also be defined as the threshold above which stable pitting can occur. On the other hand, CCT is defined as the minimum temperature below which crevice corrosion cannot occur, thus the threshold temperature value above which crevice can occur. It interesting to note that for a given steel, CCT is always lower than CPT. This occurs because crevice, as a corrosive phenomenon, is able to initiate more easily than pitting [2, 6, 13, 15]. As mentioned above, there are tests in accordance with specific standards for the determination of CPT and CCT values. One of the practices for determining these two values is given by ASTM G48, which involves the use of a corrosive environment consisting in a solution with 10% ferric chloride in distilled $H_2O$ [5]. Finally, CPCC is defined as the threshold value of chloride concentration below which pitting does not initiate. From laboratory test results, the CPCC for stainless steels turns out to be function of chemical composition and, therefore, PREN. Not only that, operating conditions in terms of temperature and pH also play a key role in determining its value. Equation 3.11 gives an empirical relationship that can be used to determine the CPCC [2, 6, 13, 15].

**Table 3.5** Calculated values of pitting resistance equivalent for the steel grades given in Table 2.2 together with CPT and CCT values from literature sources

| Group | Most common name | EN | UNS | PRE Equation 3.7 | PREn Equation 3.8 | PREn* Modified Eq. 3.8 | PREw Equation 3.9 | CPT (°C) | CCT (°C) |
|---|---|---|---|---|---|---|---|---|---|
| High-performance austenitic stainless steels | AISI 904L | 1.4539 | N08904 | 35.85 | 35.85 | 35.85 | – | 40 | 15 |
| | AISI 904LN | – | – | 34.85 | 36.37 | 37.70 | – | – | – |
| 6Mo superaustenitics | 20Mo-6 | – | N08026 | 43.31 | 45.39 | 47.21 | – | – | – |
| | 1925 hMo | 1.4529 | N08925 | 42.28 | 44.68 | 46.78 | – | – | – |
| | NAS 254N | – | S32053 | 41.15 | 44.35 | 47.15 | – | – | – |
| | 25-6Mo 1925 hMo | 1.4529 | N08926 | 41.45 | 44.65 | 47.45 | – | 70 | 30 |
| | Uranus SB8 | 1.4537 | N08932 | 42.16 | 45.52 | 48.46 | – | – | – |
| | 254 SMO | 1.4547 | S31254 | 40.63 | 43.83 | 46.63 | – | 73 | 38 |
| | Nicrofer 3127 hMo | 1.4562 | N08031 | 48.45 | 51.65 | 54.45 | – | – | – |
| | AL-6XN® | – | N08367 | 42.45 | 45.89 | 48.90 | – | 78 | 43 |
| High-N superaustenitics | Uranus B66 | 1.4659 | S31266 | 43.80 | 51.40 | 58.05 | 54.70 | – | – |
| | 654 SMO | 1.4652 | S32654 | 49.75 | 57.75 | 64.75 | – | 105 | 75 |
| | Nirosta® 4565S | 1.4565 | S34565 | 38.53 | 46.53 | 53.53 | – | 90 | 50 |

The chemical composition taken as reference for the calculation is the average composition for each steel according to the data given in Table 2.2. PREn* is the value obtained from Eq. 3.8 where the N coefficient is 30 instead of 16 and, as mentioned in the main text [6, 15, 22].

$$\log[CPCC] \cong \left[\frac{PREN}{9} - \frac{7-pH}{5} - \frac{T-25}{50}\right] \tag{3.11}$$

where

- $CPCC$ is the critical pitting chloride concentration.
- $PREN$ is the pitting resistance equivalent for the considered steel.
- $pH$ is the pH value associated with the solution in which the material is placed.
- $T$ is the temperature expressed in °C [23].

In conclusion, comparing the effectiveness of the indicators described so far, it is possible to confirm that the PREN index is a somehow "predictive" practical indicator of corrosion resistance useful for material selection, while free corrosion potential, pitting potential, CPT, CCT and CPCC are values that can be measured according to specific standards. As a consequence of this, a rule of thumb states that steels with PREN < 18 can be used in environments with low chloride content or in particular corrosive conditions of operational discontinuity, cathodic protection, absence of oxygen or other oxidizing spices, high pH (e.g., in concrete as concrete-reinforcing elements) or in the presence of favorable galvanic couplings. Steels that contain Mo (e.g., AISI 316) and having a PREN > 25 can be used in environments such as brackish waters with chloride concentrations up to 1 g/L when the temperature does not exceed 30–40 °C. However, these steels are prone to pitting corrosion when deployed in seawater. In case of need to obtain the maximum performance in terms of corrosion resistance even in chloride-rich environments, it is necessary to use stainless steels with PREN > 45. Alloys capable of reaching such values are for sure hyperduplex (see Sect. 1.2.7) and superaustenitics. Finally, as additional information, practical experience has also shown that the numerical value of PREN is often approximately equal to the CCT expressed in °C in natural seawater or in ferric chloride solutions [6, 13, 21].

### *3.3.12 Performance in Specific Conditions*

As discussed up to this point, SASSs are a family of alloys developed to show outstanding corrosion resistance in a variety of environments thanks to a chemical composition developed just for this purpose. Moreover, since the main responsible for corrosion resistance is Cr, reference is often made to the positive interaction that other alloying elements have with it. Among all of them, the positive effect that Mo and N have and how this results in better pitting resistance performance in chloride-containing environments when both Cr, Mo and N concentrations are raised is well known. However, there can be considerable differences between grades belonging to the family of SASSs. This depends on the environment in which they are placed in service and the forms of corrosion that they might face. In order to understand what has just been stated, the upcoming sections will briefly analyze the behavior of SASSs as a function of specific environmental conditions.

#### 3.3.12.1 Natural Waters and Brines

Increase of pitting, crevice and SCC resistance has been one of the most powerful driving forces for the development of high-performance stainless steels and, consequently, of SASSs. Because SASSs are virtually immune to pitting in natural seawater in normal conditions, the focus in this section will be on crevice. All stainless steels behave similarly in terms of the variables that influence crevice susceptibility, and SASSs are no exception. However, precautions and best practices to ensure adequate resistance are often more rigorous when dealing with SASSs than with standard grades. This is relatively easily explained through the fact that the environments where SASSs operate are more corrosive, and this could lead to more severe crevice corrosion phenomena [5]. Common construction materials are subjected to corrosion in water in the moment when the cathodic process can occur. Oxygen reduction is the most common cathodic process, followed by acid conditions, presence of oxidizing spices such as chlorides, and finally the activity of bacteria (MIC will be described in Sect. 3.3.12.6). The most common consequences of corrosion include the reduction of the service life of components, obstruction of small pipes due to corrosion products and alteration of the quality of the water transported or stored within the parts subjected to corrosion. There are various types of water according to chemical composition, resistivity and pH, and a classification can be made as follows [13]:

- Distilled or demineralized water with very low amount of dissolved salts and a high resistivity.
- Freshwater, which is characterized by a low salinity value (below 2 g/L). Freshwater can be classified according to its origin in: rainwater, surface waters (lakes, ponds, rivers) or groundwater percolating through the soil.
- Drinking water.
- Industrial waters.
- Brackish water and seawater. Brackish water is characterized by a salinity exceeding 2 g/L and seawater is the most abundant brackish water, with a salinity in the 34–36 g/L range.
- Formation water. This type of water is associated with oil or multiphase hydrocarbons extracted from a reservoir and is characterized by a very high ionic content, greater than that of seawater (from 3 to 4 times). Not only that, the pH is usually slightly acidic. Formation waters are rich in $CO_2$ and $H_2S$ and practically oxygen-free. Under these conditions the main cathodic process is H evolution.

There are several factors influencing the corrosion of metals and alloys in water, most important of them are: dissolved oxygen (or the cathodic reactant), pH, temperature, resistivity, presence of bacteria, hardness and scaling power [13].

Much of the information related to corrosion resistance in water comes from exposure tests. These tests involve exposing, for example to seawater, coupons with controlled-size crevices and correlating the results with the composition of the steel or a laboratory parameter such as CCT. For example, tests have shown that in natural seawater at $T_{amb}$ crevice does not initiate if the CCT of the material is at least 35 °C. With reference to Table 3.5, all steels with CCT higher than 35 °C can be considered suitable for use in natural seawater at near-$T_{amb}$ for applications such as condenser piping and tubing [5]. AISI 904L is not a SASS, but a high-performance austenitic grade according to what was explained in Chap. 2. With this material, for example, a rule of thumb dictated by field experience sets the limit to the chloride concentration in brackish or freshwater with a maximum value of 5000 ppm (see Fig. 3.12). This limit, combined with the 35 °C criterion for CCT, can be used to develop a serviceability guideline for the selection of SASSs that perform better than AISI 904L for the specific use. This concept is highlighted in Fig. 3.12, where the chloride concentration and the CCT are related for a standard austenitic grade (AISI 316), AISI 904L and 254 SMO. It is evident that the serviceability of 254 SMO is better than that of AISI 904L because the material has higher CCT and the ability to be employed with higher chloride concentrations. This type of representation become useful for the material selection phase in designing applications such as heat exchanger tubing when the composition of the cooling water is known [5].

There are some non-critical applications that require the use of large sections where localized corrosion initiation can be tolerated. In these cases, an estimate of the rate of crevice or pitting corrosion is useful so that failure can be anticipated. However, the value turns out to be difficult to be determined in service because of variables such as those related to component geometry. In this context, Fig. 3.13 provides an estimate of the crevice corrosion rate as a function of CCT (ASTM G48) from data obtained from tests lasting even several years. It can be seen that grades with CCT > 35 °C are virtually immune to this type of corrosion in seawater at near-$T_{amb}$ [5].

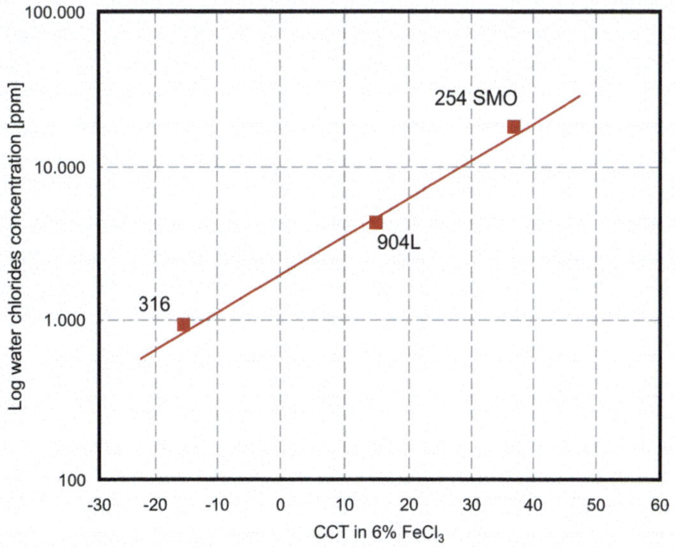


**Fig. 3.12** Serviceability guidelines for stainless steels based on alloy CCT and chlorides concentration (redrawn from [5])

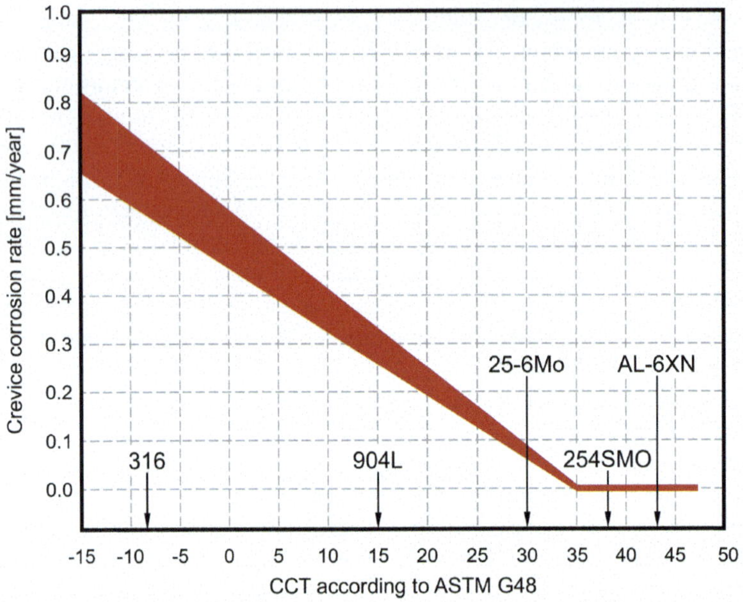


**Fig. 3.13** Estimation of the crevice corrosion rates for stainless steel in near-$T_{amb}$ seawater as a function of the CCT (redrawn from [5])

In the case of critical service conditions, such as thin wall tubing, where the available distance for corrosion propagation would be very small, the tolerability criterion just suggested would lapse and the use of a SASS grade with higher corrosion resistance performance would be necessary from the beginning [5]. As already mentioned, temperature plays a key role in determining the aggressiveness of the corrosive environment. If the temperature increases, the aggressiveness would do the same. Considering natural seawater (see composition in Table 3.6), corrosivity has a peak around 40 °C. Above this value there is a decline caused by a decrease in oxygen solubility and a polarization peak due to biological activity, or due to oxygen solubility product-temperature relations for other constituents. For applications involving clear seawater cooling with metal temperature close to $T_{amb}$, 6Mo SASSs (see Table 2.2) can resist pitting and crevice with moderate crevice attacks [5].

However, the 6Mo SASSs would be subject to localized corrosive attacks in seawater at higher temperature and higher chloride concentration. For example, 6Mo SASSs are not suitable for heat exchangers handling seawater because these heat exchangers often operate at high temperature and have gaskets with a design that can involve interstices that lead to severe crevice corrosion. High-N SASSs such as 654 SMO or Nirosta® 4565S are preferred for these applications. Figure 3.14 shows the effect of temperature and chloride content on the possibility of having crevice corrosion initiation for a one-year laboratory exposure test to oxygenated (i.e., aerated) synthetic sea salt solutions acidified to pH = 2 to simulate the corrosivity

**Table 3.6** Concentration of the main chemical spices in seawater [13, 24]

| Ion or molecule | Molar mass | Concentration (mmol/L) | Concentration (ppm) |
|---|---|---|---|
| $Ca^{2+}$ | 40.08 | 10.28 | 412.02 |
| $K^+$ | 39.10 | 10.21 | 399.21 |
| $Mg^{2+}$ | 24.31 | 53.08 | 1290.37 |
| $Na^+$ | 22.99 | 468.50 | 10770.82 |
| $Sr^{2+}$ | 87.62 | 0.09 | 7.89 |
| $Br^-$ | 79.91 | 0.842 | 67.28 |
| $Cl^-$ | 35.45 | 545.90 | 19352.16 |
| $F^-$ | 19.00 | 0.068 | 1.29 |
| $HCO_3{}^-$ | 60.988 | 2.30 | 140.27 |
| $SO_4{}^{2-}$ | 96.02 | 28.23 | 2710.64 |
| $B(OH)_3$ | 61.804 | 0.416 | 25.71 |

of natural seawater. The influence of temperature is evident, as is the superiority of SASSs over standard austenitics. The superiority of SASSs is such that in the case of exposure to NaCl solutions at 90 °C with pH and aeration variations, the 654 SMO is able to show no pitting, crevice or other forms of corrosion regardless of pH and aeration [5].

Finally, the fact that SASSs are able to resist crevice corrosion in seawater at $T_{amb}$ much better than standard austenitic grades and even under lower pH conditions, for higher chloride concentrations and for tighter crevice gaps is summarized in Fig. 3.15. This representation completes the discussion on crevice resistance by also considering a geometric factor, crevice gap width. This aspect becomes particularly

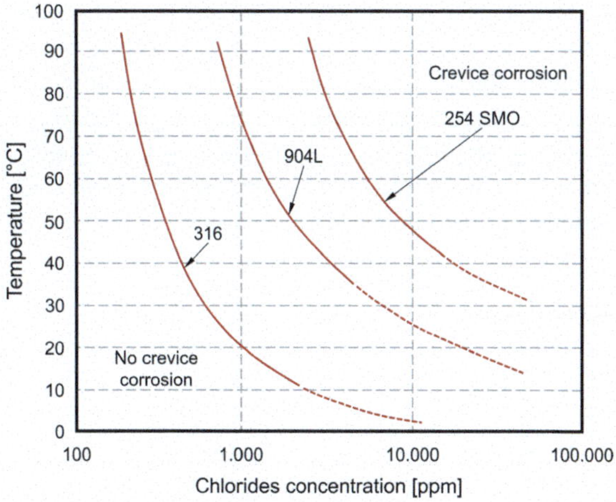


**Fig. 3.14** Effect of temperature and chloride on the initiation of crevice corrosion in aerated synthetic sea salt solutions at pH = 2 (redrawn from [5])

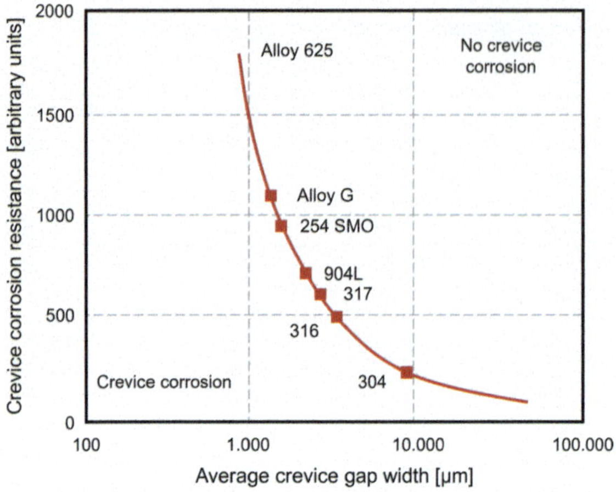


**Fig. 3.15** Behavior in terms of crevice corrosion resistance of several stainless steels and two Ni-based alloys (Alloy 625 and Alloy G) as a function of the gap width in ambient temperature seawater (redrawn from [5])

important in joints producing very tight interstices such in the case of compression fittings, or threaded connections. Looking at the representation from another point of view, it can be said that SASSs do not need wide gaps to avoid crevice corrosion, as is the case with AISI 304 or AISI 316. They can therefore be used in more critical contexts [5].

#### 3.3.12.2 Acids

Several applications involve the contact of stainless steels, and thus SASSs, with acids of various type. Many examples can be found in the chemical or in pulp and paper industry. The possibility for industrial plant components to come into contact with acids requires the need to make appropriate selection of materials according to their resistance to corrosion when placed in contacts with these substances. A description of the behavior of SASSs when in contact with various acids, both inorganic and organic, will be briefly outlined below. The inorganic acids that will be considered are:

- Hydrochloric acid (HCl).
- Hydrofluoric acid (HF).
- Phosphoric acid ($H_3PO_4$).
- Sulfuric acid ($H_2SO_4$).
- Sulfurous acid ($H_2SO_3$).

Concerning organic acids, the behavior of SASSs will be analyzed with reference to interactions with:

- Acetic acid ($CH_3COOH$)
- Fatty acids.
- Formic acid.

Hydrochloric acid is a strong and very reducing acid. Solutions containing it, even the weakest ones, lead to corrosion of standard grades once they get in contact with them. Any increase in temperature leads to increased corrosion rate and rapid H evolution. When a stainless steel component comes into contact with HCl, irrespectively from the grade, it is always good practice for the surface to be thoroughly rinsed afterward. SASSs show better corrosion resistance than standard grades when in contact with HCl thanks to the high concentration of Cr, Ni and Mo that allows for higher stability of the passive film. Of course, their chemical composition rich in Cr and Mo makes them more resistant to chloride pitting. Despite this, however, even SASSs cannot be used to handle concentrated solutions [5].

As can be clearly seen from Fig. 3.16, even members of the best-performing class of the high-N SASSs such as 654 SMO are not able to provide useful resistance at $T_{amb}$ with solutions approaching the 10 wt% concentration. As can be seen from Fig. 3.16, SASSs can be candidates for handling dilute solutions at moderate temperatures. In fact, the corrosion rate increases significantly with temperature. Evidence of this lays in the fact that at boiling temperature the isocorrosion curves clearly show that solutions with concentrations higher than about 1 wt% cannot be handled by 654 SMO without leading to its corrosion [5].

Hydrofluoric acid is also a strong acid and is considered as particularly dangerous for stainless steels as it can dissolve the protective $Cr_2O_3$-based passive film, as HF is able to dissolve ceramic materials. Standard austenitic grades are not considered suitable for use in contact with solutions containing HF. In fact, a 0.1 wt% solution

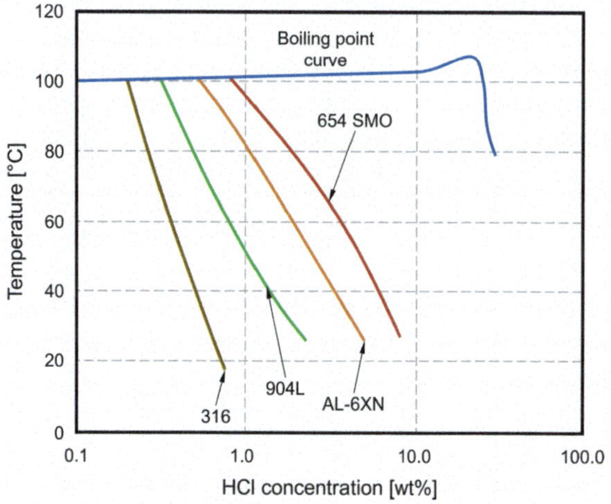


**Fig. 3.16** 0.1 mm per year isocorrosion curves in pure HCl acid solutions for various types of austenitic stainless steels (redrawn from [5])

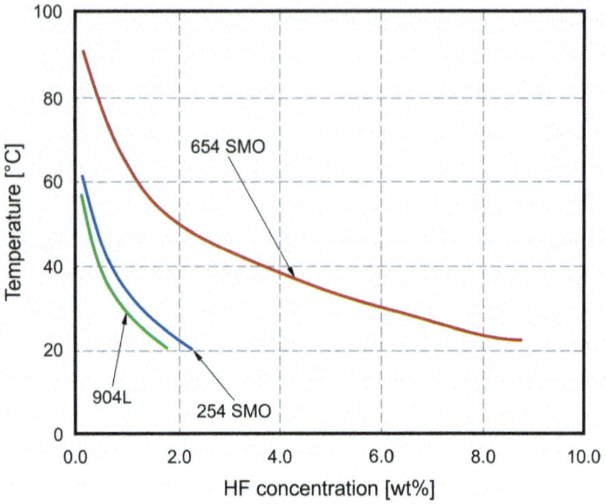


**Fig. 3.17** 0.1 mm per year isocorrosion curves in pure HF acid solutions for various austenitic stainless steels (redrawn from [5])

at 60 °C is capable of producing a corrosion rate of more than 0.6 mm per year when in contact with AISI 304. SASSs can offer acceptable performance at moderate temperatures and relatively low concentrations, especially when solutions are aerated or when oxidizing salts are present and in mixed solutions with $H_2SO_4$. As shown in Fig. 3.17, grades characterized by high Cr, Ni and Cu content can provide acceptable corrosion resistance in dilute solutions. Once again, the best performing grade turns out to be 654 SMO [5].

Pure $H_3PO_4$ solutions are less aggressive than $H_2SO_4$ solutions and have relatively low oxidizing power. High temperatures and concentrations lead to high corrosion rates, and the possible presence of ions that influence the oxidizing potential or that may lead to pitting initiation (e.g., chloride and fluoride ions). A low C content in the steel is essential because $H_3PO_4$ can produce intergranular attack of GBs that have been subjected to sensitization. As it can be seen from Fig. 3.18, at high temperatures and concentrations, good corrosion resistance performance can be obtained from 6Mo or high-N SASSs. In the moment when $H_3PO_4$ acid solutions contain chloride or fluoride ions, as is the case in $H_3PO_4$ production process, the best performance is achieved by SASSs with higher Cr and Mo contents [5].

The first stainless steels defined as high-performance grades were those developed for sulfuric acid service. Sulfuric solutions can have variable oxidizing power depending on the composition of the specific sulfuric acid environment. Pure solutions with mid-range concentrations at high temperature lead to weakly oxidizing conditions and high general corrosion rates. Aeration and presence of oxidizing ions (e.g., chromates, cupric, ferric, nitrate) lead to an increase in the oxidizing potential of dilute solutions and generally allow the material to maintain the passive behavior over a wider compositional and temperature range. The presence of chloride ions or other halides can lead to the onset of pitting corrosion. The presence of halide ions is a factor that cannot be underestimated since the behavior of steel can vary considerably between a pure $H_2SO_4$ solution and one where some halides are present (see

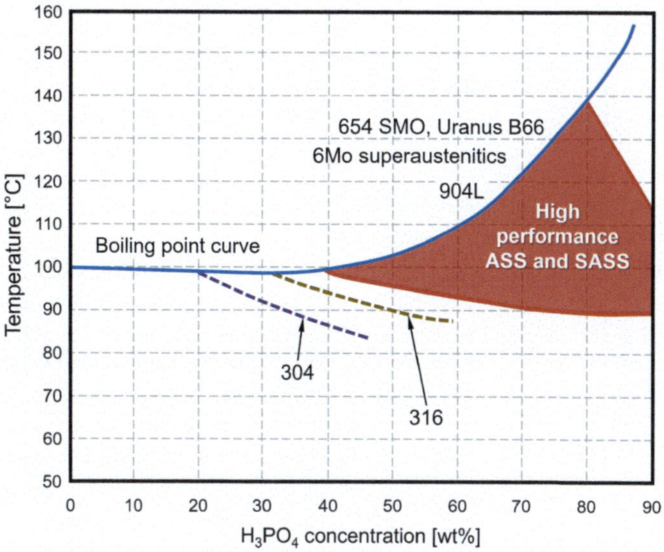


**Fig. 3.18** 0.1 mm per year isocorrosion curves in pure $H_3PO_4$ acid solutions for various types of austenitic stainless steels. ASS stands for austenitic stainless steel (redrawn from [5])

Fig. 3.19 in comparison with Fig. 3.20). SASSs such as 20Mo-6, which was develop to achieve better pitting resistance, offer good corrosion resistance in the mid-acid compositional range. Figure 3.19 shows how 654 SMO manages to perform better than AISI 904L up to concentrations close to 50 wt%. However, the performance is not that good in the mid- to high-acid concentration range because the lower Ni content compared to that of AISI 904L (see Table 2.2) makes 654 SMO less resistant against corrosion in non-aerated $H_2SO_4$ acid solutions [5].

In dilute solutions containing chloride ions or other halide ions, the best performance can be obtained from those grades characterized by higher pitting resistance due to their higher Mo content. As it can be seen from Fig. 3.20, an increasing chloride ions concentration reduces the corrosion resistance of all grades. 654 SMO turns out to show interesting behavior, even with a chlorides concentration of 2.000 ppm [5].

AL-6XN® also performs better than AISI 904L at the same chloride ions concentration. These differences also result in different application possibilities. In fact, SASSs (both 6Mo and high-N) can be successfully employed in applications where combustion product acid condensates must be handled. The reason why lays in the fact that combustion product acid condensates often contain chlorides, and, according to Fig. 3.20, this makes AISI 904L unsuitable for the specific application [5]. Applications involving the condensation of flue gases containing $SO_2$ can bring the material into contact with $H_2SO_3$ (sulfurous acid). This acid is a weak acid and, by itself, is not particularly corrosive as it induces pitting for AISI 304 and can be successfully handled with AISI 316. What has been just stated is valid as long as $H_2SO_4$ and

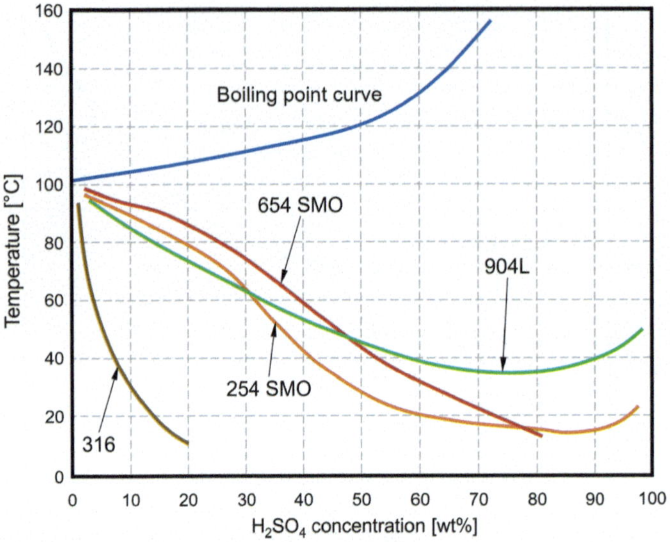


**Fig. 3.19** 0.1 mm per year isocorrosion curves in non-aerated $H_2SO_4$ solutions for various types of austenitic stainless steels (redrawn from [5])

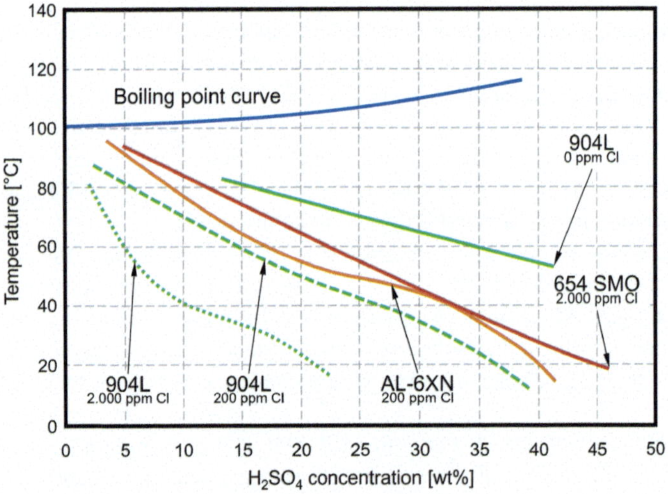


**Fig. 3.20** 0.1 mm per year isocorrosion curves in non-aerated $H_2SO_4$-chloride solutions for various types of austenitic stainless steels (redrawn from [5])

chloride or fluoride ions are not present. In any case, because flue gases may contain halide ions and become very acidic, high-performance austenitic grades such as AISI 904L or those alloys belonging to the 6Mo-superaustenitics group may offer better corrosion resistance and safety than standard grades (see Sect. 3.3.12.4) [5].

Acetic acid is a very corrosive organic acid, second only to formic acid. Acetic acid can become highly reducing at high concentrations and therefore very corrosive, especially for materials with passive behavior such as SASSs. Acetic acid becomes very corrosive also in case of high temperature, especially when chloride ions are present. Usually, for moderate temperatures AISI 304 can provide good corrosion resistance in acetic acid for any concentration. AISI 316, on the other hand, can provide better performance and resist even up to atmospheric boiling point temperature. However, due to their higher Mo content, SASSs can provide better performance at high temperatures and when chloride ions and other contaminants are present. Specifically, members of both 6Mo-superaustenitics and high-N groups are capable of markedly better performance in terms of pitting and SCC resistance. A non-negligible aspect concerning acetic acid regards its production process. In fact, it involves an oxidation process that produces other substances, including formic acid. The resulting mix of acids is particularly aggressive and, as it can be inferred from Fig. 3.21, a low corrosion rate is provided by 254 SMO, while AISI 316L shows a much higher rate also for lower concentrations. Figure 3.21 also shows the behavior of 2205 standard duplex for comparison (chemical composition in Table 1.11) [5].

Concerning industrial fatty acids, these substances are usually not dangerous from a corrosive point of view, and even standard grades can come into contact with them without any special problems. The situation could become more complicated as temperature is increased because corrosivity does the same, as does the possibility of crevice and pitting corrosion initiation. This case may occur, for example, during recovery and refining of tall oil at high-end process temperatures. In this case a

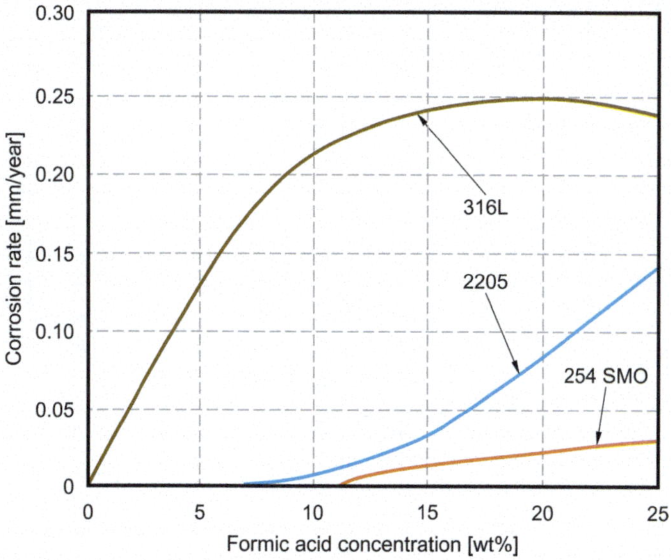


**Fig. 3.21** Corrosion rates in boiling mixtures of 50 wt% acetic acid and variable content of formic acid for three stainless steels (redrawn from [5])

standard grade such as AISI 316 is inadequate, however, an acceptable corrosion rate can also be achieved with AISI 317. It is therefore obvious that for higher-performance alloys such as SASSs fatty acids are not problematic. As a consequence of this, the use of SASSs may be unjustified, or necessary only where operating and environmental conditions would set the stage for the use of Ni-based alloys [5].

Finally, as far as acids and organic acids are concerned, the last substance to be discussed is formic acid. Among organic acids, formic acid is the most aggressive and reacts significantly with many metals and alloys, especially for what concerns warm or hot solutions. The corrosivity of formic acid is influenced by the presence of oxidizing substances, and in fact the performance of stainless steels in contact with it improves markedly with aeration and in presence of oxidizing agents. Taking AISI 316 as reference, all SASSs perform better in contact with formic acid, particularly at high temperature (e.g., boiling solutions). Notwithstanding their better properties when compared to standard austenitics, SASSs are not the best choice for corrosion resistance in formic acid. Taking the corrosion rate in boiling formic acid solutions as a benchmark parameter (see Fig. 3.22), superferritics and superduplex grades manage to provide much lower corrosion rates than a 6Mo-superaustenitic like AL-6XN® [5].

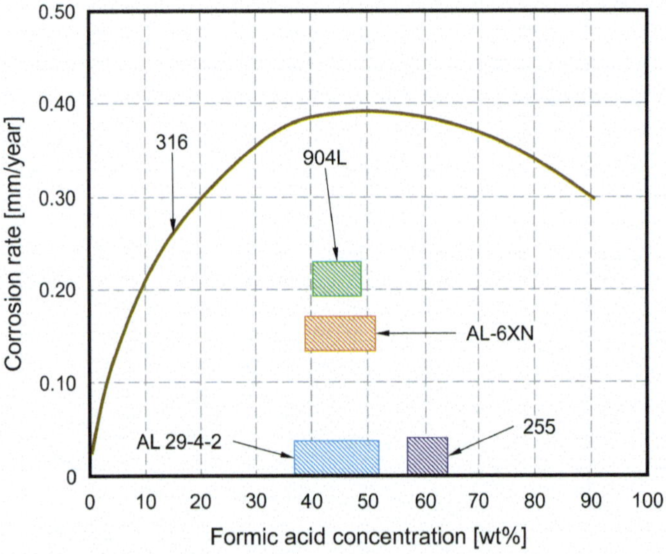


**Fig. 3.22** Corrosion rates in boiling formic acid solutions. AL 29-4-2 is a superferritic stainless steel and 255 is a superduplex (redrawn from [5])

### 3.3.12.3 Alkalies and Alkaline Salts

Regarding resistance to strong alkalies such as NaOH or KOH, both carbon steels and standard stainless steels, and consequently SASSs, are able to resist against corrosion at $T_{amb}$ up to moderate temperatures. Hence, it turns out that weak alkalies and alkaline salts such as $Na_2CO_3$ are not dangerous from the point of view of corrosion up to the boiling temperature. In contrast, other compounds such as NaOCl (sodium hypochlorite), which is an oxidizing salt, can be corrosive to many metals, including stainless steels [5]. Regarding direct contact with NaOH at high temperature, SASSs definitely perform better than standard austenitics. However, the best performance is attributed to some ferritic grades. From Fig. 3.23 it can be seen that SASSs have a corrosion rate in pure boiling NaOH that is almost an order of magnitude lower than standard austenitics. High-purity ferritics (i.e., those having a very low Ni content), on the other hand, can achieve a corrosion rate that is an order of magnitude lower even than SASSs [5].

In the moment when strong alkalies contain impurities, such as oxidizing salts, high-performance stainless steels such as SASSs and superferritics show outstanding corrosion resistance. This happens in particular because of their high Cr content. An alkaline environment in which contaminants of oxidizing nature are present is paper pulp mill white liquor (used in the first stage of the Kraft process during which hemicellulose and lignin are separated from cellulose fiber). In these cases, SASSs definitely offer better performance than standard grades, especially with regard to resistance against general corrosion and SCC. Corrosion rates are somewhat higher in strong alkalies containing impurities than for pure alkalies solutions, but still within acceptable ranges. Figure 3.24 represents the behavior of four different materials in a simulated environment involving the presence of NaCl, $NaClO_3$ and $Na_2SO_4$

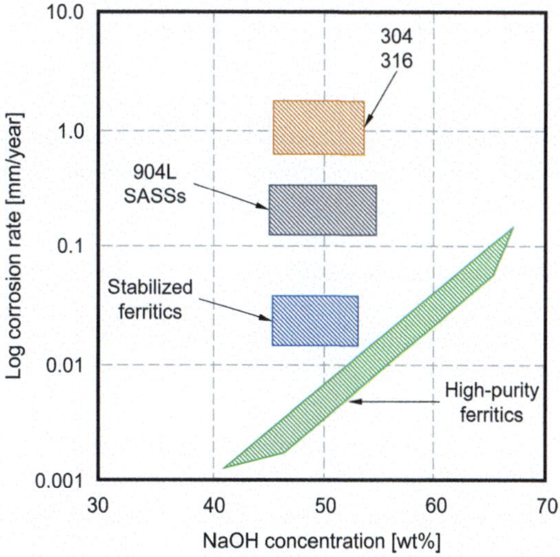


**Fig. 3.23** Corrosion of stainless steels in boiling reagent grade NaOH solutions (redrawn from [5])

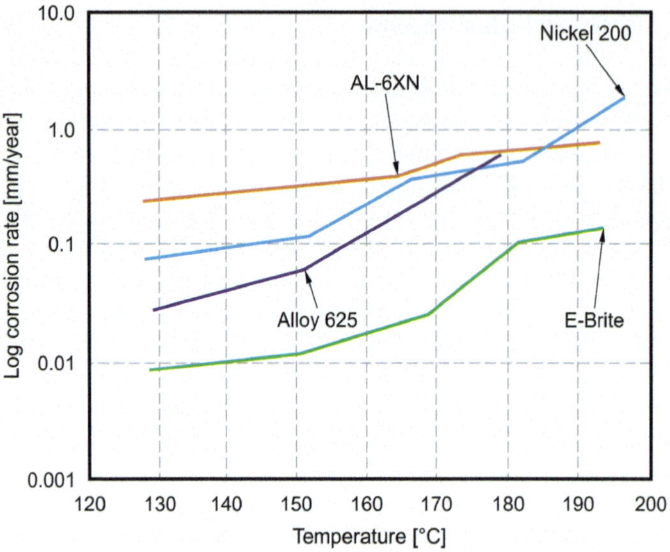


**Fig. 3.24** Comparison between corrosion rate of high-performance stainless steels with a Ni-based superalloy and Nickel 200 in a simulated caustic evaporator liquid with composition 43 wt% NaOH, 7 wt% NaCl, 0.45 wt% $Na_2SO_4$ and 0.15 wt% $NaClO_3$ (redrawn from [5])

(sodium sulfate) and reproducing the conditions of a caustic evaporator liquid. As it can be seen from Fig. 3.24, it is a member of the superferritic family (E-Brite®, see Sect. 1.2.7) that shows the lowest corrosion rate for any temperature. In order to give a more complete picture of the performance level in the specific environment, Alloy 625 is also included in the plot [5]. This material belongs to the family of Ni-based superalloys and features a Cr content of 21.5%, Mo content of 9% and Fe content of 5% [12]. It is interesting to note that E-Brite® performs better than both the AL-6XN® and Alloy 625, but also that AL-6XN® offers virtually the same performance as Alloy 625 at 180 °C. Not only that, the behavior of Nickel 200 (99% pure Ni) is also reported as an additional benchmark material. In particular, it is interesting to note that E-Brite® performs better even than Nickel 200, but more interestingly, at 190 °C it is AL-6XN® that does better than Nickel 200 [5]. Based on the data just discussed, it is evident that superferritics are better performing materials than SASSs when it comes to being in contact with pure or contaminated strong alkalies. However, in some situations SASSs are able to offer comparable and in some cases superior performance than Ni alloys.

#### 3.3.12.4 Halide-Containing Environments and Flue Gas Condensates

In terms of localized corrosion, the aggressiveness of halide-containing environments is determined by temperature, pH, type of halide, and oxidizing power of the oxidant. In near-neutral solutions, bromide is the most aggressive halide, followed by chloride. Iodide and fluoride are relatively harmless in near-neutral environments. In acidic solutions, however, fluoride can be very aggressive [5]. Strong oxidants can take the potential of stainless steel above that of pitting or crevice for any halide. High temperature and low pH can contribute to the increase in corrosivity of the environment. An example where such conditions can be encountered is the pulp bleaching step in papermaking. A further example is chlorination for fouling control in seawater-cooled heat exchangers (fouling consists of micro- and macroorganisms growing on the submerged part of a structure, or even inside plants) [5, 12]. In order to better explain the effect of chlorination, Fig. 3.25 highlights the change in potential vs Ag/AgCl standard electrode for 254 SMO exposed to natural seawater with different chlorination conditions. As it can be easily seen, even 0.1 ppm of continuous chlorination can produce a positive shift in the corrosion potential [5].

Field experience shows that SASSs and superferritics can be used with continuous chlorination at $T_{amb}$ with Cl levels at least as high as 1 ppm. For severe high-T crevices, such as conditions that flanges might face, 654 SMO equals the performance of Ni alloys and has been successfully employed in flange connections in critical seawater handling systems [5]. As explained in Chap. 2, one of the possible applications of SASSs, particularly high-N grades, is in the fabrication of components for FGD plants. The possibility for pitting and crevice to occur increases with decreasing pH and increasing T, halide concentration and, especially, with reducing conditions that can also promote generalized corrosion. Such conditions can be encountered in FGD plants. Fuels containing high levels of Cl or S can produce some of the most

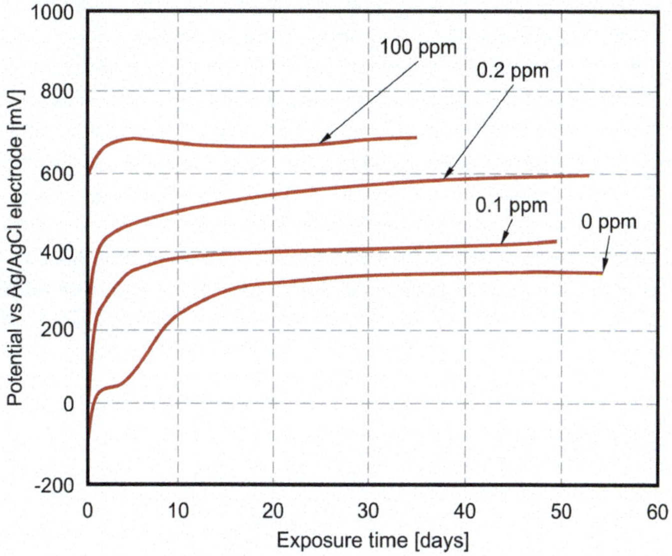


**Fig. 3.25** Variation of the open circuit potential of 254 SMO exposed to natural seawater as a function of the chlorination (redrawn from [5])

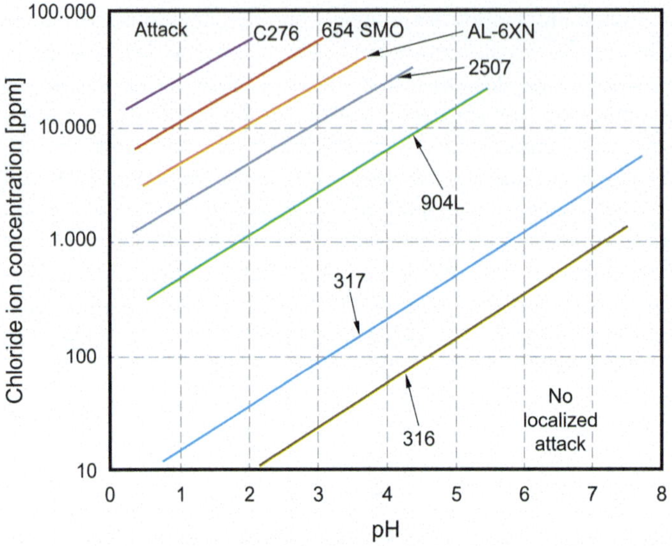


**Fig. 3.26** Approximate service limits for several stainless steels and the C276 Ni-based alloy in flue gas condensates and acidic brines at moderate temperatures (60–80 °C) (redrawn from [5])

corrosive combustion products. The most common examples are fuel oils, municipal waste, and sulfur coals. Condensate and scrubbing liquors become acidic with $SO_2$ and reducing with HCl and $SO_3$. The characteristics of the gas-handling system in terms of design and method of operation strongly influence the severity of the corrosive environment that may be generated. As a general indication, washed walls and adsorbent-treated liquors are less corrosive, while most corrosive conditions are produced by raw condensates and recycling. For more information regarding the structure and conditions present in FGD systems, the reader can refer to ASTM STP 837 where a generic schematic of the FGD system and operating zones can also be found. In any case, the approximate behavior of various stainless steels together with the C276 Ni-based alloy is illustrated in Fig. 3.26 regarding the service limits in flue-gas condensates and acid brines at moderate temperatures (60–80 °C). From this representation it is possible to see that SASSs can perform better than some superduplex grades (i.e., 2507) [5].

#### 3.3.12.5 Sour Oils and Gas Environments

In case $H2_S$ was also addedo to waters with high chloride contents, the corrosivity of the environment to the material would grow non-negligibly. Such conditions are often found in oil and gas production, and the presence of $CO_2$ or intentionally introduced acidifiers can also add to them. Such conditions increase the likelihood of pitting, crevice and SCC and even generalized corrosion as the corrosivity of the

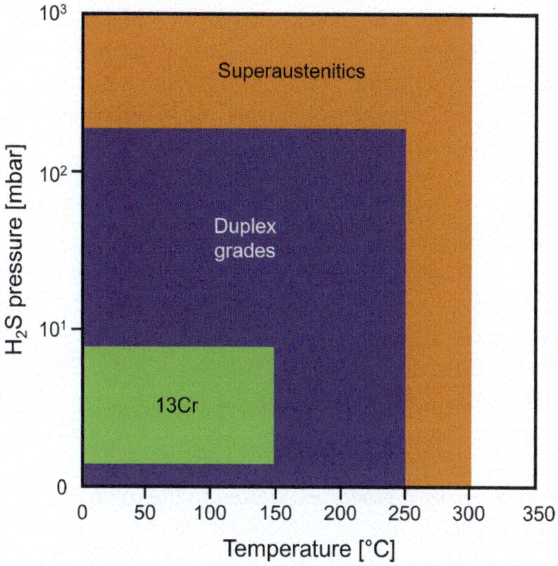


**Fig. 3.27** Application range in terms of temperature and partial pressure of $H_2S$ for corrosion resistant stainless steels in sour environments containing 50 g/L NaCl (redrawn from [5])

environment increases. At relatively low $H_2S$ levels, standard grades manage to offer adequate corrosion resistance. However, as temperature, chloride concentration and $H_2S$ partial pressure increase, along with a lowering of pH, high-performance stainless steels become an obvious choice. SASSs generally outperform duplex grades in terms of resistance to $H_2S$-assisted SCC and this can be seen in Fig. 3.27. The superiority results in an increase of both maximum partial pressure and temperature values for the same Cl salts content in the environment [5].

Regarding behavior in environments containing high H concentrations, standard austenitics, and even more SASSs, are reported to have high resistance to such operating conditions characterized by high partial pressure of H. For this reason, they are often used for handling H over a wide range of temperatures and pressures [5].

#### 3.3.12.6 Microbes-Containing Environments

As introduced in Sect. 3.3.9, under some circumstances microbiological activity can lead to the occurrence of localized corrosion for standard austenitic stainless steels where it would not have occurred in the absence of microorganisms. As mentioned, stagnant water conditions can lead, along with the presence of an adequate microbial population, to the occurrence of MIC [5]. Some of the industrial applications where MIC can take place are nuclear power plants service water and emergency cooling systems. Evidence of MIC-related failures in such applications have been reported in the literature for coated C-steel components and also for piping in AISI 304 and AISI 316. Compared with other forms of corrosion that do not involve microorganisms, MIC is characterized by less literature data [5]. MIC does not actually provide for any new form of corrosion than those already introduced in this chapter. Consequently, it can be said that the resistance of stainless steels to this type of corrosion is influenced by the same parameters considered for the others, e.g., the content of alloying elements that increase resistance to crevice and pitting [17]. As a consequence of this, and from what emerges at the present date, it would seem plausible that the corrosion resistance of SASSs is greater than that of standard austenitics. For this reason, SASSs, and in particular those in the 6Mo group, have been extensively used in environments described above due to their higher corrosion resistance ($E_{pit}$ value in particular) [5]. Not only that, the use of SASSs, SDSSs and HDSSs in marine environments where MIC is likely to occur has become a common practice [17]. As mentioned, there are various types of microorganisms that can lead to MIC. A detailed analysis of the consequences of the action of each of these is beyond the scope of this textbook, and the reader is referred to the references at the end of this chapter. However, in order to offer a description that can be a general guideline in understanding the mechanism behind MIC, an interesting and fairly representative case could be that of one of the microorganisms mentioned in Sect. 3.3.9: *Pseudomonas Aeruginosa.* The influence of its presence has been studied in the marine environment with regard to the possible corrosion of one of the members of the high-N group of SASSs: 654 SMO. Tests showed that the presence of the *biofilm* produced by the bacteria accelerates the corrosion of the 654 SMO. This was also verified by the detection of a negative shift in the open circuit potential, a decrease in polarization resistance and an increase in the current density. The detected corrosion mechanism was pitting, and it would appear that the biofilm catalyzes the formation of $CrO_3$, thus damaging the passive film and compromising the resistance due to a damage of the $Cr_2O_3$-rich passive layer. The formation of the *biofilm* can destroy the passive layer or impair its reformation on the metal surface. High concentrations of C, N and O were detected underneath the *biofilm* compared with those characteristics of the surface of SASSs. Not only that, Cr, Fe and Ni were detected underneath the *biofilm* at corrosion products. This could suggest that the *biofilm* is involved in the loss of these elements and consequently participates in the corrosion process. The current density in the presence of *Pseudomonas Aeruginosa* undergoes a significant increase. The increase in current density, together with the shift in potential, suggests that the anodic dissolution reaction of the metal is allowed. There is a decrease in the $E_{pit}$ value, indicating that pitting should be the corrosive phenomenon most likely to occur. It is interesting to see that the tests conducted did not show substantial changes in pH due to the presence of the bacterium. As a result of this, it could be assumed that the acceleration of corrosion is not caused by the organic acids secreted by the microorganisms. Compared with a standard duplex like 2205, 654 SMO performed much better in the environment where *Pseudomonas Aeruginosa* was present because the average pit depth detected was about 5 times lower. However, the material is not totally immune to MIC, and the acceleration of the corrosive process seems to be mainly caused by the formation of $CrO_3$ and subsequent damage of the surface passive film on 654 SMO [18].

#### 3.3.12.7 Stress Corrosion Cracking

As already specified within Sect. 3.3.7, SCC is one of the most insidious forms of mechanochemical corrosion. For what concerns stainless steels, SCC involves either cathodically-controlled H cracking or anodically-controlled cracking in presence of specific ionic spices, usually chlorides. Other halides can cause cracking like chlorides, but their presence is less common and their effect depends on other variables in the environment such as oxidizing potential and acidity. The influence of the presence of specific cations in halogen salts is mainly on the pH and hydrolyzation, where more acidic salts are more aggressive. For example, NaCl is fairly neutral and is generally less aggressive than salts containing Ca or Mg cations [5].

SASSs perform better against SCC than standard austenitics due to their higher Cr, Ni and Mo content and are therefore less susceptible to this type of degenerative process. For example, for water and brine environments, the grades having Ni > 18% are characterized by a substantial increase in SCC resistance when compared to standard austenitics (see Table 2.2). This is evident from Fig. 3.28 where it is also clear how duplex grades perform better than standard austenitics in the most common chloride environments [4, 5]. Of course, it is not possible to define the performance of SASSs against SCC in an absolute way since this is a function of the type of environment where conditions for SCC may occur, as well as the stress field and geometry of the specific component. AISI 904L has often been employed in applications where AISI 316 would be inadequate, and 6Mo-superaustenitics, as well as the high-N group, could be used in even more aggressive environments where SCC is likely to occur [5]. In order to give a classification of materials according to SCC resistance that could also be useful in design, laboratory tests can be employed. However, laboratory tests could have limitations due to the many system variables involved in the phenomenon. These include, as partly mentioned above, the potential (determined by oxygen concentration), the actual stress pattern, ion concentration, and the possibility of evaporation. Usually, an application where it is possible to have significant incidence of all these factors is constituted by heat exchangers subjected to localized boiling and hot surfaces covered by insulation. Among the different tests useful for simulating such applications is DET (drop evaporation test) according to ISO 15324:2000 standard for evaluating the SCC resistance of metals and alloys. DET involves holding a specimen under constant load in a horizontal configuration and heating it by resistance heating to the test temperature. A salt solution is then dropped onto the center of the gauge length at a constant rate until the specimen fails. The output data of this test is usually the time until failure. However, other indicators related, for instance, to a threshold load value above which crack initiation occurs can also be obtained. At the end, DET is basically a ranking test [5, 25]. As mentioned in Sect. 3.3.7, SCC is dangerous because it can occur at load levels well below the yield strength of the material. An additional way to classify resistance to SCC would be

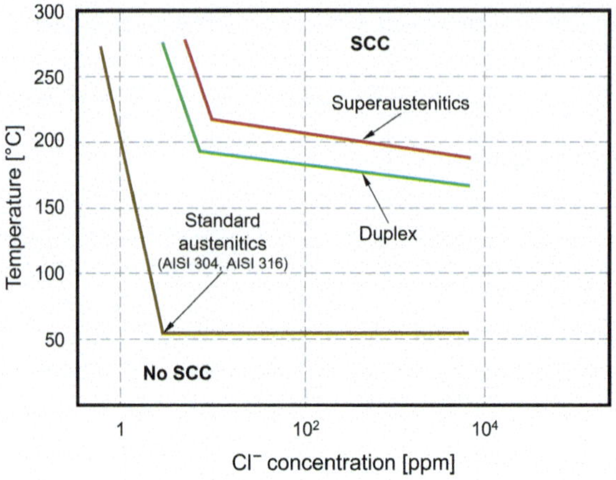


**Fig. 3.28** Stress corrosion cracking resistance as a function of temperature and $Cl^-$ concentration (redrawn from [4])

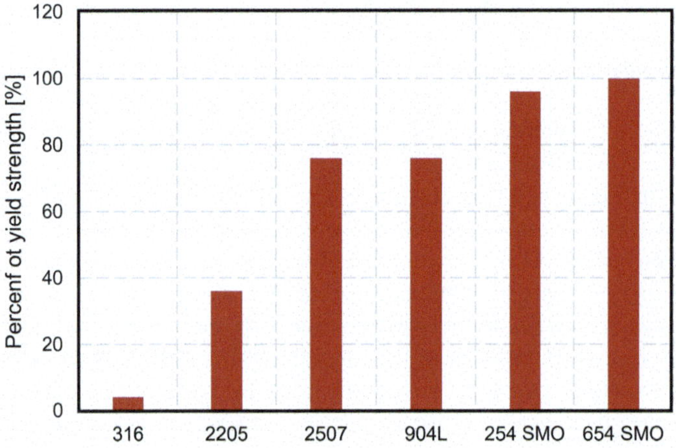


**Fig. 3.29** Stress corrosion cracking resistance expressed as percentage of the yield strength at which crack initiation takes place during DET with NaCl solution at 120 °C (redrawn from [5])

to check the stress value at which crack initiation takes place. The higher this value turns out to be, the greater the resistance.

In this context, Fig. 3.29 shows the performance of various stainless steels in terms of percentage of yield strength at which crack initiation takes place during DET with NaCl solution at 120 °C. The lower the associated value, the worse the SCC resistance behavior. From the analysis of the data in Fig. 3.29, it is evident that SASSs are superior to standard austenitics, and even to standard duplex (2205) and superduplex (2507) stainless steels [5].

## References


1. Groover, M.P.: Fundamentals of Modern Manufacturing—Materials, Processes and Systems, 7th ed. Wiley (2019)
2. Schweitzer, P.A.: Metallic Materials—Physical, Mechanical and Corrosion Properties. Marcel Dekker Inc (2003)
3. Liu, Y., et al.: Predicting the density of molten alloys using computational thermodynamics. Calphad **68**, 101690 (2020)
4. Capelli, F.: L'acciaio INOX. Centro Inox Servizi Srl (2021)
5. Kovach, C.W.: High-Performance Stainless Steels. The Nickel Institute (2000)
6. McGuire, M.F.: Stainless Steels for Design Engineers. ASM International, Materials Park, Ohio (2008)
7. Rossiter, P.L.: The Electrical Resistivity of Metals and Alloys. Cambridge University Press (1991)
8. Hosford, W.F.: Mechanical Behavior of Materials. Cambridge University Press (2010)
9. Straffelini, G.: Ductility and Formability of Metals—A Metallurgical Engineering Perspective. Academic Press (2023)
10. Masamura, T., et al.: Work-hardening mechanism in high-nitrogen austenitic stainless steel. Mater. Trans. **61**(4), 678–684 (2020)
11. McCafferty, E.: Introduction to Corrosion Science. Springer (2010)
12. Gialanella, S., Malandruccolo, A.: Aerospace Alloys. Springer (2020)
13. Pedeferri, P.: Corrosion Science and Engineering. Springer (2018)
14. Schweitzer, P. A.: Fundamentals of Metallic Corrosion—Atmospheric and Media Corrosion of Metals, 2nd ed. CRC Press (2007)
15. IMOA: Practical Guidelines for the Fabrication of Austenitic Stainless Steels, 2nd ed. International Molybdenum Association (IMOA), London (2020)
16. Schibicheski Kurelo, B., et al.: Improved saline corrosion and hydrogen embrittlement resistance of superaustenitic stainless steel by PIII nitriding. Technology **18**, 1717–1731 (2022)
17. Loto, C.: Microbiological corrosion: mechanism, control and impact—a review. Int. J. Manuf. Technol. **92**(9–12), 4241–4252
18. Li, H., et al.: Microbiologically influenced corrosion behavior of S32654 super austenitic stainless steel in the presence of marine Pseudomonas aeruginosa biofilm. J. Mater. Sci. Technol. **33**(12), 1596–1603
19. Li, H., et al.: A new insight into high-temperature oxidation mechanism of super-austenitic stainless steel S32654 in air. J. Alloy. Compd. **686**, 326–338 (2016)
20. Farrar, J.C.M.: The Alloy Tree: A Guide to Low-Alloy Steels, Stainless Steels and Nickel-Base Alloys. Woodhead Publishing Limited (2004)
21. ASM International: ASM Handbook Volume 13B—Corrosion: Materials. ASM International, Materials Park, Ohio (2005)
22. Schweitzer, P. A.: Fundamentals of Metallic Corrosion—Atmospheric and Media Corrosion of Metals, 2nd ed. CRC Press (2006)
23. Lazzari, L.: Engineering Tools for Corrosion—Design and Diagnosis. Woodhead Publishing (2017)
24. Housecroft, C., Sharpe, A.: Inorganic Chemistry, 3rd ed. Pearson—Prentice Hall (2008)
25. Steinsmo, U., Drugli, J.M.: Assessment of susceptibility to chloride stress corrosion cracking of highly alloyed stainless steels. Part 1: Drop evaporation test method. Paper presented at the National Association of Corrosion Engineers (NACE) International Annual Conference and Corrosion Show, New Orleans (LA), 09 March 1997


## *Further Reading*


Ashby, M.F.: Materials Selection in Mechanical Design, 4th ed. Butterworth-Heinemann (2011)

# Chapter 4
# Manufacturing, Processing and Uses of Superaustenitic Stainless Steels

**Abstract** The final building block related to the understanding of superaustenitic stainless steels concerns the techniques by which these materials can be produced, manufactured, processed and brought to market. It is known how the material influences the production process dedicated to it, and also how in some cases it is the process that determines whether the material can be produced or not. The manufacturing and processing characteristics of each material then force the choice of the most suitable process for it, the exclusion of its production by certain techniques, and also the development of technologies that can process it in a way that has never been done before. In this context, the knowledge of the properties of the material is as fundamental and important as knowing the aspects of each technological process to be employed in the production cycle. Manufacturing and processing possibilities strongly influence the type of product that can be realized and, in combination with material characteristics and associated cost, contribute to the determination of the reference markets.

**Keywords** Manufacturing · Processing · Uses · Melting · Casting · Hot-working · Cold-working · Welding · Heat treatment · Machining

## 4.1 Introduction

Chapter 1 was structured in such a way as to introduce the reader to the world of stainless steels. A first part of historical contextualization was provided in order to come to a better understanding of the path that led to the development of these alloys and, subsequently, of SASSs. In fact, in addition to the development of new materials, one of the major driving forces for materials science in recent decades has been the improvement of the performance of existing materials. Moreover, performance implementation is actually the overlap of many improvements in materials, processes and design [1]. SASSs are certainly one of the cases where these concepts have come to fruition, given the needs that led to their development. The mechanism that enables stainless steels to achieve higher corrosion resistance in many environments was also



A. Malandruccolo, *Superaustenitic Stainless Steels*, Engineering Materials,
https://doi.org/10.1007/978-3-031-68744-0_4

explained in Chap. 1, and the main families were introduced. Finally, a generalized overview of possible uses was addressed in order to complete the initial picture of the world of stainless steels. This paved the way for Chap. 2, where, starting from the chemistry, specific aspects of SASSs were analyzed. Using the chemical composition as a reference, the types of SASSs and the individual members of this class of alloys were defined. Again, referring to chemical analysis, important aspects concerning the solidification of these alloys were addressed. Finally, the possible secondary phases that can be detected in these steels were described. In turn, Chap. 3 used what was previously explained to set forth the properties of SASSs. In sequence, physical, mechanical and corrosion resistance-related features were described. This was enriched with a brief description of corrosive phenomena and the response of SASSs in particular environments, a summary description of which was also provided. The journey through the universe of SASSs must necessarily conclude with the processes that lead to the manufacture of products with the associated possibilities for their use. In fact, technological properties and aptitudes of a material can only be adequately defined if its chemical nature and resulting properties (i.e., physical, mechanical, corrosion resistance) are understood. Etymologically, the word *manufacture* originates from the combination of two Latin words: *manus* (hand) and *factus* (make). The resulting word means *made by hand*. This makes perfect sense because, in origin, human hands were the first tool employed to shape materials. From the perspective of the technological meaning, the term manufacturing can be defined as the application of physical, chemical and mechanical processes aimed at altering the geometry, properties and/or appearance of a given material to make parts or products (assembly of parts into products is also included). Still from the technological point of view, in order to properly complete manufacturing, a precise combination of machinery, tools, power, environment and labor is required. On the other hand, from an economic perspective, manufacturing is the transformation of materials into products resulting from the process and characterized by increased value given by the employed processes themselves and/or assembly operations. Of course, the evolution of humankind has led to the introduction of new technologies that now make it possible to replace human hands and accomplish manufacturing through automated, computer-controlled equipment and machinery [2]. The onset of artificial intelligence (AI) is perhaps about to set the pace for further development in manufacturing, eventually making it possible the development of even more autonomous manufacturing systems that can manage themselves independently of human input.

Processes employed in manufacturing are able to convert materials from one form to another and add value to them. As depicted in Fig. 1.1, the history of humankind is linked to the ability to work with tools and materials. In the ancient past there has been the transition from the Stone Age to the Iron Age passing through the Bronze Age; more recently there has been the so-called Steel Age. Finally, although ferrous materials continue to dominate the world of manufacturing in many fields, humankind has entered an era where molecular engineering products may dominate the stage [1]. As with materials, processes have also seen considerable development over the past 50 years, and new types are constantly being developed to handle the new materials. This has been done in order to achieve more efficient handling and use, and with the goal of minimizing waste as well [1]. The processes must be compatible with the chosen material, and this can lead to two different conditions. The first happens when the material defines the process, and so the process must be developed or chosen to match the characteristics of the material in order to process it properly to achieve the desired result. On the other hand, the second condition happens when the process dominates and a material match must be identified [3].

In order to put SASSs into the context of manufacturing, this chapter aims to analyze the processes that enable the transformation of these materials. Obviously, this last chapter does not have the goal of describing in detail the technological processes that will be discussed with regard of SASSs. It will be taken for granted that the reader is already familiar with the general aspects of the processes presented, and only the most relevant for processing SASSs will be outlined. For this purpose, the main stages that can compose the manufacturing process related to products made of SASSs will be covered. The discussion will begin with the melting and refining processes. After this, the main casting processes by which SASSs can be solidified will be covered. Both hot working and cold working operations will be presented and, immediately after, most important aspects of heat treatments will be covered. Before reaching the conclusion of the path, the processes of machining and joining by welding will be illustrated. Finally, some considerations regarding the use and markets of SASSs will be presented.

## 4.2 Melting, Refining and Casting

As with virtually all metal alloys, the production of stainless steels begins with smelting processes and can have two different starting points in terms of *raw material*: ore or scrap. When reference is made to Fe-rich ores as feedstock, the initial processing route is said to be that of the *extractive metallurgy* or *primary metallurgy* of steel and this involves the use of a blast furnace for the conversion of ore to *pig iron*. Pig iron does not possess a chemical composition that can be ascribed to that of steel and therefore requires further refining operations by means of converters in order to achieve the desired chemistry [4]. Market data and those related to the annual production of stainless steels and market share of SASSs will be presented in Sect. 4.8, however, it should be stated already that as far as the world production of stainless steel is concerned, almost all of it has scrap as its starting material. As a result, the focus of this section will be on the production process that has scrap as feedstock. In addition to this, metal alloys are usually classified into two main categories: cast and wrought. This distinction is made depending on whether a casting process or further deformation processing is sufficient to obtain the final product, or semi-finished product to be processed, respectively. SASSs can be produced in either cast or wrought form. Up to this point in the text, reference has always been made to the nomenclature of SASSs in their wrought form, this is because the market of cast products turns out to be much narrower than that of wrought products. In addition to this, as far as cast products are concerned, even the most common trade names are different from those mentioned starting from Chap. 2. For example, the 254 SMO becomes CK3MCuN and the AL-6XN® takes the name CN3MN. In this context, regarding the nomenclature and compositions used for the fabrications of castings in SASSs, the reference is usually the ASTM A744/A744M-21a standard. This standard groups precisely the specifications for the production of castings, Fe–Cr–Ni, corrosion resistant alloys for severe service; hence, SASSs. As already mentioned, since the market for cast products is much smaller than that for wrought products, the discussion from here on will be based on the latter type. Nonetheless, with regard to possible problems related to the manufacture of cast products, useful references may be those presented in Chap. 2 regarding peculiarities and problems related to solidification. Furthermore, again with regard to the properties of SASSs, in case of interest in casting processes for obtaining finished products, the properties of specific heat, thermal conductivity and expansion coefficient are presented in Sect. 3.2. As a result of what has just been explained, the typical manufacturing process from scrap of wrought SASSs products involves a melting stage, a refining and finally a casting stage. A schematic representation of this process is provided in Fig. 4.1.

The exact processing route to which the material should be subjected is a function of the availability of equipment and facilities within the factory and, more incidentally, the required quality of the final product. The processes depicted in Fig. 4.1 will be briefly presented in the following in order to understand the main features and implications in the production of SASSs.

### *4.2.1 EAF Melting*

EAF stands for electric arc furnace and it is one of the most important plants in steel production factories, and is nowadays employed by more than 80% of steel producers. Regarding the rest, slightly less than 20% of the melting processes from scrap are carried out using induction furnaces; the remaining are other less common melting practices. The EAF consists of three main parts: a tiltable internally refractory-lined vessel, a moving roof, and electrodes. The roof is movable to allow the furnace to be opened in order to load it with the scrap. The EAF charge consists of scrap and other substances such as ferroalloys and slag-forming compounds:

- Scrap: internal waste material of the production plant such as nonconforming material, chips, grinding powder etc. In addition to this there are machining industry scraps, selected scrap, and demolition of products and systems that contain material with similar chemical composition to the final alloy to be produced.
- Ferroalloys such as low-C FeNiCr, low-C FeMn and low-C FeCr.
- Slag forming substances such as CaO, $CaF_2$, MgO, $Na_2SiF_8$.

Typically, the EAF is loaded in more than one step because the scrap has high volume due to limits in its compaction resulting in much empty space. Each charge is melted before the next is loaded in the furnace. The temperature increase required

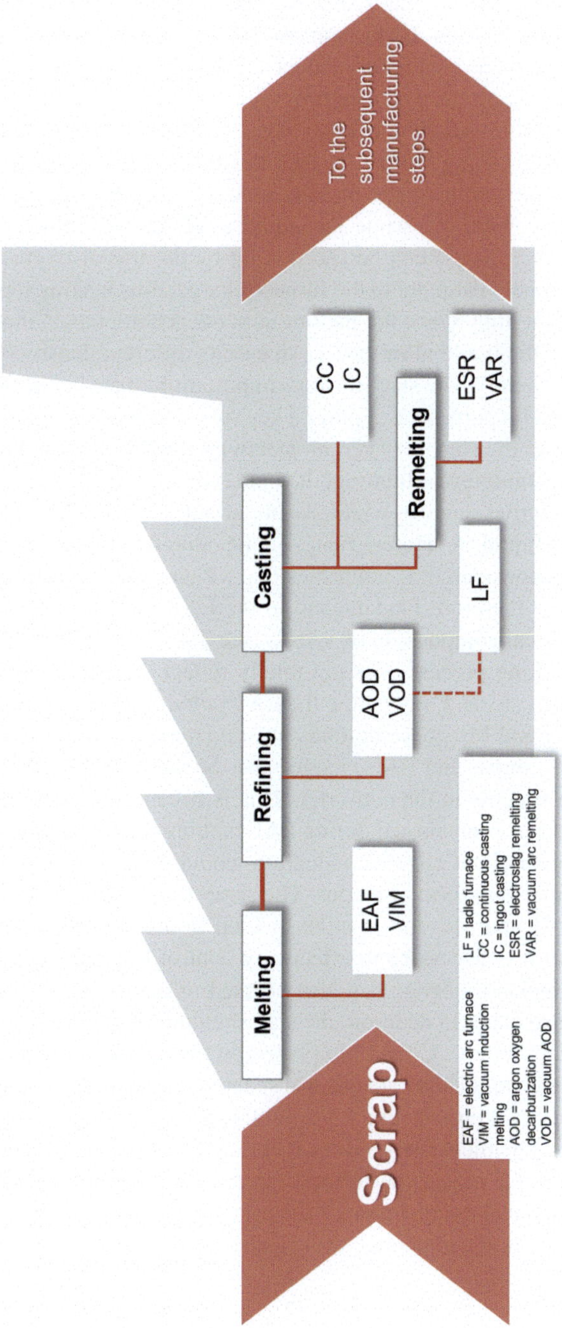


**Fig. 4.1** Typical melting, refining and casting sequence and processes employed in the production of wrought SASSs

for melting is achieved by energy transfer between the graphite-based electrodes and the charge due to an electric arc that sparks between them as a consequence of the high voltage difference. Usually, EAFs employed in steel production are alternating current furnaces with three electrodes. Direct current furnaces are less common and feature only one electrode. The electrodes are mounted vertically through the movable roof and are able to float above the scrap, or the molten bath, according to the specific step of the melting process. In fact, the distance between the electrodes and the charge/melt varies during the melting process. The temperature of the arc reaches values around 3600 °C, while the liquid steel reaches values even close to 1800 °C [4, 5]. The process can be speeded up by the use of oxyfuel burners, which provide additional heat input to the furnace charge, thus making it possible to anticipate complete melting. Once the melting process is complete, a liquid slag is formed and floats on the liquid alloy surface due to its different density (the slag is an oxide-based compound). The slag is a very important by-product of the melting process and is formed on purpose as it plays a key role in capturing impurities from the bath. Not only that, the slag has a certain reactivity toward the liquid bath and is able to regulate, by promoting or limiting, the oxidation of alloying elements in the melt. Excessively reactive slag can withdraw important alloying elements, such as Cr and Mo, from the liquid. A saturated slag, on the other hand, may jeopardize the removal of elements considered as impurities (e.g., P and S). A proper calibration of the slag features is therefore fundamental to the EAF melting process. In light of the complex chemical composition of SASSs (see Table 2.2), rich in Cr and Mo, the output of the melting process does not totally reflect the final composition of this family of stainless steels. C content in the EAF output must be limited as much as possible, while Cr and Mo concentrations must be maximized to limit additions in the converter. Of course, the EAF charge must be calibrated according to the subsequent refining process in the converter. This is because even the charge will have specific Cr and Mo contents and during EAF melting the high temperature and the impossibility to carry out effective reduction would result in some removal of these important elements in form of oxides. The scrap for the production of SASSs obviously contains Cr and Mo, and in order to keep as much of these elements in the bath as possible, it is necessary to effectively control the temperature to limit their oxidation (Ellingham diagram can also be used in this context to understand the stability field of oxides). In addition, the slag should have a reactivity such that it does not excessively remove Cr and Mo present in the liquid. During the melting process, samples of the liquid alloy are taken at specific moments during the process. These small amounts of liquid are quickly solidified and sent to the quality control laboratory to perform chemical analysis in real time, giving feedback to the operators on the effectiveness of the process performed. Once the melting process is complete and the expected chemical composition of the alloy at this stage in the production has been achieved, the liquid is poured into a ladle and sent to the converter for the refining process.

### 4.2.2 *VIM Melting*

VIM stands for vacuum induction melting and it is a melting process involving the use of a furnace consisting of a melting vessel placed in a steel shell kept under vacuum thanks to a high pumping rate vacuum system. Melting of the charge is achieved thanks to induction heating obtained by a magnetic field that induces a current into the metal charge. The system devoted to heating is made of water-cooled Cu-based induction coils wrapped around the melting vessel. One of the key parameters to be carefully adjusted during the process is the induction coil frequency, to be varied according to the mass and composition of the alloy charge. The driving force for the development of VIM technology has been the need for producing materials with better controlled mechanical properties as a result of a particular compositional cleanliness. Compared to air melting (i.e., EAF), VIM can provide a considerable reduction in the oxygen content of the liquid. In addition, other impurities such as Pb and Sb can also be removed through evaporation promoted by the low-pressure processing environment. In addition, compared to the EAF + AOD process sequence, VIM allows for a more accurate control of the final chemical composition of the alloy [4]. The characteristics featured by alloys produced by the VIM process also allow for better behavior with regard to corrosion resistance. In fact, corrosion resistance is also affected by microcleanliness and the presence of impurities in the steel (improved microcleanliness and lower impurity content improve corrosion resistance). However, VIM also has disadvantages directly resulting from the inherent limitations associated with the principle on which it is based:

- Process and equipment are more expansive than air melting.
- Elements with high vapor pressure (e.g., Mn) are difficult to be controlled from the compositional point of view.
- The chemical composition of the alloy resulting from the melting process is essentially fixed and determined by the composition of the charging material. This results in the need to use only scrap with certified chemical composition and in line with the final product, or material with the correct chemical composition in case VIM is employed as a secondary purification process. Accordingly, compositional changes such as C, Si and S content reduction, cannot be implemented.

In light of the above information, VIM is hardly used as a melting process for SASSs. Its use as a secondary purification process is more common, such as in the production of 254 SMO, which can employ VIM to achieve better cleanliness values by limiting the presence of impurities in order to further improve corrosion resistance.

### 4.2.3 *Refining*

When refining is mentioned in the context of steel production, it refers to those processes whose task is to modify the chemical composition of the liquid alloy in order to set it within the values required for the specific steel type. Among the refining processes for stainless steels, the leading role is played by the AOD (argon oxygen decarburization). As mentioned in Chap. 1, and emphasized in more detail in Chap. 2, the very birth of stainless steels is inextricably linked with the AOD process. AOD has made possible the production of alloys with high Cr, N, Ni and Mo content; key elements in the definition of the chemistry of SASSs. The refining process is conducted in a processing vessel consisting of a refractory-lined steel shell mounted on a tiltable trunnion ring. The shape of the AOD converter is an evolution of that of the older Bessemer and looks quite similar, but the specific geometry in terms of height and width are typical of each manufacturer. The converter is equipped with tuyeres in the lower sidewalls, whose job is to inject Ar, N and O into the liquid steel that has previously been received from the ladle. In order to increase the reaction efficiency of oxygen with liquid, and to make the process faster, many modern converters are equipped with a top lance for oxygen blowing even from above as is the case with the LD converter. These AOD converters are usually referred to as OTB (oxygen top blowing). The AOD converter has no heat supply systems, and the heat needed to maintain the steel in the liquid state is provided by the exothermic oxidation reactions that occur in the converter as a result of oxygen injection. Even in the case of the AOD converter, temperature control is essential to avoid excessive oxidation of the liquid and to prevent the slag from undergoing changes in its reactivity. Temperature control is carried out by carefully adjusting the amount of injected oxygen, by the addition of material with a lower enthalpy content into the liquid (e.g., alloying elements, ferroalloys and solid steel scrap), and by the inert gases injected from the tuyeres. Typically, the refining process in the AOD converter consists of two main steps: decarburization and reduction. The decarburization phase usually consists of three steps in which the ratio between the injected gases varies and where, of course, oxygen is always employed. Usually, in the first step a mix of $N_2$ and oxygen in a 3:1 ratio is injected, in the second step a 1:1 mixture of Ar and oxygen is instead employed. During the third, the last oxidation step with Ar and oxygen in a 1:3 proportion, the C content is decreased down to the very low levels needed by the composition of SASSs (see Table 2.2). It is important to note how proportion between the employed gases and blowing time are dependent on the condition of the liquid and the typical fabrication practices of each manufacturer. In fact, even during the AOD process, the chemical composition of the melt is checked at regular intervals in the same way as described for EAF. Once C oxidation is completed, oxygen blowing is stopped and only Ar is used, which does not react with the liquid and has the sole function of stirring. This is the reduction stage in which, in addition to stirring operated by Ar, slag reduction occurs thanks to the reduction mixture released within the converter [4]. The reduction mixture contains elements that are more likely to bond with oxygen than Cr and Mo (e.g., Al, Mg and Si). Proper slag reduction is critical in order to allow the recovery of the alloying elements and to ensure the expected chemical analysis of the material at the end of the process. As mentioned in Chap. 2, one of the most difficult aspects in the production of SASSs is keeping high N concentrations in the liquid. This is achieved by continuous injection of N and by high Cr content in the liquid, which is capable of retaining N in it. In addition to this, the reactivity of the slag throughout the process must be such that excessive removal of Cr and Mo is avoided, and therefore it is important that the slag is properly saturated during the oxidation step. Temperature control is also fundamental since, as already mentioned, an excessive increase could lead to a variation of the slag activity and to the loss of some alloying elements from the liquid. In case the reduction step is successfully completed, it is possible to move on to the next step, which can be either delivering the heat to casting processes or an intermediate processing in a ladle furnace (LF).

In case the intention is to limit Cr and Mo loss during refining and obtain a higher C removal rate, one option could be to carry out the conversion process under vacuum. In this case the process can be either VOCD (vacuum oxygen decarburization converter) or VOD (vacuum oxygen decarburization). The difference between the two consists in the fact that the former is performed in a converter, while with the latter the conversion is carried out in a ladle. The conversion process does not differ substantially from that in air with the AOD converter, however since VOD and VOCD operate in vacuum environment, thorough control of gas blowing is paramount. Thanks to the very low CO partial pressure, the C removal rate is higher than that of AOD. In addition to this, S levels in the range of 10–30 ppm can be achieved for any type of steel. This latter aspect, in view of the extreme sensitivity of SASSs to the presence of S (reduction in high-T ductility, see Sect. 4.3.1), makes vacuum conversion processes particularly attractive if a premium quality product must be obtained from the produced SASS heat [4].

As mentioned above, a ladle furnace (LF) could be used after the conversion processes, if the factory is provided with one of it. The LF can be regarded as an intermediate post-conversion and pre-casting station. Basically, the LF implies placing the ladle containing the liquid steel inside a station equipped with a movable roof with three electrodes that can be lowered onto the ladle in a configuration that resembles that of the EAF. The difference with the EAF is first of all related to the fact that the electrodes are smaller in size (i.e., diameter) and are not used to melt, but at most for regulate the temperature of the liquid by operating heating of the heat. The LF is also provided with additional equipment such as a wire injection system, an electromagnetic stirrer, and a gas blowing system from the bottom of the ladle. As mentioned in Chap. 2 and as it will be explained in Sect. 4.5, SASSs do not have versions with improved workability and therefore do not need special treatments during LF processing (e.g., spheroidization of inclusions or increase in S content). Concerning SASSs, the main functions of the LF are related to eventual temperature control (arc heating or cooling by Ar bubbling), chemical and thermal homogenization before casting thanks to the stirring action performed by the electromagnetic stirrer.

### 4.2.4 Casting Practices

As mentioned above and as it will be explained in Sect. 4.8, since the market share of castings is very small, the casting processes that will be addressed will be those related to the production of semi-finished products intended to be further processed through other processes (e.g., forging, hot rolling, etc.). Thus, in this context, reference will be made to techniques useful for the production of semiproducts that can have long products (e.g., bars, wire rods, wires) or flat products (e.g., sheets) as output. Generally speaking, casting processes can be defined as procedures that allow the steel heat to solidify into useful forms for subsequent production steps. In light of previously stated information, the casting techniques will be ingot casting (IC) and continuous casting (CC). IC is a casting process that employs non-expendable molds for the production of ingots of various sizes and shapes. Ingots can be defined as large discrete castings weighting from less than 1t up to around 300t or more (the weight of an entire heat). Ingot molds are usually made of cast iron and are tapered at the top or bottom in order to allow the removal (*stripping*) of the ingot once solidification is complete. The shape of the ingots is dependent on their mass and segregative tendency of the alloy. As seen in detail in Chap. 2, SASSs have high tendency to segregate some alloying elements and therefore both aspects will have to take this phenomenon into account with the aim of minimizing it. Ingots of relatively low mass (up to usually around 4–5 t) tend to have a square cross-section. For large weights, and therefore larger volume of liquid to be solidified, the shape is polygonal with a higher number of sides the greater the mass in order to promote a greater heat flow out of the ingot, i.e., higher cooling rate. There are two main types of ingot casting: top pouring (direct teeming) or bottom pouring (uphill teeming). Of the two processes, bottom pouring is practically the one that dominates the stainless steel ingot production. The process involves teeming through a trumpet that feeds a runner system capable of filling the ingot molds from the bottom until they are full. The trumpet is required to funnel the steel from the ladle to a sprue plate that has separate channels to direct the steel to the molds. Both trumpet and runners are refractory lined. Thanks to this specific architecture, lower turbulence in the liquid, better surface and internal quality of the ingots can be achieved. One of the major advantages related to bottom pouring is the possibility to fill multiple ingots simultaneously. The main disadvantages related to this process consist mainly in the higher cost compared with top pouring and the greater complexity of setting up the equipment prior to casting. Concerning SASSs, due to their high segregative tendency already explained in Chap. 2, severe ingot segregation of Mo can be said as an unavoidable phenomenon. Once the ingot has solidified, it can be subjected to forging or hot rolling processes depending on the final product to be obtained. In this context, it is necessary to carefully calibrate the reheating of the ingots prior to processing. The soaking time in the furnace should not be excessive because of possible reactions of the surface with the atmosphere of the reheating furnace (more detail will be provided in Sect. 4.3). In addition, the process parameters must also be carefully defined in terms of strain and strain rate. This is because incorrect values could lead to incipient melting between the grains, thus causing defects in the processed product. In fact, ingots tend to have higher segregation passing from the surface to the center, and this leads to a different structure and composition along the cross-section. More segregated regions with lower melting temperature can be found toward the center of the ingot, and any melting of them during plastic deformation processes can lead to internal defects in the produced material.

On the other side, continuous casting (CC) is a casting process developed with the purpose of bypassing conventional ingot casting, thus overcoming the inherent limitations of the discrete process, in order to cast a product that can be directly rolled on finishing mills. This holds for plants that are built in such a way to have the CC machine in series with a rolling mill (integrated mill). In case the layout of the production plant does not allow the continuous casting machine to be in series with the rolling mill, continuous casting process still allows for a higher production rate than ingot casting (mini-mills). This is because the CC machine can process virtually an unlimited number of heats in sequence due to the feeding mechanism that provides the ability to alternate multiple ladles filled with liquid steel in sequence. The general sequence of operations for CC implies the delivery of the liquid steel to the casting machine from the converter or LF. The liquid steel is poured from the ladle into an intermediate buffer distributor (tundish). After this, the liquid flows from the tundish into a Cu-based water-cooled mold system that may consist of one, two or more molds that will produce the material on the resulting *casting lines*. Once the liquid gets in contact with the Cu mold, solidification starts thanks to the heat removal. The material is subjected to a continuous withdrawal action from the mold and further heat is extracted thanks to the water spraying system located immediately after the Cu mold. Once the required length is reached, the material is cut by mechanical shears or thermal cutting and evacuated from the CC machine. Concerning stainless steels, the typical product of the CC process are square-section billets or rectangular-section slabs. SASSs can be readily continuously cast, but the high Mo content makes the process more difficult than standard austenitics. Flat products made of SASSs are usually produced employing slab CC. Subsequently, the slab is hot-rolled in order to obtain the desired flat product that will then be processed through heat treatments and surface conditioning, as it will be presented in the upcoming sections. Of course, even in the case of continuous casting, the issue of segregation of SASSs still remains. As explained in Chap. 2, Cr and Mo segregate in the liquid present in the IDRs during solidification and can lead to the formation of eutectics. For the material produced by means of CC, this segregation is particularly pronounced in the center of the material. During subsequent processing steps, e.g., hot rolling, IDRs can be subjected to various deleterious phenomena [6]. One of these phenomena is the development of internal defects if the process parameters (temperature, strain and strain rate) were such that they fell in the region of the processing map of the material where it is unable to dissipate heat effectively (see Sect. 4.3). Under this condition, the accumulated heat could lead to the melting of the most segregated areas and, due to the differential flow that would occur between solid and liquid part of the material during hot rolling, internal decohesion and cavities could be formed. On the other hand, IDRs rich in solute elements and possible intermetallics become elongated and aligned along the rolling direction, giving rise to the so-called *chemical composition bands.* Once cooling is completed, bands with high alloying elements content and depleted neighboring regions give rise to the phenomenon of *microstructural banding* (MSB). Although comparative studies on the evolution of MSBs and the consequences of their presence in SASSs are not yet systematically reported, there is already some common evidence of the effects of their presence. Studies to date show that MSBs are mainly made of σ phase particles and that their presence markedly deteriorates mechanical properties in the short transverse direction. Other studies have analyzed the spatial distribution, interconnectivity, and morphology of precipitates in MSBs and the presence of internal voids associated with these precipitates. The results seem to suggest that virtually all coarse precipitate particles (σ phase) have an associated void and that the presence of the σ-void combination inevitably has important implications for the mechanics of the material. This result may in part provide further explanation of the reasons related to defect formation (e.g., cracking) during high-T deformation, in addition to what will be presented in Sect. 4.3 [6].

As reported in Sect. 2.4.3, one possibility to reduce segregation during solidification of steels with complex chemical composition such as SASSs could be to employ a process named twin-roll strip casting (TRSC). TRSC is one of the so-called near-net shape casting (NNSC) type casting processes. These processes make it possible to obtain products with dimensions and geometries very close to those of the final product. Because of this feature, post-casting processes can be limited, saving considerable time and cost. In fact, regarding strip casting processes specifically, this process is defined as the casting of molten metal into a metal strip with a thickness in the order of several millimeters in a single processing step. These strips can be subsequently hot rolled into thin strips of desired product thickness. Since this process involves integrating casting and rolling into a single process, the typical slab reheating process usually employed in the production of strip coils is eliminated or limited, thus leading to a significative advantage in terms of process improvement. Similar to the continuous casting process explained previously in this section, the strip caster employs a tundish, a liquid transfer system from the tundish through a nozzle and a solidification system (mold) that allows the formation of the strip. In case of the TRSC process, the mold consists of two counter-rotating water-cooled rolls that form the strip by solidifying the material directly from the sump in which the nozzle is immersed. In fact, the defining feature of strip casters is the so-called *moving mold* design, where the mold surface (i.e., roll surfaces in this case) moves along with the material under solidification, thus reducing the relative speed difference between the two surfaces [7]. The rapid solidification of a thin product, combined with the compressive stress operated by the two rolls on the material, positively affect the quality of the material by greatly reducing segregation as explained in Sect. 2.4.3. The limitation of segregation in the IDRs during solidification is also expected to lead to a reduction/elimination of MSBs, with the obvious advantages in product quality, properties and homogeneity [8]. Once the strip is formed, it is typically straightened by pinch rolls, rolled and then cooled and coiled into a coil strip. This type of TRSC plant layout is depicted in Fig. 4.2.

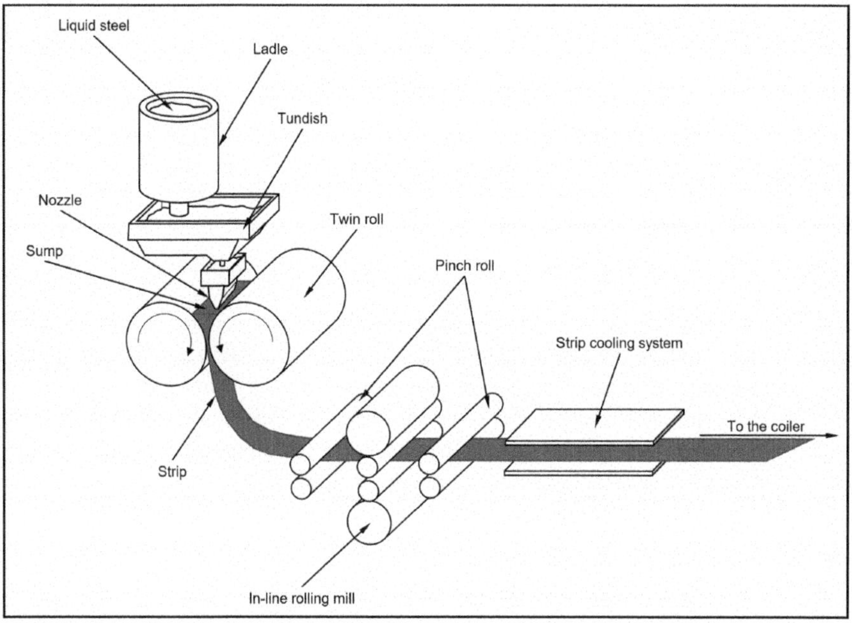


**Fig. 4.2** Schematic overview of a commercial-scale TRSC caster (redrawn from [7])

The original idea of directly producing metal alloy strips from liquid is practically as old as the steel industry in that the first concept was developed directly by Henry Bessemer in 1846 and patented in 1856 with the first example of a twin-roll caster. On the other hand, as for industrial-level deployment in the modern era, many R&D projects concerning strip casting were started worldwide from 1980 onward. The most popular design was, and still is, however, very similar to the twin-roll caster design initially envisioned by Bessemer. The development of this technology led to the successful commercialization of some processes, and beginning in the late 1980s, stainless steel strips also began to be produced by TRSC. However, the results with the first pilot plants were not satisfactory in comparison with the traditional manufacturing process (rolling process starting from slabs) because the surface quality of the strips did not match that of the conventional counterparts. Thanks to the effort put into the development of this technology, since the mid-1990s stainless steels have also been produced by means of TRSC [7]. As seen in Sect. 2.4.3, the TRSC process has the potential to greatly limit the segregation issues typical of traditional solidification products made of SASS [8].

### *4.2.5 Remelting Processes*

There are two main types of remelting processes used for metal refinement: electroslag remelting (ESR) and vacuum arc remelting (VAR). Although the two technologies are different, the common denominator for them is that an electrode made of the material to be refined is melted and subsequently resolidified. In both cases, the plant has a vertical-type structure that serves to take advantage of gravity with regard to the resolidification process. One of the parts of the electrode, namely the working face, is heated to the melting point, and the drops of liquid metal that fall from the face itself are collected in a crucible on the bottom and are rapidly resolidified. ESR was patented in 1945 and remelting is accomplished in atmosphere thanks to Joule heating. Liquid drops are passed through a slag with a chemical composition designed such that it is reactive toward certain undesired elements and impurities in the steel. The slag then operates a filtering action for the liquid alloy, which resolidifies in the part below after passing through the slag itself [4]. The main advantages of the ESR process can be summarized as:

- Reduction in the inclusion content.
- Higher compositional homogeneity and reduced segregation.
- Controlled solidification microstructure.

Regarding the drawbacks associated with the use of ESR, the most evident ones are related to the cost associated with equipment and process, and the know-how needed to properly accomplish the remelting process. On the other hand, VAR was developed in the late 1950s specifically for the aerospace industry. Remelting takes place in the VAR furnace thanks to an electric arc that melts the electrode in a vacuum environment. In this case, the metal droplets do not pass through a slag, but quickly resolidify in a vacuum environment. VAR is usually employed for the refinement and purity improvement of air-melted alloys produced via EAF or VIM. Alloys produced thanks to VAR are characterized by improved mechanical properties and corrosion resistance [4]. The main advantages related to the use of VAR are:

- Porosity elimination.
- Absence of macrosegregation thanks to micro-fusion and layer-by-layer solidification of the VAR ingot.
- Removal of dissolved gaseous elements such as $N_2$, $O_2$, CO and H that can be trapped in the material during conventional casting operations.
- Less formation of oxides and nitrides in the solidified material.
- Controlled solidification microstructure.

Even in the case of VAR, the main drawbacks are related to the high initial cost of the plant and to the high know-how needed to properly run the plant [4]. SASSs are also produced through remelting processes, both ESR and VAR. In both cases, the main target is to reduce segregation and consequently obtain products with higher homogeneity and improved properties, with a focus on a better corrosion resistance for those applications that involve particularly aggressive environments (see Sect. 3.3). As a result, remelting processes for SASSs are used to obtain premium products that can maximize the corrosion resistance properties of these alloys and achieve even better performance in this regard. A direct consequence of what stated above is related to the fact that ESR and VAR products are more expensive. Moreover, out of the two technologies, the one implying the greatest increase in price (and also in performance) is VAR.

## 4.3 Metal Forming

When talking about metal forming operations, reference is made to a family of manufacturing processes in which the shape of metal workpieces is changed by plastic deformation of the workpieces themselves. Metal forming operations involve the use of special equipment such as presses and hammers in case of forging, rolling mills in case of metal rolling, bending machines in case of bending, and so on. Deformation is caused by stresses applied to the material by means of special tools, which are different depending on the plastic deformation process (e.g., dies in case of forging, rolls in case of metal rolling etc.). Obviously, in order to achieve permanent deformation, the stresses applied to the material must exceed the yield strength value (see Table 3.3 for room temperature yield strength values of SASSs). Metal forming operations associated with the greatest volume in terms of production output generally involve the application of compressive stresses, however, in other cases tensile or bending stresses can also be employed. As it is well known, material properties are affected by temperature. An increase in temperature generally results in a decrease in strength and an increase in deformability, i.e., less effort is required to cause the plastic deformation of the workpiece. The difference in material behavior when temperature changes lead to the distinction between three different families of processes depending on the temperature itself:

- Hot working when $0.5T_m < T < 0.75T_m$, where T is the process temperature and $T_m$ is the melting temperature of the material. The process occurs above the recrystallization temperature and has the main advantages consisting in a significant alteration of the workpiece shape, lower force and power required for the same deformation when compared with processes carried out at lower temperatures. Last, but not least, materials that fracture during cold working operations may be successfully processed by means of hot working operations. Main disadvantages of hot working operations consist of: lower dimensional accuracy, higher total energy required (i.e., thermal energy contribution for heating the workpiece), shorter tool life, poorer surface finishing, development of surface scale and oxides due to interaction with high-T atmosphere in reheating furnaces and cooling media during operations.
- Warm working when $0.3T_m < T < 0.5T_m$, where T is the process temperature and $T_m$ is the melting temperature of the material. The process occurs above room temperature, but below the recrystallization temperature of the material. Warm

working operations have intermediate characteristics between hot working and cold working operations.
- Cold working when the process is carried out at room temperature, or slightly above. The main advantages of cold working lay in a greater accuracy than hot working, better surface finishing, increase in properties due to strain hardening and no need for heating processes. The microstructure becomes deformed (i.e., elongated grains) and acquires a certain degree of directionality with regard to properties, which in some cases can be an advantage, and in others give rise to problems. Cold working operations require higher forces and power, higher care of the surface prior to deformation and make it possible to accomplish deformation with limited extent due to lower deformability and possible strain hardening of the material [2].

A fundamental understanding of workability is essential for the development of a science-based technique in manufacturing processes and for achieving processing conditions in which the material can be shaped without the onset of fracture. In the upcoming discussion about SASSs, reference will be made to the two most common families of processes for metal forming: hot working and cold working. Warm working for SASSs, and for austenitics in general, is often employed for the deformation of thick plates in order to exploit the decrease in strength with increasing temperature; without, however, causing precipitation or excessively compromising the surface of the material. In fact, since at warm working temperatures SASSs do not undergo precipitation of deleterious secondary phases (see Sect. 2.4.6), the material can be processed under temperature conditions that do not even cause excessive oxidation of the surface. As a result of this, post-processing operations used to improve surface finish can also be less severe [9].

### *4.3.1 Hot Working*

Generally, austenitic stainless steels have excellent formability characteristics for what concerns forging, hot rolling and extrusion processes. They also turn out to have higher high-T strength than carbon steels. As a result of this and what is explained in Chap. 3 regarding mechanical properties, SASSs are much more difficult to be deformed than plain carbon steels and are characterized by even greater hot strength than standard austenitic grades [9]. This results in the need for plants and equipment that can apply higher stress values on the material and, consequently, withstand the reaction of the material during high-T plastic deformation processes. The processing cycle for hot working basically involves three steps: heating in reheating furnace, extraction from the furnace and processing, and final cooling.

It is important that heating in the reheating furnace is done properly and that the material is homogeneous over the entire section. Moreover, again for comparison with carbon steels, the reheating cycle must be carried out with special care and for longer time because SASSs have lower thermal conductivity than carbon steels and standard austenitic grades (see Table 3.2) [9]. However, as explained in Sect. 3.3.10, SASSs show poor corrosion resistance at typical hot working temperatures reported in Table 4.1. The decrease in corrosion resistance, and the resulting degradation of the surface due to the reaction with the environment, is mainly due to the transformation of the protective $Cr_2O_3$ into the non-protecting and volatile $CrO_3$ at temperatures above 950 °C. In addition to this, $MoO_3$ formation induces further surface degradation. Furthermore, in case $MoO_3$ remains on the surface in liquid form, this would accelerate the corrosive process. These reactions at the interface between the material and the environment in the reheating furnace occur in case the atmosphere is oxidizing. The problem resulting from the high-T corrosion process described above, and with more detail in Sect. 3.3.10, is twofold. In fact, on one hand the formation of corrosion products in form of a thick scale on the surface of the material takes place. In case homogenous detachment of the scale occurs during the process, the oxide may act as a dry lubricant during hot working operations by remaining between the tool and the workpiece, thus affecting the friction between them. However, the scale may not spall off completely and remain imprinted on the surface causing a defect called *wrinkling*. The imprinted scale would then be found on the product at the end of the process. In case the imprinted scale detaches during the various steps of the deformation process, it may still leave marks on the surface and constitute a defect that might not be acceptable by quality control. On the other hand, if the scale completely detaches or is removed before plastic deformation, the exposed surface would have a different composition than the bulk. This condition is caused by the reaction of some elements with the atmosphere, which causes the formation of corrosion products and the consequent formation of a depleted surface layer. This would result in a different material interacting with the tools used to perform the plastic deformation. In hot working practice, this condition causes different surface plasticity and the formation of surface defects. These defects are caused by the different plasticity of the surface compared to the bulk and a processing condition that would be optimized for the material with the correct chemical composition not altered by the depletion of certain elements. The defects that develop on the surface are the result of the material being unable to undergo processing and failing under the action of the applied stresses. Defects that may form on the surface in such conditions are typically cracks, slivers and surface decohesion. Obviously, proper heating cycles of SASSs take into account issues related to excessive soaking time in the furnace. The time must be calibrated to achieve thermal homogenization without damaging the surface. However, there are conditions under which it is unavoidable that the time in the furnace will be longer than expected. Some of the common cases may be:

- Problems in the hot working process of the previous workpiece that extend its processing cycle time.
- Plant downtime due to equipment failure.
- Overnight furnace residence in case of plants that do not work three shifts in case the furnace is not evacuated at the end of the last shift.

**Table 4.1** Hot working temperature range for SASSs

| Group | Most common name | EN | UNS | Hot working range (°C) |
|---|---|---|---|---|
| High-performance austenitic stainless steels | AISI 904L | 1.4539 | N08904 | 850–1150 |
| | AISI 904LN | – | – | – |
| 6Mo superaustenitics | 20Mo-6 | – | N08026 | 980–1150 |
| | 1925 hMo | 1.4529 | N08925 | 900–1200 |
| | NAS 254N | – | S32053 | – |
| | 1925 hMo<br>25-6Mo | 1.4529 | N08926 | 900–1200<br>900–1200 |
| | Uranus SB8 | 1.4537 | N08932 | – |
| | 254 SMO | 1.4547 | S31254 | 1000–1200 |
| | Nicrofer 3127 hMo | 1.4562 | N08031 | 1050–1200 |
| | AL-6XN® | – | N08367 | 850–1150 |
| High-N superaustenitics | Uranus B66 | 1.4659 | S31266 | 1150–1180 |
| | 654 SMO | 1.4652 | S32654 | 1000–1200 |
| | Nirosta® 4565S | 1.4565 | S34565 | 900–1200 |

Note that for each grade the optimal processing temperature depends on the process type (e.g., forging, hot rolling etc.), the characteristics of the product to be deformed (i.e., ingots, billets, etc.), and the characteristics of the plant/equipment that has to process it [9, 10]

In order to limit or avoid issues related to excessive time in the furnace, some strategies can be:

- Have well-established manufacturing instructions that allow processing in specific time frames and minimize the risk of extending the duration of the hot working process.
- Perform adequate maintenance to facilities and equipment in order to minimize the risk of failure and resulting downtime.
- In case it is impossible to avoid excessive time in the furnace, consider the possibility of producing the material with special precautions anyway. Alternatively, consider evacuating the furnace and removing part of the surface, for example, by grinding.
- Avoid conditions that lead to limitation in the $MoO_3$ volatilization.
- Limit exposure to high-T to what is stricly necessary.
- Perform the reheating cycle in a furnace with controlled atmosphere (i.e., non-oxidizing conditions).

Regarding material deformability, it is very important to maintain a uniform temperature throughout the entire workpiece. Colder regions, which appear darker on visual inspection, will be characterized by much greater resistance to deformation. This can lead to geometric irregularities after deformation, excessive stresses experienced by the tooling or the formation of defects on the surface. In fact, colder regions may crack or tear because they do not have sufficient plasticity to withstand deformation [9]. In addition to this, thermal inhomogeneities in the workpiece can also lead to issues related to the microstructure at the end of the production process. In fact, assuming a non-negligible temperature difference between the surface in contact with the tooling and the core of the workpiece, the lower temperature surface layer could be subjected to work hardening when undergoing the application of the loads needed for accomplishing the plastic deformation. This condition would lead to differences in internal stresses between surface and inner part of the workpiece. Immediately after the hot working process, and once complete cooling is accomplished, this situation does not lead to noticeable differences. The phenomenon emerges after solution annealing treatment. In fact, it is known that solution annealing causes a change in the grain size with a growth that is directly proportional to the treatment temperature and soaking time. Work hardened regions react differently and with greater and faster grain growth than regions that have not experienced this phenomenon. Given the fact that grain growth is a thermally activated phenomenon that requires a certain energy barrier to be overcome, the work-hardened regions are able to reach the energy condition for grain growth earlier than the others because work hardening itself gives an energy contribution. This allows the region itself to reach the condition for grain growth sooner with the same temperature and time, compared to regions that have not undergone work hardening. After the solution annealing treatment, the metallographic analysis of the microstructure shows a material with enlarged grains in the regions that have undergone work hardening. This condition affects both corrosion resistance and cold-workability. In particular, for cold working, in agreement with what is stated by the Hall–Petch equation and Eq. 3. 1, grain size affects yield strength. The presence of extended regions with different grain size results in different deformability also due to different associated yield strength values that have to be overcome, and this affects the final result due to differential material flow and unbalanced behavior during plastic deformation. The incidence of this phenomenon can be such that for certain applications it can cause product nonconformity. Thermal homogeneity during reheating is therefore critical. In addition to this, however, temperature control during the hot working process is also critical. The balance between heat generated in the material as a consequence of plastic deformation and heat removal from the surface is extremely important. Therefore, any equipment/tooling cooling system employing water or other cooling media must be calibrated with appropriate flow rates without excess.

As mentioned, temperature control during the process must also take into account the heat produced by the material during the deformation process, and the ability of the material to dissipate it. As it is well known from the theory related to the study of hot working operations, among the fundamental parameters for the design of the process, the strain rate also plays a key role. A useful tool to better design the hot working process, and one that takes into consideration the two aspects just mentioned, are the so-called *processing maps*. Processing maps are explicit representations of the response of a material in terms of microstructural mechanisms to the imposed process parameters. Processing maps are obtained as a consequence of the superimposition of instability maps and power dissipation maps. Instability maps are log strain rate versus temperature representations based on the determination of the condition for microstructural stability according to a dimensionless parameter $\xi$ (*xi*). By greatly simplifying, when the value of $\xi$ is negative, metallurgical instability during plastic deformation occurs and phenomena like adiabatic shear bands and flow localization take place. On the other hand, power dissipation maps are again log strain rate versus temperature plots that consider the partitioning of power between heat generated from plastic deformation and microstructural changes. They express the efficiency of energy dissipation during the process also taking the strain rate sensitivity of the material in consideration. Consequently, for a given applied strain it is possible to look at a processing map that provides a range of optimal combination of temperature and strain rate that allows the material to deform without instability issues and with the best possible heat dissipation [11]. An example of processing map is given in Fig. 4.3, where the plot for 654 SMO at a true strain of 1.0 is represented. Processing maps can be obtained by means of hot compression test in the desired temperature interval and with a specific strain rate range [12]. The strain value is kept constant, this means that each processing map is related to one specific strain condition. As a consequence of this, complete description of the material at various strain values requires extensive experimental activity and is very time consuming in the moment when literature data are not available. For more in-depth discussion on the topic of processing maps and instability phenomena such as adiabatic shear bands, the reader is referred to the texts used as references for this section and also to the further readings at the end of the chapter.

Having discussed the topic of processing maps, it is clear that the correct process temperature is a key parameter. Increasing the temperature of the material indiscriminately in the belief of increasing its ductility is a rather serious mistake with austenitic stainless steels of any type, including SASSs. In fact, SASSs experience a drastic drop in ductility when the temperature is too high and approaches the melting temperature as shown in Fig. 4.4. Furthermore, at temperatures in excess of 1200 °C, ductility can also have a drastic drop due to the negative effects of trace elements such as O, P and S at the GB. The possible presence of these elements leads to severe problems, thus producers put special effort into the melting and refining stages in order to minimize their presence. For this reason, process windows in terms of temperature should always be respected for both minimum and maximum values. In addition to this, again referring to Fig. 4.4, the process window for high ductility values in terms of temperature is narrower than that of standard austenitics [10, 13].

The high strain rate ductility of SASSs is reduced when compared to standard austenitics due to the low self-diffusion rates in austenite and high N content [13]. Finally, in view of the process temperatures (see Table 4.1), SASSs must be solution annealed in order to ensure a homogeneous microstructure and the absence/elimination of deleterious secondary phases that could form during slow cooling after the hot working process. Failing the solution annealing cycle could lead to the negative consequences related to the precipitation of secondary phases that were described in Sect. 2.4.6. Furthermore, given the chemical composition of SASSs and the critical issues related to the formation of secondary phases, even though the hot working temperature never falls below the typical solution annealing temperature (see

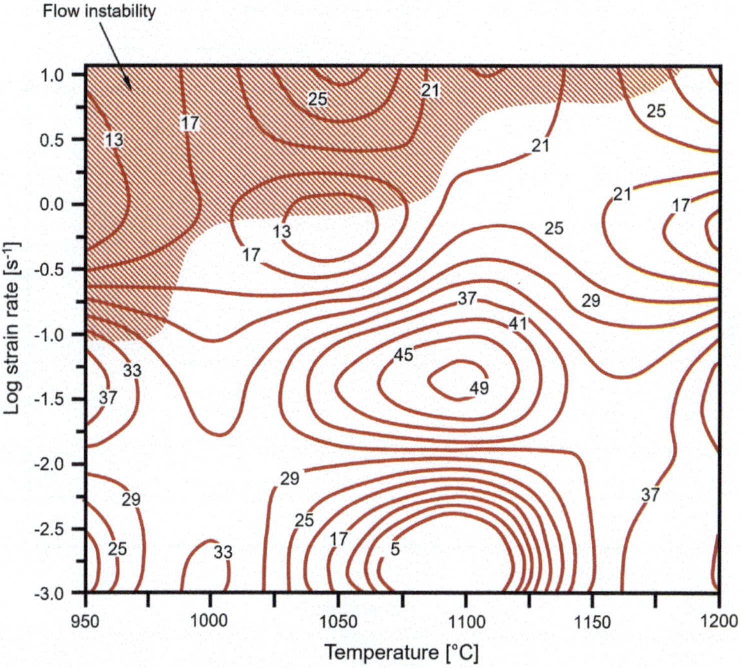


**Fig. 4.3** Processing map of 654 SMO at a true strain of 1.0. Contour numbers represent the percent efficiency of power dissipation (redrawn from [12])

Table 4.2), these steels are unlikely to undergo the so-called *process annealing*. This process would involve exploiting the hot working temperature for solution annealing and rapid cooling at the end of processing [9]. However, there are often plant and equipment limitations to the possibility of rapid cooling since a water-cooling system, for example, would be required. Furthermore, for complex geometries it is quite difficult to ensure that process anneal is achieved throughout the entire workpiece because of the non-uniform cooling that occurs during hot working.

### 4.3.2 Cold Working

The austenitic stainless steel family is characterized by high ductility at room temperature. This property results in excellent formability through cold working operations, from bending to deep drawing and even stretch forming. SASSs can be cold worked through virtually all traditional methods, however they will be substantially more demanding than standard grades because, as mentioned within Sect. 3.2, SASSs are characterized by higher mechanical properties than standard grades, and this obviously requires more power for the machines and equipment employed in cold deformation operations. The effect will be seen particularly on required power, forming

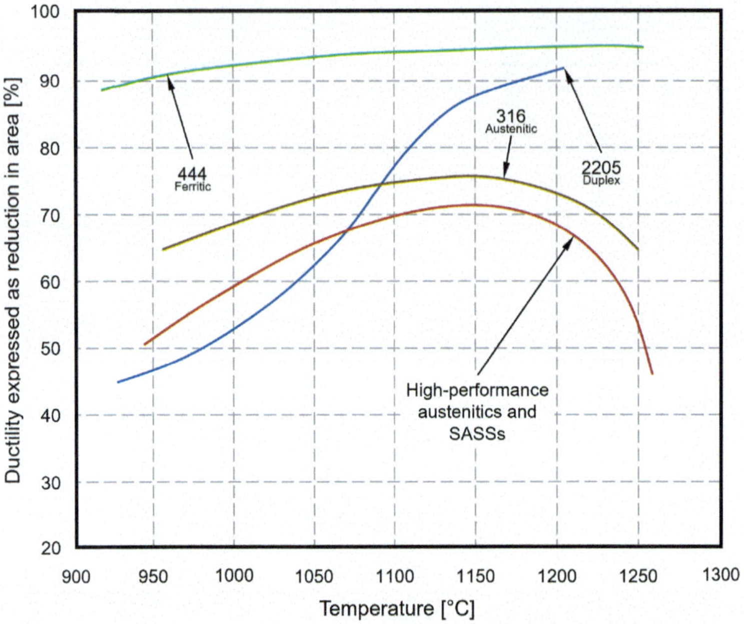


**Fig. 4.4** Hot ductility of SASSs in comparison with the other stainless steels families (redrawn from [5])

**Table 4.2** Solution annealing temperatures and ranges for the steel grades given in Table 2.2 [9, 10]

| Group | Most common name | EN | UNS | Solution annealing (°C) |
|---|---|---|---|---|
| High-performance austenitic stainless steels | AISI 904L | 1.4539 | N08904 | 1095 |
| | AISI 904LN | – | – | – |
| 6Mo superaustenitics | 20Mo-6 | – | N08026 | 1150–1200 |
| | 1925 hMo | 1.4529 | N08925 | 1120–1200 |
| | NAS 254N | – | S32053 | – |
| | 1925 hMo<br>25-6Mo | 1.4529 | N08926 | 1120<br>1120 |
| | Uranus SB8 | 1.4537 | N08932 | – |
| | 254 SMO | 1.4547 | S31254 | 1150 |
| | Nicrofer 3127 hMo | 1.4562 | N08031 | 1150–1180 |
| | AL-6XN® | – | N08367 | 1105 |
| High-N superaustenitics | Uranus B66 | 1.4659 | S31266 | 1150–1180 |
| | 654 SMO | 1.4652 | S32654 | 1150–1200 |
| | Nirosta® 4565S | 1.4565 | S34565 | 1120–1170 |

equipment loading and process-related lubrication requirements. In addition, the work hardening of SASSs is also considerable, and this, combined with higher mechanical properties, also results in higher springback of these steels. As a consequence of this, taking bending operations as reference, non-negligible overbending is almost always necessary to obtain the desired profile [9]. In case of considering a 180° free bending, SASSs have different practical production criteria than standard austenitics. Specifically, with regard to the bending radius, a standard austenitic in the annealed condition implies a minimum bending radius of 0.5t; being *t* the thickness of the material thickness. In case of SASSs, the criterion usually implies a minimum bending radius of 1.0t, which is exactly double. Considering a 90° bending process, with a standard austenitic grade the springback is around 10°, meaning that for the material to achieve the 90° bend, it must be bent at about 100°. In case of the members of the SASSs family, the bending angle must be increased to a value around 103° to achieve a 90° bend [10].

The different average behavior in terms of the effects of cold working performed on SASSs in comparison with the AISI 316 standard austenitic and also high-performance ferritic grades is shown within Fig. 4.5. What can be immediately noticed is that SASSs exhibit similar trend to standard austenitics as cold working percentage increases, but with a shift toward higher stress values for the yield strength, and toward lower elongation values for what concerns ductility. The cold working response of SASSs is basically amplified and this is easily explained by what is reported in Sect. 3.2. In fact, this behavior can be justified through the higher mechanical properties in the annealed condition and also to the higher strain hardening. The latter characteristic, as already stated, is strongly influenced by the percentage of N and increases proportionally to it. Regarding the comparison with the high-performance ferritic grades, it can be seen that the behavior is different in the initial phase due to the fact that ferrite has an initial high yield strength and an initial high work hardening rate [13]. In addition to bending, cold rolling is also employed in the production of flat products made of SASSs. Typically, cold rolling of these steels involves the use of cluster-type rolling mills. These mills have small-diameter working rolls that are supported by multiple layers of the so-called backed rolls. Such architecture is very suitable for cold rolling performed on wide sheets of stainless steel allowing for reaching close tolerances and improved surface finishing [5].

With regard to processing, specifically rolling of sheets or long products (both hot and cold rolling), materials tend to lose isotropy and instead to be characterized by some anisotropy. This occurs because stainless steel products are usually manufactured through casting processes followed by one or more steps of rolling. This induces a directionality of the properties as a consequence of the deformation of grains, in particular when cold-rolling is concerned. This anisotropy can also lead to problems in subsequent processing, one of the cases is the sheet metal bending. In fact, for good practice, the sheet metal bending process should have the bending axis not parallel to the sheet metal rolling direction. This type of anisotropy is often measured with the so-called $\Delta r$, calculated as follows [5]:

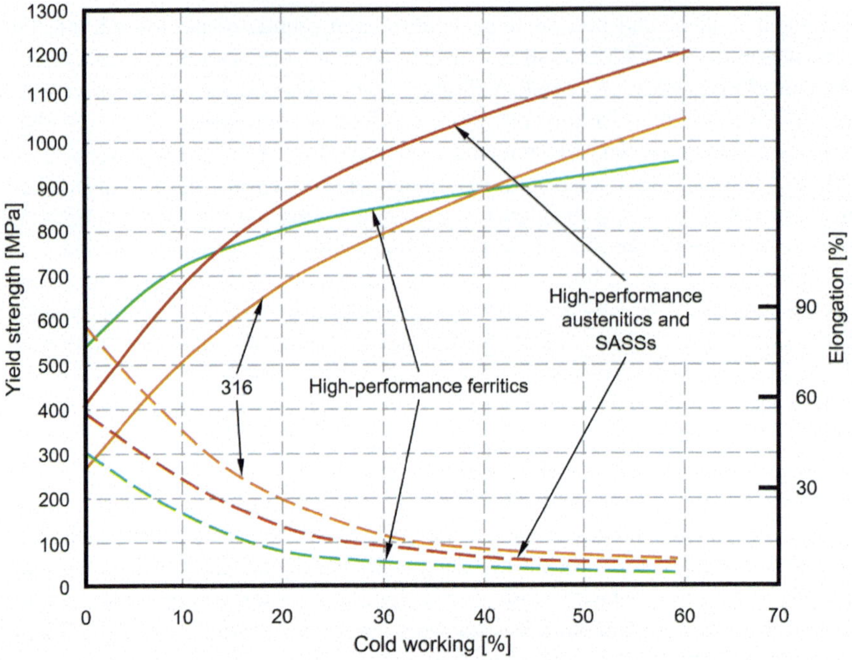


**Fig. 4.5** Comparison of the effect of cold work on yield strength and elongation of SASSs, AISI 316 standard austenitic and high-performance ferritics (redrawn from [13])

$$\Delta r = \frac{r_T - 2r_{45} + r_L}{2} \tag{4.1}$$

where

- $\Delta r$ is the planar anisotropy.
- $r_T$ is the plastic strain ratio in transverse direction with respect to the rolling direction.
- $r_{45}$ is the plastic strain ratio at 45° with respect to the rolling direction.
- $r_L$ is the plastic strain ratio in longitudinal direction with respect to the rolling direction.

The plastic strain ratio $r$ is an expression of the ability of a sheet metal to resist thinning or thickening when the material is subjected to tensile or compressive stresses. Typically, when the material is characterized by a minimal reduction in area when subjected to these stresses, this attitude is considered positive and is indicated by a high $r$ value. A material exhibits greater anisotropy the greater the difference in the $r$ value in different directions calculated according to Eq. 4.1. With reference to Fig. 4.6, it can be seen that SASSs such as 254 SMO and 654 SMO do not exhibit very high $r$ values, as in the case of with the ferritic AISI 441. However, they are better or equally performing than the lean duplex UNS S32101 (see Sect. 1.2.6).

Compared with the ferritic instead, the $\Delta r$ value is lower because the difference in value between the three directions is smaller. This implies less anisotropy and less tendency to develop defects such as earing. In any case, the cold-formability of SASSs is also good for processes such as stretch forming. In this case, 654 SMO undergoes just such a process for the manufacture of heat exchanger components. Specifically, this grade is employed for the fabrication of heat exchanger plates obtained by stretch forming [5].

## 4.4 Heat Treatments and Post-processing

Heat treatments are technological processes that can be defined as a set of controlled steps of heating, soaking and cooling of a solid material in order to obtain changes in microstructure, properties and, in some cases, formation of specific secondary phases with positive effects on the material. Concerning stainless steels, there are four possible heat treatments these alloys may undergo: annealing, quenching, tempering and aging [4]. As explained in previous chapters, SASSs are alloys that cannot be strengthened by precipitation of secondary reinforcing phases like PHSS. Furthermore, they do not exhibit martensitic transformation upon cooling, even with quenching or in presence of strain. Given what stated above, aging, quenching, and consequently tempering, are not applicable to these alloys. Consequently, the only heat treatment that can be employed for SASSs is annealing. There are several types of annealing, specifically:

- Diffusion annealing.
- Normalization.
- Softening.
- Solution annealing.
- Spheroidization.
- Stabilization annealing.
- Stress relief annealing [4].

Among those listed, SASSs are subjected to only two: solution annealing and stress relief annealing. Solution annealing is used for dissolution of particular phases that have precipitated during high-temperature processes or during slow cooling, again from high temperature. The deleterious phases that can precipitate in SASSs are carbides and intermetallics and were discussed to a greater extent in Chap. 2, in particular in Sect. 2.4.6. During solution annealing, the material is heated to a temperature higher than the solvus temperature of secondary phases and held at this temperature for a time sufficiently long for dissolving these compounds, but not excessively long to cause excessive grain growth. Solution annealing is often referred to as full annealing because it implies an optimal metallurgical condition to be established for the material: complete dissolution of the secondary phases and homogenization. As a result of this, the material is characterized by maximum corrosion resistance.

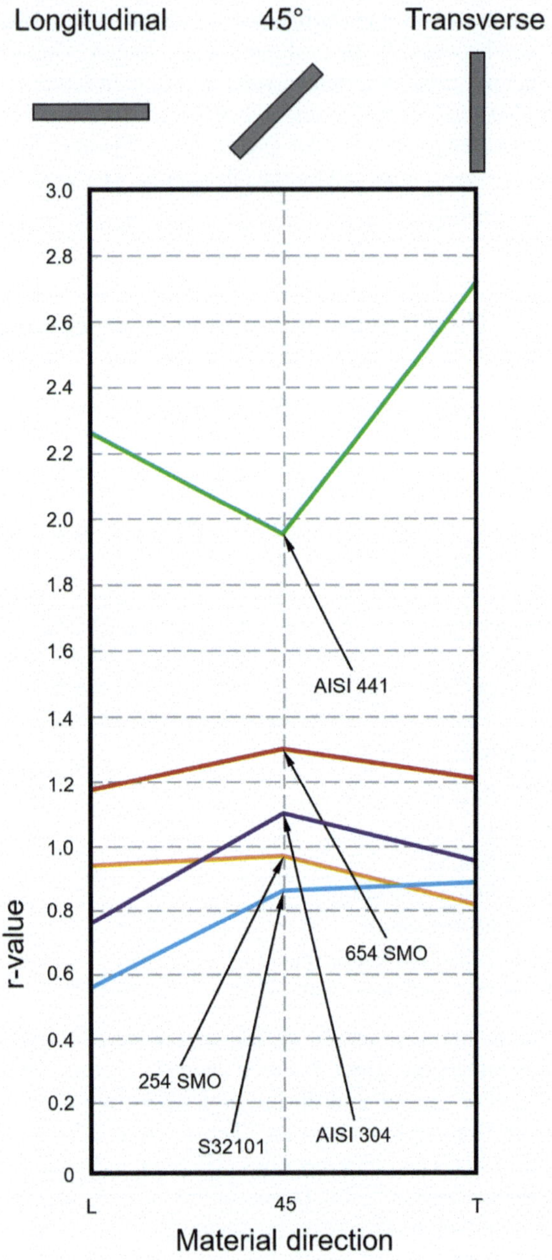


**Fig. 4.6** Typical *r* values for 254 SMO and 654 SMO in comparison with standard AISI 304, AISI 441 ferritic grade and UNS S32101 lean duplex (redrawn from [5])

Typical temperatures for solution annealing of SASSs are given in Table 4.2. A limitation for soaking time and for annealing temperature is also related to the need to avoid excessive oxide scale formation. A too thick scale would be complicated to be removed by post-processing and, even if removed, could jeopardize the surface due to excessive irregularities and roughness. Usually, the atmosphere in heat treatment is oxidizing due to air and combustion gases. Air annealing results in the formation of surface scales that must be removed to restore the corrosion resistance of the material. Prevention of scale formation can be achieved by performing the solution annealing treatment in protective atmosphere saturated with Ar, He, cracked ammonia or vacuum. Once protective atmosphere annealing is performed, no scale is produced and no post-processing for surface cleaning is needed [9].

In addition to this, as explained in Sect. 3.3.10, SASSs at high temperatures in oxidizing atmospheres are characterized by rapid scale formation due to the high Mo content. The $MoO_3$ formed can leave the surface of the material after volatilization, or accumulate in liquid form, thus accelerating the corrosion process in case volatilization is inhibited [4, 9]. In order to limit the problems related to these phenomena, some precautions taken by manufacturers are:

- Avoid conditions that lead to the inhibition of $MoO_3$ volatilization.
- Avoid re-annealing material with thick scales.
- Limit exposure to high-T to what is necessary.
- Perform the heat treatment at the minimum temperature that will allow solution annealing to be effective.

After the soaking time is over, the material is rapidly cooled in order to avoid re-precipitation [4, 9]. Generally , solution annealing is carried out in continuous furnaces (typical of wire and wire-rod coils) composed of several zones of increasing temperature, in batch furnaces (typical of single wire and wire-rod coils, or forgings), in bogie heart furnaces or in chamber furnaces. In principle, there is no specific type of heat treatment furnace for solution annealing. The fundamental requirement is that the furnace must be capable of reaching the solution annealing temperature for the alloy together with the ability of ensuring the complete thermal homogenization of the material. The furnace choice is usually made according to the type of product that needs to be processed. For example, continuous furnaces are often used in integrated steel mills plants, while bogie heart and chamber furnaces are usually employed for long products such as bars, billets or forgings. Rapid cooling usually takes place by dipping the material in water. To avoid issues of limited heat exchange between the cooling water and the material, water stirring is essential. This can be accomplished by the water re-circulation circuit, by mechanical stirring, or by both methods at the same time. Water stirring is needed to prevent the formation of a gas layer adherent to the material surface resulting from the evaporation of water itself when it comes into contact with the high-T surface. The gas layer separates the surface from additional liquid that could remove heat, but because of this cannot reach the surface, thus limiting heat extraction.

On the other hand, in the case of stress relief, heat treatment is not carried out at high temperature like solution annealing. The heat treatment temperature is highly dependent on the material's past history in terms of deformation and accumulated internal stresses. The phenomena leading to stress relief are basically two: recrystallization and recovery [4]. Stress relief aims to eliminate or significantly reduce internal stresses resulting from cold working processes as well as from welding operations. Generally, for austenitic stainless steels, stress relief annealing has two different ways to be carried out:

- In the 200–450 °C temperature range in order to partially reduce the accumulated internal stresses.
- At temperatures above 850 °C. In this case, however, soaking time and cooling must be calibrated in order to avoid the precipitation of deleterious secondary phases [14].

In many cases, the surface of the material before annealing is characterized by an oxide formed during previous high-T processes. There are cases where the surface could be contaminated with grease, forming lubricants, crayon marks, oil and other substances. The high temperatures typical of solution annealing can lead to the burning of contaminants into the surface. This phenomenon modifies the surface and, usually, the only way to remove contaminants turns out to be surface grinding. In case contaminants are rich in C, it can diffuse into the surface altering its chemical composition and leading to a condition that promotes sensitization (see Sect. 3.3.5). In case the surface is contaminated with low-melting-point metals such as Cu, Pb and Zn, the resulting issues can become critical. This is because a grain boundary penetration of these contaminants can take place. This phenomenon can lead to the so-called liquid–metal-embrittlement. Removal of these kind of contaminants is paramount before annealing, but also prior to welding operations [9].

As mentioned above, in case annealing is carried out in air, scale formation results in a Cr-rich surface oxide and in a Cr-depleted underlying layer that is also characterized by a lower corrosion resistance. Both the surface oxide and the depleted layer must be removed to restore the original corrosion resistance of the alloy surface. The most common methods for accomplishing this are pickling and shot blasting [9]. Pickling is a surface treatment for improving the quality of a metal or alloy that implies the removal of impurities and scales by means of the cleaning action of an acidic solution, the so-called *pickle liquor*. The most common technique for pickling steels is by immersion, but in principle the process could also be carried out with gels, pastes and sprays. The acids used for pickling steels with more than 6% alloying elements are generally HCl, HF, $HNO_3$, $H_3PO_4$, and $H_2SO_4$; either alone or in solutions implying their combination. Acidic solutions employed in pickling clean the surface through a process of generalized corrosion and resulting dissolution of the surface scale. $HNO_3$ is particularly important for the pickling process of stainless steels because it operates a selective dissolution of Fe from the surface. This leads to a reduction in the Fe content of the surface and a consequent enrichment in Cr. For this reason, $HNO_3$ immersion is usually one of the last steps in pickling. Once the material is exposed again to the atmosphere, a stable, adherent and protective Cr-rich passive layer forms on the surface of the material. Stainless steel manufacturers usually collaborate with acid solution manufacturers to develop solutions with specific proportions calibrated to the composition of the alloys produced in order to achieve the best result on the surface. In fact, the chemical composition of the steel subjected to pickling strongly influences the result. The reason lays in the fact that pickling is a cleaning process implying the generalized corrosion of a layer of material, therefore its corrosion resistance has an influence on the process itself. For high corrosion-resistant alloys such as SASSs, the material removal rate related to pickling will certainly be lower than for standard austenitics such as AISI 304L and 316L. These aspects have already been extensively addressed in Sect. 3.3 for what concerns the corrosion resistance of SASSs in comparison to standard grades. However, to complete the information, Fig. 4.7 shows the pickling-related material removal rate for different stainless steels immersed in a solution that simulates the pickling process. It is evident that the better corrosion resistance of SASSs is reflected in a lower material removal rate and, consequently, in a process that must be optimized in the duration and composition of the *pickle liquor*. This is necessary to achieve a qualitatively compliant surface with maximum corrosion resistance after repassivation.

On the other hand, shot blasting implies the propelling of a high-pressure stream of abrasive material against the surface for removing contaminants, and thus scale. Shot blasting is the term used in case a metal-based abrasive media is employed. The process leads also to a slight improvement of surface mechanical properties as a result of residual compressive stresses and strain hardening of a thin layer. Shot blasting equipment typically consists of a chamber in which the abrasive mean and air are mixed. This mixture is then accelerated by means of special pumps and travels through a nozzle to direct the particles toward the surface of the material. In addition to this, shot blasting can be accomplished also manually, in this case for safety

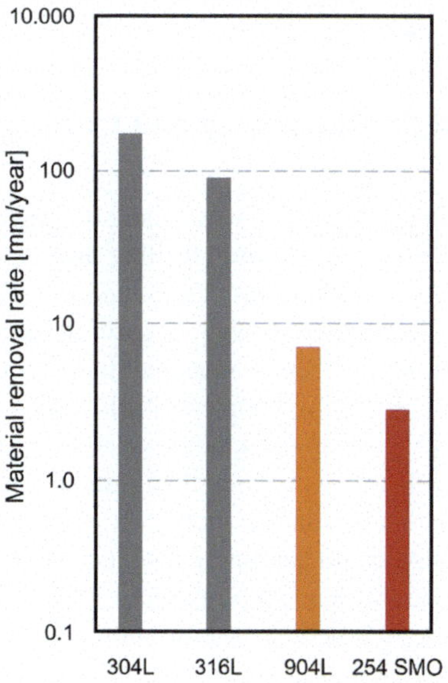


**Fig. 4.7** Material removal rate of various stainless steels in a simulated *pickle liquor* (pickling solution) containing 10% $HNO_3$ and 3% HF at 21 °C [9]

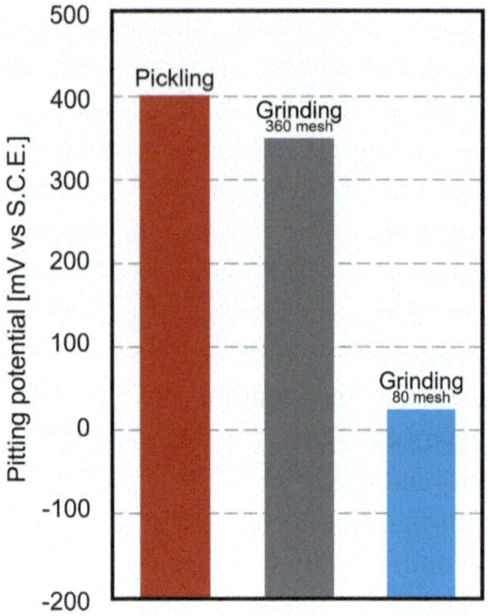


**Fig. 4.8** Effect of the cleaning method on the corrosion resistance expressed by means of the pitting potential ($E_{pit}$). A higher pitting potential expresses better corrosion resistance [5]

reasons the operator must wear special protective suits. Besides being different in principle, the two main cleaning methods also differ in the achievable result in terms of corrosion resistance of the resulting surface. In fact, shot blasting removes the surface scale, but because of the slight surface work hardening and the increased surface roughness it leads to lower corrosion resistance of the material. In fact, the lower the surface roughness, the better the corrosion resistance. This difference in corrosion resistance can be quantified in a different pitting potential value ($E_{pit}$). This can be seen in Fig. 4.8, where various surface cleaning methods are compared. It is evident that among those presented, the one able to maximize the corrosion resistance is pickling.

## 4.5 Machining and Cutting

In the present context, under the definition of machining operations fall the processes of turning, milling, drilling and threading. In this section, more emphasis will be given to these processes, and in the final part some remarks on cutting operations will be given as well. Regarding machining, in general, austenitic stainless steels are more difficult to machine than carbon steels. In addition to this, the machinability of austenitic stainless steels varies greatly from grade to grade and is lower for SASSs than for standard austenitic grades. The characteristics that make SASSs more difficult to machine than standard austenitics, and consequently than carbon steels, lie mainly in the properties expressed in Chap. 3: high ductility, high strength and high strain hardening tendency, and low thermal conductivity [10, 15]. The physical and mechanical properties of SASSs have implications for machining operations as they cause high specific cutting pressure, formation of built-up edge (BUE), long and serrated chip generation, increased tool wear rate, and low quality of the machine surface. Furthermore, the forces exchanged during machining of SASSs tend to have greater fluctuations than those characteristic of carbon steels mainly due to fluctuations in chip thickness. Finally, the tool wear rate is amplified by the low thermal conductivity, which obviously implies an increase in temperature at the interface between the workpiece and the tool [15]. This leads to a general rise in temperature in the workpiece due to the heat generated, which, however, is not effectively dissipated through it due to the lower thermal conductivity of these steels compared to ferritics and to standard austenitics (see Table 3.2) [10]. However, despite the fact that SASSs present greater challenges than standard austenitics, these steels can be successfully machined by suitably employing typical stainless steel machining guidelines: rigid setups, powerful machines, proper cooling, positive feeds and adequate depth of cut underneath the prior work-hardened layer. The latter point means that the depth of cut must be high enough to cut below the layer that has undergone strain hardening as a result of the compressive stresses exerted by the tool on the surface during the previous cutting pass. If this was not the case, the tool would be machining a work-hardened material with negative consequences on surface quality [10].

Moving back for a moment to a more general view, within the family of austenitic stainless steels there are grades with improved machinability called free-machining grades. This is the case, for example, of modified AISI 304 or AISI 316 which have been subjected to a special processing starting from AOD together with advanced ladle treatments (which, for steelmakers that possess it, could be done in a ladle furnace, also known as LF). The goal of these treatments is to control the composition, size, shape, and distribution of oxide inclusions and sulfides in order to achieve maximum machinability as a result of an increased ability of the microstructure to break the chip during cutting operations. In addition to this, high-S austenitics such as AISI 303 (see Table 1.7) are also part of the free-machining family. They feature a minimum S content of 0.15% up to a maximum of 0.3–0.4%. Although in most cases S is considered a detrimental impurity in steels, its presence greatly increases machinability due to the formation of sulfides that promote this property. In this context, SASSs, on the other hand, are shown to be characterized by a poorer machinability also for chemistry-related reasons (see Fig. 4.9). In fact, according to Table 2.2, they do not contain S for improving machinability. The high Mo content negatively affects the interaction with the tool, causing it to be subjected to more wear. Furthermore, the high N content is detrimental because of the positive effects on mechanical properties (strength, toughness and strain hardening), which, however, are counterproductive with regard to machinability [10]. The guidelines that should be employed for processing austenitic stainless steels and that must necessarily be carefully followed for SASSs can be summarized as follows:

- Employ machine tools characterized by high power, rigidity and extremely strong, rigid workpiece and tooling setups able to provide a depth cut below the strain-hardened layer produced by prior passes.
- Employ tools characterized by small tool nose radius, in any case not larger than that necessary for adequate tool life.

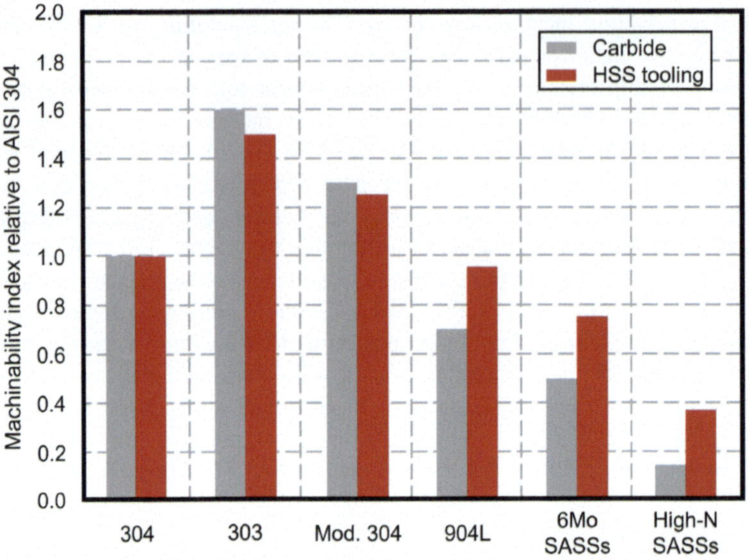


**Fig. 4.9** Machinability index of various stainless steels taking AISI 304 as reference (redrawn from [10])

- Feed rate and depth of cut must be high enough to cut below the layer that has undergone strain hardening.
- Cutting speed must be adequate to limit the extension of the work-hardened layer.
- Tools should be replaced or sharpened at regular intervals before they become dull. In fact, a dull edge leads to the formation of a thicker work-hardened layer.
- The use of coolant/lubricant flow directed at the cutting region is essential. Emulsions with extreme pressure additives have particularly positive effects.
- The quality of the tools must be high, whether they are made of high-speed steels (HSSs) or coated carbide inserts [10].

As explained extensively in previous chapters, stainless steels have a wide range of properties and consequently an equally wide range of machinability resulting from these properties. Defining a single machinability index that gives an explanation of this property of stainless steels is difficult, because it is not possible to effectively represent all grades and the related processing conditions. However, the indices take value when comparing different grades taking one of them as reference, as represented in Fig. 4.9. The figure represents the machinability index declined according to tool type and taking AISI 304 as reference for the comparison with the other grades. It is possible to see that among the steels in Fig. 4.9, SASSs are those with the lowest machinability index, and this is not surprising when considering the abovementioned information [10]. Obviously, the index does not refer to a particular machining operation like turning, milling, etc., but considers machinability in an overall manner and thus averaging among possible machining operations.

Generally, in case of surface finishing operations, the depth of cut is reduced if compared with roughing operations. However, based on what has been stated above, finishing passes for SASSs should be as deep as possible in order to cut below the surface layer that has undergone work hardening from the previous pass [13]. Finally, the machinability characteristics of SASSs closely resemble those of solid solution strengthened Ni-based alloys in the annealed condition. This group includes alloys such as Incoloy 800, Inconel 600, and Monel K-500. According to this, as a first approximation, for the definition of machining parameters for SASSs a good starting point could be those of the abovementioned Ni alloys [13].

In terms of cutting operations, SASSs can be cut by means of the most common cutting techniques employed for the other stainless steels such as:

- Laser cutting.
- Plasma cutting.
- Punching and shearing.
- Saw cutting.
- Water jet.

Laser cutting refers to a family of processes in which a laser focuses on the surface of the workpiece. Thanks to the high energy, the laser beam melts and vaporizes small portions of the material in a controlled manner. Laser cutting operations are classified as high power density operations, meaning that they allow for very rapid temperature increases with minimal heat input to the workpiece. This feature reduces heating-related issues such as those related to secondary phase precipitation (see Sect. 2.4.6). Regarding laser cutting of SASSs, there are small differences from standard grades [9, 10]. Plasma cutting involves the use of a plasma beam as cutting tool in order to perform rapid cutting of various types of materials. Compared with laser cutting, the material removal rate is higher; however, the heat input is greater and this could be dangerous for the characteristics of the cutting region for materials such as SASSs in terms of precipitation of deleterious secondary phases. Furthermore, for cutting of SASSs and stainless steels in general, plasma cutting requires high-purity N. This is because even a slight oxygen contamination would cause oxidation of Cr and deplete the cutting zone, thus reducing the corrosion resistance [9, 10]. Regarding punching and shearing operations, the high ductility and strain hardening tendency of SASSs make these steels more complicated to cut than standard grades. This implies the use of machines with higher power than those used for carbon steels. The setup becomes especially important and requires rigid frames, sharp blades, and small blade clearances [9, 10]. For what concerns saw cutting operations, SASSs can resist cutting with any machine except very high-powered machines due to their combination of high strength and high strain hardening tendency. In the event that a machine with particularly high power cannot be employed and a positive feed cannot be maintained, the choice of another cutting process is mandatory for accomplishing the cutting operation [9, 10]. Water jet cutting is a group of cutting processes in which the tool is a high-pressure water jet, with or without the addition of abrasive media. In case the process employs the use of a water jet only, the operation is referred to as water-jet machining (WJM). On the other hand, when abrasive media is added the process is named abrasive water-jet machining (AWJM). Water jet cutting operations allow any material to be cut with reduced burr, without heat input, and absence of workpiece deformation. SASSs can be readily cut by water jet operations without any particular process issues [9].

## 4.6 Welding

Welding can be defined as a technological process of permanent joining in which two or more parts are brought to coalescence on their contact surfaces by an appropriate application of heat and/or pressure, with or without the use of a filler material (used to promote coalescence). The weldability of a material is a complex technological property that is influenced by:

- Processing history. In this case, residual stresses and thermal treatments performed on the material have influence on the weldability. In fact, austenitic stainless steels, and also precipitation hardening alloys, are welded in the solution annealed condition.
- Surface conditions. Surface moisture can lead to porosity in the fusion zone after solidification as a consequence of the resulting gas evolution. Surface preparation and cleaning prior to welding are paramount, as well as the nature and properties of surface oxides or passive layers.
- Chemical composition. Weldability of steels decreases with increasing C content and varies as a function of the presence of oxide-forming elements (e.g., Al, Cr, Si).
- Physical properties: thermal expansion, thermal conductivity and electrical resistivity.
- Cleanliness of the steel. High content of impurities (e.g., oxides and inclusions) in the steel reduce the weldability of the material.

In this context, the analysis of the physical properties of SASSs and their consequences on welding processes turns out to be of primary importance for understanding the behavior of these materials. Austenitic stainless steels are characterized by a higher coefficient of thermal expansion than, for example, ferritic grades (see Table 3.2). This has an influence on the behavior of the material after welding and, in case of austenitics and SASSs, leads to a more pronounced tendency to undergo distortion. On the other hand, as far as thermal conductivity is concerned, again referring to Table 3.2, values associated with SASSs are significantly lower than those of ferritics, and lower even than those of standard austenitics. A low thermal conductivity leads to a limited heat dissipation capacity. This implies that the heat generated by welding process will tend to concentrate near the weld joint causing a more pronounced thermal gradient. This, in combination with the coefficient of thermal expansion, causes a greater tendency for the development of shrinkage stresses and post-weld distortion. The lower thermal conductivity with the resulting limitations in heat removal also leads to the possibility of achieving higher temperature in the heat-affected zone (HAZ), which is also maintained for longer time. This will in turn promote the precipitation of carbides and intermetallics that are deleterious to the materials, as extensively explained in Chap. 2. Finally, with regard to electrical resistivity, SASSs are characterized by a higher electrical resistivity than ferritics and standard austenitic grades, as shown in Table 3.1. Higher electrical resistivity leads to a reduction in the current required for heat generation. This, in combination with a lower melting temperature than other stainless steels, results in a lower value of current required to weld this family of alloys [9].

Regarding filler metals, the most common ones used to weld standard austenitics like AISI 304L and AISI 316L have a chemical composition designed to form between 5 and 10% δ ferrite because, as explained in Sect. 2.4.6, the presence of δ ferrite greatly improves resistance against hot cracking. δ ferrite better accommodates shrinkage stresses and has higher solubility for P, S and for impurities that can lead to embrittlement of austenite. The filler metals for welding SASSs do not contain or form δ ferrite, which is the reason why these steels are prone to hot cracking. In case δ ferrite is formed, this would only be a consequence of the chemical composition of the base metal as described in Sect. 2.4.6. In order to minimize the problem of hot cracking, filler metals for SASSs have extremely low S and P contents. In addition to this, heat input is also carefully controlled and limited. In fact, any practice that would increase the size of the weld pool (e.g., excessive heat input, excessive weaving) should be avoided when welding SASSs. Furthermore, large weld pools can also lead to grain growth in the weld and in the HAZ. A material with large grains is characterized by a smaller GB surface area, which leads to increased impurity concentrations and possible reduction in corrosion resistance when compared to fine-grained microstructures [9].

As already seen in Sect. 2.4.6, SASSs are very prone to the formation of secondary phases among which the most common and deleterious is the σ phase. In the present context, regarding precipitation characteristics, temperature ranges and consequences, the reader is invited to refer to what is reported in Sect. 2.4.6. In addition to this, SASSs feature a very low C content, even lower than that of the standard L-grades. Hence, the issue related to carbides formation is of less concern, although carbides can still form as seen in Sect. 2.4.6. As seen in Sect. 2.4, SASSs are particularly affected by Mo segregation during solidification, and this phenomenon also affects the material with regard of welding processes. The phenomenon of Mo microsegregation can lead to the formation of regions of low content of this alloying element, with values even lower than 4% (partly already seen in Sect. 2.4.4). The formation of these areas leads to an inhomogeneous material with lower corrosion resistance characteristics. In order to tackle the problem of Mo depletion, SASSs need to be welded with over-alloyed filler metals for compensating microsegregation effects. Usually, SASSs are welded with Ni-based filler with high amount of Mo. For example, a metal filler with at least 9% Mo is used to successfully weld members of the 6Mo SASSs family. This would result in the possibility of achieving a Mo concentration of at least 6% in the region of material that solidifies first, thus preserving good corrosion resistance of the weld region. In case post-weld annealing treatment is not possible, autogenous welding of SASSs, i.e., without filler material, is prohibited. Autogenous welds can be accomplished only if the resulting assembly is subsequently subjected to a solution annealing treatment to homogenize the weld and reduce microsegregation in order to preserve corrosion resistance [9]. Regarding the welding techniques for SASSs, as well as stainless steels in general, shielding must be employed in order to prevent oxidation of the weld and HAZ. This is because oxidation during welding has deleterious effects as it compromises not only corrosion resistance, but also mechanical properties. In order to maximize corrosion resistance, it is preferable to shield the backside of the weld as well. Ar, N, and mixtures of N and $H_2$ in 90–10% ratio are used as backing gases [9]. The most commonly used techniques used for welding SASSs are:

- Gas tungsten arc welding (GTAW). GTAW technique is also known as TIG (tungsten inert gas) or WIG (wolfram inert gas) welding. The process employs the use of a non-consumable W electrode and inert gas to accomplish welding. GTAW can be performed either with, or without filler metal.
- Gas metal arc welding (GMAW). GMAW uses a bare metal electrode shielded by a protective gas. GMAW was first introduced in the 1940s for the welding of Al alloys with the name MIG (metal inert gas) welding [2].
- Flux-cored arc welding (FCAW). With this process a flexible continuous consumable filler metal tubing electrode containing flux and other substances (e.g., alloying elements, deoxidizers) is employed to weld the material. Shielding can be provided by the flux core containing gas-generating compounds, or externally supplied in form of gas. This latter version is called gas-shielded FCAW [2].

Table 4.3 shows some of the shielding gas mixtures for the abovementioned welding operations that can be employed, together with some useful notes.

**Table 4.3** Shielding gases and mixtures for GTAW, GMAW and FCAW welding processes [9]

| Process | Shielding gas | Notes |
|---|---|---|
| GTAW manual | 99.95% Ar | Good penetration, high quality welds, expansive, low gas flow rate |
| | 99.99% Ar | |
| GTAW automated | He | High heat input, deep penetration |
| | Ar + He | Improved wettability and bead contour |
| | Ar + $H_2$ | Improved penetration and welding speed |
| | Ar + He + $H_2$ | |
| | Ar + $N_2$ | Limitation of N loss in N-alloyed grades |
| GMAW spray arc | 99.95% Ar | – |
| GMAW short arc and pulsed arc | 99.95% Ar | – |
| | Ar + 30% He + 1% $O_2$ | For 6Mo SASSs and AISI 904L |
| FCAW | Ar + 20–25% $CO_2$ | Horizontal welding |
| | 100% $CO_2$ | Vertical welding |

## 4.7 Additive Manufacturing

In modern manufacturing, the group of technologies known as additive manufacturing (AM) is gaining market acceptance because of the undisputed advantages it can bring. In fact, this technology has emerged in recent years due to the possibility of producing solid products with shapes and characteristics that cannot be achieved by traditional machining and plastic deformation processes. Typically, AM involves manufacturing the product by sequentially overlapping layers of materials that are joined or solidified together resulting in the desired component. This type of process is in sharp contrast to machining operations because they involve removing material from a block or sheet in order to produce the desired shape. For this reason, traditional machining techniques are classified as subtractive manufacturing processes. From an historical point of view, the first examples of bringing AM technologies to the market date back to 1988 and the processes were referred to as rapid prototyping (RP). In fact, the original use was not related to the production of working and functional finished components, but rather to the manufacturing of prototypes with new design, typically in the automotive sector. The term 3D printing was then applied to processes in the 1980s and finally the name additive manufacturing was formally adopted from January 2009 as a result of decisions by one of the ASTM technical committees [16]. The adoption of the new terminology brought not only a change in the name, but also in the paradigm related to AM. In fact, AM techniques are now used for the production of finished components employed in several products and industries such as automotive, aerospace, biomedical. The major challenges in this context lay mainly in verifying the equivalence, or superiority, of the same components made, however, using AM techniques. From the perspective of materials applicability, the challenges and critical issues are also far from trivial. Currently, all major families of materials (i.e., metals, ceramics, polymers and composites) can be manufactured through AM techniques. However, not all members of these families are currently suitable for use in the manufacturing of AM products. This last consideration is based on two main themes. The first concerns the optimization of existing materials and the adoption of process parameters that make them feasible with current AM techniques. The second regards the development of new materials better suited to existing techniques, or the development of new technologies better suited to materials currently on the market. For this reason, the AM world is one of the most vibrant technological areas in this era. Furthermore, even as this text is being written, a new technique or material may be developed and make the information that will be displayed obsolete [17].

Obviously, detailed analysis of AM processes is beyond the scope of this text, and the reader is therefore encouraged to refer to the references and further reading at the end of this chapter. However, basic concepts will be laid out to enable understanding of the applicability or non-applicability of AM to SASSs. To summarize the main features of AM and give a brief description of the main aspects, reference will be made to Fig. 4.10 where six key aspects for the interpretation of the technology are highlighted in comparison with traditional CNC machining. This comparison makes particular sense because AM shares something with CNC machining: both are based on computer-supported technology. Moving clockwise through the figure, the first parameter is velocity. CNC machining processes tend to be faster than AM processes in terms of realizing the single feature (e.g., chamfers, holes, etc.) or simply external geometries. By means of multi-axis and multi-tool machine tools, CNC machining can be accomplished in a short time. Furthermore, the speed at which material can be removed is much greater than the speed at which the same volume can be added. However, AM allows multiple of the same component to be made in a single manufacturing process. In addition to this, AM can make a product in a single production stage, whereas CNC often requires many steps, a long setup of the machine, and specific planning of the operations. Hence, in some respects the velocity could be greater. Thus, considering the single physical interaction of the part material, CNC is definitely faster and AM slower (full line in Fig. 4.10). If, on the other hand, speed is considered as referring to the whole production process of a part, AM may be faster because the part could be ready immediately after production (dashed line in Fig. 4.10). With CNC, more machines, more machining and finishing operation may be needed, which, in series, could take weeks [17]. As far as complexity is concerned, in this case AM can guarantee a much higher degree than CNC machining as the possibilities in this regard have fewer limitations than with traditional processes. Cavities, cones, cuboids, internal channels, and undercuts can be easily manufactured because the section of the workpiece to be produced is composed of overlapping layers. The possibilities here are only limited by the geometry that can be imparted to the layer compatibly with the technology and material used so that the functionality is retained. It could almost be said that geometry is more limited by the imagination of the designer than by the possibility of the process. Regarding accuracy, the most common AM technologies currently have dimensional accuracy on the order of tens of $\mu$m. Subtractive processes can guarantee results of at least an order of magnitude lower. This is mainly due to the way the part is produced: layer overlapping. Furthermore, AM machines also have different resolution along orthogonal axes being the vertical one characterized by the lower value when compared to the other two [17]. Moreover, even from the point of view of surface roughness there are differences between the two processes, being CNC machining the one capable of reaching the lowest roughness values. Parts made in AM are typically characterized by higher surface roughness and by a stepped appearance on oblique surfaces, the so-called *stair-stepping effect* caused by the overlapping of the layers. Often, to achieve specific surface roughness values, AM parts must be machined by conventional processes. This is true in a variety of technological fields and particularly in one where additive techniques are rapidly gaining ground: the biomedical sector. AM parts can be used as a replacement for human body parts, or as a scaffold to allow tissues to grow. In particular, in the case of parts that must integrate with bone, roughness must be tailored in order to account maximization of osseointegration and also all durability requirements of the specific part. For example, the surface roughness of dental implants is higher than that of ball joints. The prosthesis could therefore be produced through AM with a high roughness that could be maintained in service, or reduced by machining processes such as micromilling [17–19]. Discussion regarding geometry is similar to that made for complexity. AM is superior to CNC machining in that it can decompose a complex 3D problem into a series of simple bidimensional sections with a certain thickness. In this way, the three-dimensional connection between surfaces is lost, and continuity is solely defined by how close successive layers are to each other. Since this cannot be done easily with CNC machining, the surface must be generated in three-dimensional space. As a consequence of this, the realization of freeform surfaces can be a challenge since complex geometries may not be feasible even with 5-axis interpolated control, or greater [17]. Regarding programming, in the case of AM, no special knowledge is required because the profile generation is done automatically by the software. The result will then be a function of the accuracy of the software. On the other hand, CNC machining requires skilled labor, and inaccurate programming could not only generate a nonconforming part, but also damage the machine tool itself. In terms of material, as mentioned above, all families can be employed in AM with some exceptions for specific members. However, there still remain some critical issues related to the anisotropy of the produced parts due to the preferential direction of layer addition. Furthermore, even density of the parts still does not reach 100%, and the presence of porosity has obvious implications on the achievable properties. It is true that CNC machining is particularly indicated for metals and relatively hard materials; however, the parts produced are more homogeneous and with a quality that is better predictable than their AM counterparts [17].

Since steel still remains the most popular engineering material, it is obvious that it has catalyzed attention to the possibility of producing components through AM processes. Steel is weldable, with differences in weldability depending on chemical composition. Consequently, it is natural that these alloys have been considered as candidates for additive manufacturing. Stainless steels are currently used for the

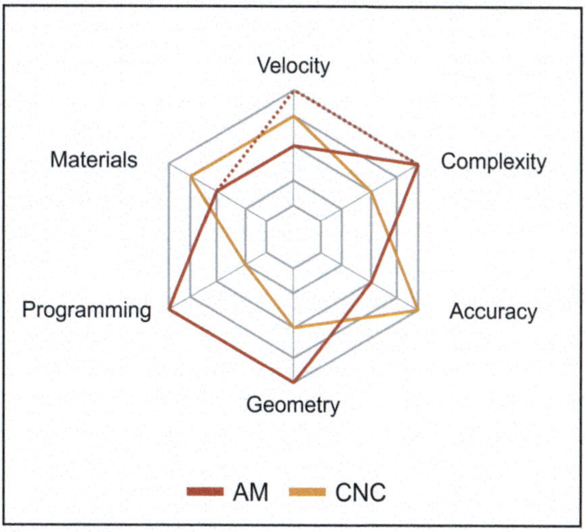


**Fig. 4.10** Comparison between traditional CNC machining technology and AM

fabrication of components produced by AM and the feedstock employed is usually in form of powder or wire based on traditional cast, wrought, and weldable grades. Challenges in this area include the formation of unexpected microstructure and phase precipitation. Both these issues are related to the very rapid solidification during the additive process followed by heating and cooling cycles due to the various manufacturing steps (see also Sect. 2.4). Other issues such as anisotropy, residual stresses caused by the near-net shape production, and porosity that can result from incomplete melting should be faced. Currently, the most common technologies for additive processing steels are binder jet (BJ), directed-energy deposition (DED), and powder bed fusion (PBF). BJ involves the deposition of liquid droplets over a bed of binder powder. This process requires postprocessing to remove the binder and, as far as stainless steels are concerned, is generally limited to AISI 316L and AISI 420 with the possibility of achieving density values between 98 and 99% with respect to bulk products. On the other hand, DED uses a concentrated energy source to generate a melt pool where feedstock is deposited. The feedstock can be wire or powder and the energy source is typically laser, although electron beam (EB) can be employed as well. The PBF process also employs a focused energy source, which can be EB or laser, with greater diffusion of the latter. In fact, for metals, PBF is also often referred to as selective laser melting (SLM). The laser beam selectively melts a layer of powder and the sequence of various heating, melting, and solidification processes creates the three-dimensional part. The cooling rates associated with SLM can be one or two orders of magnitude greater than with DED, resulting in the production of much finer microstructures. In addition to this, the AM of stainless steels is not immune to the phenomenon of microsegregation (see Chap. 2). The phenomenon, which occurs during the solidification of wrought and cast alloys, is also typical of AM, but with different genesis. In fact, although AM very rapid cooling rates limit microsegregation at an early stage, the continuous sequence of heating and cooling can promote the phenomenon. Microsegregation issues become particularly evident for those grades that may be subjected to precipitation hardening (e.g., PHSSs) and for those with chemical compositions rich in elements with a high segregative tendency, as SASSs might be [16]. Currently, there are a few dozen grades that can be employed in AM processes. Concerning stainless steels, these are mainly PHSSs and austenitic grades such as: AISI 304, AISI 316L, AISI 410, AISI 420, 15-5PH and 17-4PH [16]. At the moment of writing this textbook, there seems to be no evidence of using SASSs feedstock for AM, and even the references in the literature do not mention the application of AM to produce SASSs artifacts. There is some experimental evidence of application of particular AM techniques to AISI 904L, which is a close relative of SASSs. In addition, the technology employed is not among the three most popular mentioned before as it is wire arc additive manufacturing (WAAM) [20]. However, it is not ruled out that in the short term SASSs could also be used for the production of AM parts. Certainly, the segregative and secondary phase precipitation issues presented in Chap. 2 are among the major limiting factors in the employment of these materials. Moreover, given also their use in niche applications and low production volumes, they do not make the use of SASSs a priority in the AM landscape. Nevertheless, the time may be quite near as some materials that are in some ways similar are already used in the AM industry. In fact, the Ni-based alloy named Alloy 625, which is similar in Cr and Mo content to the best-performing SASSs, is already being used successfully in AM through BJ, DED and SLM [16]. It is likely that the more widespread use of this alloy, and the strong difference in Ni and Fe between Alloy 625 and SASSs, are among the reasons why critical production issues have already been solved for the Ni-based alloy.

## 4.8 Uses

The entire text presented so far has reported as comprehensive an overview as possible regarding SASSs. Through all the chapters and sections, the history, types, chemical characteristics, properties and manufacturing capabilities of this family of steels have been defined. The conclusion is necessarily related to the understanding of the possible intended use for these materials. However, it should be emphasized that SASSs represent a niche within the global landscape of stainless steels. This becomes evident when considering that the world steel production in 2022 was around 1,900 Mt and that about 55 Mt of this was stainless steel. Regarding product categories, in the reference period for stainless steels it is possible to say that the largest volume accounts for flat products (more than 60%), slightly more than 20 percent are long products, and the rest are castings and other products. According to some market estimates SASSs would account for no more than 0.3% of the world stainless steel production, resulting in a production around 160,000 t/year. Given the typical areas of use for SASSs, e.g., oil and gas, scrubbers, chemical industry, and desalination plants; it is enough for important orders to be taken worldwide to allow the presented values to vary significantly during the year. At time of writing of this text, the specific closing data for 2023 are not yet available. However, as far as stainless steel market is concerned, 2023 data state that world production grew by about 3 Mt. This growth was practically all due to increased production in China (+ 14.7%) and a decrease of about 6% in the rest of the world.

The demand for SASSs appears to be mainly related to specific projects, although there are stockholders and manufacturers specialized in these alloys. However, they face constant difficulties in deciding which and how many products to stock. Because of this "project-related" nature of the demand, it is difficult for it to be regular for specific grades, although some appear to be more required and popular than others. In this regard, Fig. 4.11 gives an overview of the qualitative breakdown among the different SASSs, also considering the high-performance austenitic AISI 904L for comparison. Talking in general terms, the industrial sectors in which SASSs find the most application are oil and gas (especially 6Mo), off-shore (e.g., umbilical-type applications), green energy, desalination plants, seawater storage, chemical industry related to acid processing (acetic, sulfuric, phosphoric), paper industry, food industry (e.g., in soy sauce processing). It should be noted, however, that for some applications,

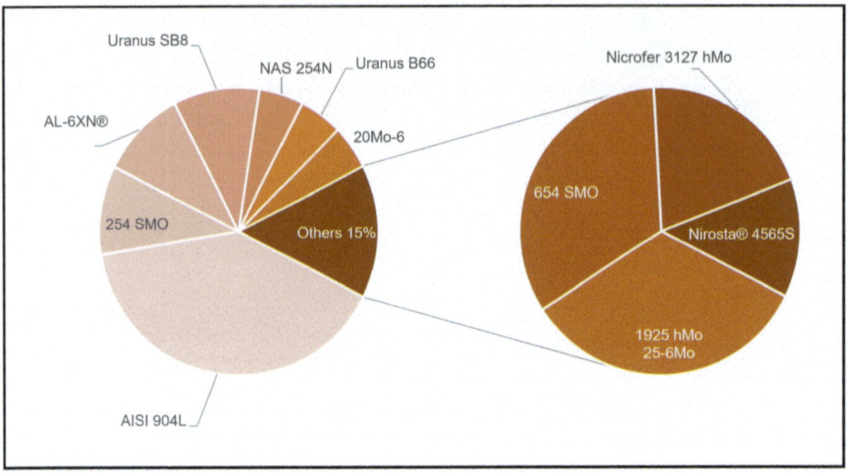


**Fig. 4.11** Qualitative breakdown among the different SASSs including the AISI 904L high-performance austenitic grade. Nirosta® 4565S corresponds to the UNS S34565

SASSs have been replaced by duplex grades. For example, when considering the scrubber industry, SASSs have recently had a lot of competition from SDSSs like 2507. Precisely with regard to the construction of scrubbers, the last two to three years had seen an increase in demand for SASSs due to new legislation to equip both cruise ships and cargo ships with these systems. This was expected to increase the market share of SASSs. However, partly because of the competition with SDSSs and partly because of the total non-application of the new regulations, the amount of SASSs employed had a sharp decline. This brought the market values back into the average of previous years. In order to provide the reader with information on the last aspect of relevance to the knowledge of SASSs, Table 4.4 gives the main sectors, applications and examples.

**Table 4.4** Applications of SASSs in modern industry [5, 13, 18, 21, 22]

| Industrial sector | Example of applications | Example of employed grades |
|---|---|---|
| Chemical industry | Bolting<br>Components of plants for phosphoric acid production via wet digestion process<br>Digestor vessels<br>Distributor systems<br>Flanges<br>Process equipment for all but strong reducing and hot sulfuric acid<br>Pump shafts<br>Pumps<br>Rotors<br>Tanks<br>Valves<br>Vaporizers | 254 SMO<br>AL-6XN®<br>25-6Mo<br>Uranus B66<br>UNS S34565<br>3127hMo |
| Energy production | Components of geothermal energy production equipment | Uranus B66 |
| Flue gas treatment and pollution control | Condenser tubing<br>FGD adsorbers and piping, also operating at high chloride levels<br>Power plant flue-gas scrubbers | AL-6XN®<br>NAS 254N<br>25-6Mo<br>3127hMo<br>654 SMO<br>Uranus B66<br>UNS S34565 |
| Food industry | Food processing equipment<br>Soy sauce fermentation tanks<br>Storage vessels | AL-6XN®<br>NAS 254N |
| Marine | Marine scrubbers<br>Ship diesel scrubbers | 3127hMo<br>25-6Mo |
| Medical | Centrifugal separators<br>Reaction tanks | NAS 254N |
| Mineral mining industry | Components of high-pressure acid leach plants for digestion of laterite ores | 3127hMo |
| Nuclear | Condenser<br>tubing<br>water<br>piping | 654 SMO<br>AL-6XN® Service- |
| Oil and gas and offshore | Bolting<br>Components of off-shore platforms<br>Firewater systems<br>Flanges<br>Pipework<br>Pump shafts<br>Pumps<br>Rotors<br>Umbilical systems<br>Valves | 254 SMO<br>25-6Mo<br>3127hMo<br>654 SMO<br>Uranus B66<br>UNS S34565 |

(continued)

**Table 4.4** (continued)

| Industrial sector | Example of applications | Example of employed grades |
|---|---|---|
| Paper industry | Digestor vessels<br>Flanges<br>Highly oxidizing paper bleach processing equipment<br>Paper bleach equipment operating at moderate chloride, temperature and pH conditions<br>Pumps<br>Rotors<br>Tanks<br>Valves | 254 SMO<br>25-6Mo<br>654 SMO<br>Uranus B66 |
| Seawater treatment | Bolting<br>Chillers<br>Condenser tubes<br>Desalination plants<br>Flanges<br>Heat exchangers/evaporators handling hot seawater and brines<br>Piping systems<br>Pressurized systems components<br>Pumps<br>Rotors<br>Tanks<br>Valves | 254 SMO<br>25-6Mo<br>3127hMo<br>654 SMO<br>UNS S34565<br>NAS 254N |
| Steel production industry | Components of pickling plants | 3127hMo |

## References


1. Black, J.T., Kohser, R.A.: DeGarmo's Materials and Processes in Manufacturing, 13th ed. John Wiley & Sons (2019)
2. Groover, M.P.: Fundamentals of Modern Manufacturing—Materials, Processes and Systems, 7th ed. Wiley (2019)
3. Ashby, M., et al.: Materials: Engineering, Science, Processing and Design, 4th ed. Butterworth-Heinemann (2019)
4. Gialanella, S., Malandruccolo, A.: Aerospace Alloys. Springer (2020)
5. Outokumpu Oyj: Handbook of Stainless Steels. Outokumpu Oyj, Espoo, Finland (2013)
6. Hao, Y., et al.: Microstructural bandings evolution behavior and their effects on microstructure and mechanical property of super-austenitic stainless steel. Mater. Sci. Eng., A **736**, 258–268 (2018)
7. Ge, S., et al.: Progress of strip casting technology for steel; historical developments. ISIJ Int. **52**(12), 2109–2122 (2012)
8. Hao, Y., et al.: Solidification structures of Fe–Cr–Ni–Mo–N super-austenitic stainless steel processed by twin-roll strip casting and ingot casting and their segregation evolution behaviors. ISIJ Int. **58**(10), 1801–1810 (2018)
9. IMOA: Practical Guidelines for the Fabrication of High Performance Austenitic Stainless Steels. International Molybdenum Association (IMOA), London (2010)
10. IMOA: Practical Guidelines for the Fabrication of Austenitic Stainless Steels, 2nd ed. International Molybdenum Association (IMOA), London (2020)

11. Prasad, Y., et al.: A Compendium of Processing Maps. ASM International, Materials Park, Ohio (2015)
12. Pu, E., et al.: Hot deformation characteristics and processing maps of superaustenitic stainless steel S32654. Mater. Sci. Eng., A **598**, 174–182 (2014)
13. Kovach, C.W.: High-Performance Stainless Steels. The Nickel Institute (2000)
14. Capelli, F.: L'acciaio INOX. Centro Inox Servizi Srl (2021)
15. Passari, E.S., et al.: Machinability investigation of 254 SMO super austenitic stainless steel in end milling under different cutting and lubri-cooling conditions. Int. J. Adv. Manuf. Technol. **131**, 6061–6073 (2024)
16. ASM International: ASM Handbook Volume 24—Additive Manufacturing Processes. ASM International, Materials Park, Ohio (2020)
17. Gibson, I., et al.: Additive Manufacturing Technologies, 2nd ed. Springer (2015)
18. Cappellini, C., et al.: A feasibility study of promoting osseointegration surface roughness by micro-milling of Ti-6Al-4V biomedical alloy. Int. J. Adv. Manuf. Technol. **126**, 3053–3067 (2023)
19. Cappellini, C., et al.: Analysis of Ti-6Al-4V micro-milling resulting surface roughness for osteointegration enhancement. Mater. Res. Proc. **28**, 1255–1264 (2023)
20. Vishnukumar, M., et al.: Microstructural characteristics, mechanical properties and corrosion performance of superaustenitic stainless steel 904L produced by wire arc additive manufacturing. Mater. Today Comm. **35**, 105801 (2023)
21. Davidson, R., Redmond, J.: Practical Guide to Using 6Mo Austenitic Stainless Steels. NiDI Technical Series n°10 032. Nickel Development Institute (1994)
22. Farrar, J.C.M.: The Alloy Tree: A Guide to Low-Alloy Steels, Stainless Steels and Nickel-Base Alloys. Woodhead Publishing Limited (2004)

## *Further Reading*

ASM International: ASM Handbook Volume 15—Casting. ASM International, Materials Park, Ohio (2008)

Di Nicolantonio, M., et al.: Advances in Additive Manufacturing, Modeling Systems and 3D Prototyping. Springer Nature Switzerland (2019)

Kumar, J.L., et al.: 3D Printing and Additive Manufacturing Technologies. Springer Nature Singapore (2018)

Milewski, J.: Additive Manufacturing of Metals—From Fundamental Technology to Rocket Nozzles, Medical Implants, and Custom Jewelry. Springer International Publishing (2017)

Straffelini, G.: Ductility and Formability of Metals—A Metallurgical Engineering Perspective. Academic Press (2023)

# Appendices

**Appendix A** Alloying elements in steels

| Element | Symbol | $\gamma$-stabilization | Carbides formation | Main alloying effects |
|---|---|---|---|---|
| Aluminum | Al | No | No | Al is a strong de-oxidizer and can be used for this purpose during alloy refinement, either alone or in association with Mn and Si. Al can form non-soluble inclusions (e.g., $Al_2O_3$) in molten steel, both during primary melting and refining process. It is a very effective element when used as grain refiner. Its presence promotes precipitation hardening in PHSS thanks to the formation of the $\gamma$' precipitate. It is used in nitriding steels leading to the formation of very hard aluminum nitrides (1000 HV). Al is used for increasing corrosion resistance in low-carbon steels and for increasing oxidation resistance in heat-resistant steels. |
| Antimony | Sb | No | – | It is considered as an impurity and it is not intentionally added. Moreover, it is difficult to be removed during melting and in refining processes. Its presence lowers the surface quality of continuously cast billets. Sb causes temper embrittlement in alloy steels. Sb has a strong tendency to segregate, especially at grain boundaries during cooling processes |
| Beryllium | Be | – | – | The main function of beryllium is to improve the corrosion resistance of steels |
| Boron | B | No | – | Boron is usually added in very limited amounts, ranging from 0.0005 to 0.0035%. B tends to segregate at grain boundaries, thus improving hardenability of steels. Boron is a strong nitride former and this feature interferes with the hardening effect of elemental B. The addition of a specific amount of Ti to the steel protects B from nitride formation because Ti combines preferentially with N, thus avoiding boron nitride formation |

(continued)


A. Malandruccolo, *Superaustenitic Stainless Steels*, Engineering Materials,
https://doi.org/10.1007/978-3-031-68744-0

**Appendix A** (continued)

| Element | Symbol | γ-stabilization | Carbides formation | Main alloying effects |
|---|---|---|---|---|
| Calcium | Ca | – | – | Ca is a strong deoxidizer. Concerning steelmaking processes, the *SiCa treatment* can be used in order to modify inclusion morphology and type. In particular, the process helps removing alumina from the molten bath thanks to the combination of Ca and S. This process leads to CaO·x$Al_2O_3$ inclusions surrounded by liquid CaS. Thanks to this phenomenon, $Al_2O_3$ is removed from the liquid steel and goes into the slag. SiCa treatment can reduce sulfur and oxygen content in the alloy. In *killed steels*, CaS is present as spherical particles. Ca, Al and Si tend to form low melting compounds that results in better machinability of the steel. Some problems arise when CaS-based inclusions become too large, promoting the formation of fatigue-cracks |
| Carbon | C | Yes | Yes | Carbon is, with Fe of course, the base element in steels. It is inexpensive and can be easily added in various forms during steelmaking. During solidification, carbon shows a moderate tendency to macrosegregation, being more prone to segregating at microstructural defects, such as grain boundaries and dislocations. Carbon is an interstitial, both in *fcc* and in *bcc* iron, although with different solubility limits, and has a strengthening effect, enhancing both tensile strength, yield strength and hardness. This is possible through carbide dispersion strengthening and solid solution strengthening. Strengthening effect is proportional to the C content up to a threshold value (around 1% for carbon steel). Notch toughness and weldability of steels decrease at higher carbon concentrations |
| Chromium | Cr | No | Yes | Chromium is one of the most important alloying elements in steels, it is a carbide-former, and for increasing Cr/C ratios, $Cr_2C$, and, the most common, $Cr_7C_3$ and $Cr_{23}C_6$ would form. Cr-carbides increase hardness and wear resistance. Chromium carbides precipitation at the grain boundary (i.e., sensitization) lowers the corrosion resistance of stainless steels, in particular in the heat-altered zone (HAZ) after welding processes. For this reason, titanium is introduced as preferential carbide-forming element. Chromium implies an increase in mechanical properties, in hardenability, in oxidation and general corrosion resistance. Stainless steels usually contain more than 13% Cr and this leads to the formation of a thin, stable and adherent $Cr_2O_3$-based layer on their surface causing an increase in the corrosion resistance in many different environments. Cr improves in particular the pitting corrosion resistance of the alloy |

(continued)

**Appendix A** (continued)

| Element | Symbol | $\gamma$-stabilization | Carbides formation | Main alloying effects |
|---|---|---|---|---|
| Cobalt | Co | Yes | No | The use of cobalt as alloying element has been recently limited by its toxicity. As a consequence of this, Co-free alloys have been developed. By adding cobalt to carbon steel, the material undergoes a reduction in its hardenability. When combined with Cr it increases the hardenability of Cr–Mo steels. Its use is generally limited to the production of tool steels, maraging steels, creep-resistant steels, high-speed steels and for steels used in high-temperature applications. Co is a grain boundary stabilizer at high temperature and contributes to maintaining high-temperature strength and temper as well. This latter feature is the reason for the increase in useful lifetime of tool steel items. Cobalt promotes precipitation hardening, and acts as a stabilizer agent for martensite increasing the $M_s$ temperature and reducing the concentration of retained austenite in alloy steels |
| Copper | Cu | Yes | No | Copper is generally regarded as an impurity in steels, and its content is kept below 0.04%. It cannot be oxidized and removed during refining of steels and the best approach for reducing Cu content in steel is through dilution. A particular class of low-alloy steels contains an appreciable amount of Cu for increasing atmospheric corrosion resistance. This class of steels is named weathering steels and, thanks to the presence of Cu, P, Ni and Cr, they are able to form a protective layer on the surface. This family of steels is also known as COR-TEN. PH stainless steels use Cu for increasing the strength thanks to the precipitation of Cu-rich intermetallics during ageing treatment. Copper is also added to austenitic stainless steels and when exceeding 1% it causes an improvement in the corrosion resistance against HCl, $H_2SO_4$ and against stress corrosion phenomena. Furthermore, Cu improves cold workability of austenitic stainless steels that contain it |
| Hydrogen | H | – | – | Hydrogen is an interstitial atom, it is soluble in liquid steel and its presence can be very detrimental to the material. Modern steelmaking processes imply degassing stages for removing this element from liquid steel. The presence of hydrogen can lead to hydrogen-induced damage |
| Lead | Pb | – | – | Lead is insoluble in iron and it is used for increasing machinability of the material. When added to steel, it tends to form a particle dispersion throughout the material. This is particularly useful during machining, since it results both in a good lubrication and increased chip-breaking ability. However, leaded steel market has experienced a contraction in the last decades because of the environmental hazard and health risk associated with lead. Therefore, modern steelmaking processes tend to prefer the use of sulfur for the production of steels with improved machinability |

(continued)

**Appendix A** (continued)

| Element | Symbol | $\gamma$-stabilization | Carbides formation | Main alloying effects |
|---|---|---|---|---|
| Manganese | Mn | Yes | Yes | Manganese is both a $\gamma$-stabilizer element and a moderate carbide-forming element, it dissolves in $Fe_3C$ and forms alloyed cementite as well. Typical Mn-carbides are $Mn_{23}C_6$ and $Mn_{15}C_4$. Manganese is present, as substitutional, in almost all steel grades in a quantity $\geq$ 0.30%. In steelmaking processes, Mn is used both as deoxidizer agent (often in association with Al, Si or both) and for S removal, through the formation of MnS, thus avoiding the deleterious formation of FeS. Manganese enhances the steel hardenability for concentrations exceeding 0.8%. Steels with Mn concentrations larger than 2% are used for special purposes, such as railway track connectors and rock crushers. These steels belong to a special class named Hadfield manganese steels, also known as austenitic manganese steels. Mn increases the N solubility in steel and is, therefore, an important alloying element in SASSs |
| Molybdenum | Mo | No | Yes | Mo is an $\alpha$-stabilizer element. It is a strong carbide-former element, $Mo_2C$ and $Mo_6C$ being the main phases found in steels. Mo-carbides are important for steel grades used in wear-resistant applications (e.g., tool steels). Molybdenum is added for minimizing temper embrittlement, but it cannot avoid it completely. It is an effective substitutional solution hardener and can induce secondary hardening during tempering of quenched steels and in austenitic alloys at high temperatures. It acts also as grain refiner when added to fine-grained steels. Mo improves corrosion resistance, it is used both in ferritic and austenitic stainless steels, acting as a passivation-enhancing element, thus promoting a fast formation of the passivating $Cr_2O_3$ layer, also increasing its stability. Mo is also present in duplex stainless steels. Superaustenitic stainless steels have a Mo content $\geq$ 6%: the high Mo-content reduces the susceptibility to pitting corrosion, but can lead to the formation of brittle intermetallic compounds (e.g., $\chi$ and $\sigma$), which are responsible for the reduction in deformability of the alloy. Furthermore, Mo decreases the solubility of C in the alloy matrix leading to the promotion of carbides precipitation |

(continued)

**Appendix A** (continued)

| Element | Symbol | $\gamma$-stabilization | Carbides formation | Main alloying effects |
|---|---|---|---|---|
| Nickel | Ni | Yes | No | Nickel is a strong $\gamma$-stabilizer and a non-carbide-forming element and, thereby, it is a necessary alloying element in austenitic stainless steels. Similarly to Cu, due to its position in the Ellingham diagram it does not undergo oxidation during steelmaking processes, meaning that the amount of Ni is constant over time during the process. The substitutional solid solution of nickel in iron has a limited effect on increasing the alloy yield strength. Furthermore, this element improves the hardenability of steel, in particular when combined with Cr and Mo. In this condition the material undergoes also an increase in fatigue resistance and in impact toughness. It reduces the ductile to brittle transition temperature. For this reason, nickel is used as alloying element in steels for cryogenic applications. If combined with Al, it forms the intermetallic $Ni_3Al$ upon aging, on which precipitation hardening of some steels, similarly to superalloys, is based. Furthermore, Ni decreases the solubility of C in the alloy matrix leading to the promotion of carbides precipitation. In addition to this, an increased Ni content is also effective in limiting or suppressing the formation of $\delta$ ferrite formation upon solidification in those steels that, according to their chemical composition, could be prone to the phenomenon. Together with N, Ni is a fundamental alloying element in SASSs |
| Niobium | Nb | No | Yes | Niobium is a strong carbide-, nitride- and carbonitride-forming element. Niobium (named Columbium in the US literature) introduces a grain refinement effect in steels. It is used in HSLA (high-strength low-alloy) steels in order to enhance the strength by maintaining good toughness through precipitation strengthening and the formation of Nb-carbonitrides and it is also able to prevent carbide dissolution and reprecipitation in undesirable locations. For this reason, it is added to some ferritic stainless steels (e.g., AISI 436 and 437), austenitic stainless steels (e.g., AISI 347, 348 and 384) and also to 17-4 PH stainless steels. Furthermore, Nb improves creep resistance of the material |
| Nitrogen | N | Yes | No | Nitrogen is an important $\gamma$-stabilizer element, which forms, together with carbon and metals, carbonitrides. Nitrogen may also partially replace nickel as austenite stabilizer in austenitic steels, in many of which it is used and results in solution or precipitation strengthening. This is the case of AISI type 201, 202, 205, 304N, 304LN, 316N and 316LN. Nitrogen is also used in duplex stainless steels for increasing pitting corrosion resistance and strength. Along with Mo, N is a key alloying element in superaustenitic stainless steels. It provides solid solution hardening, as well as improved pitting corrosion resistance and delayed precipitation of secondary phases |
| Oxygen | O | – | – | Oxygen is considered as an impurity in steels. Usually, the oxygen content has to be lower than 60 ppm. The solubility of oxygen in iron is very low and this leads to undesired oxides formation or porosities |

(continued)

**Appendix A** (continued)

| Element | Symbol | γ-stabilization | Carbides formation | Main alloying effects |
|---|---|---|---|---|
| Phosphorus | P | No | No | Phosphorous is generally regarded as an impurity in steels and its content is usually below 0.02%. Upon increasing the P content, the material experiences a decrease in toughness and ductility. It has a strong tendency for segregation at the grain boundary and this leads to deleterious effects like temper embrittlement or cracking during hot working operations. However, for specific steel grades, a relatively high P content can have beneficial effects. Low alloy structural steels with high P-content show better atmospheric corrosion resistance. The addition of P can cause precipitation hardening and yield strength improvement in some austenitic Cr–Ni steels. *Re-phosphorized* steels are a special class of alloys with improved strength and enhanced machinability |
| Rare earth metals | | – | – | Rare earth metals are a group of 17 elements belonging to the III B group. They can have strong deoxidizing and desulfurizing effects. These elements have also a tendency to segregate at the grain boundaries. Rare earth metals have effects on shape modification and control of inclusions, in particular La and Ce are added for the sulfide shape control. Ce is also characterized by low solubility in austenite. Furthermore, Ce is able to reduce segregation in interdendritic regions and cause grain refinement of the microstructure. Finally, rare earth metals can also increase the hardenability of steels |
| Silicon | Si | No | No | During steel refinement it is used as de-oxidant, either alone or in combination with Al, Mn or both. It does not form carbides in steel, but its presence may lead to the formation of $SiO_2$ inclusions. Silicon retards the formation of $Fe_3C$ during tempering and increases yield strength, hardenability and wear resistance in heat-treated steels. It is also added to a particular family of steels called *transformer steels*. They have a silicon content around the 3% and are characterized by low core losses and high magnetic permeability. Silicon dissolves into martensite and it is able to retard martensite decomposition up to 300 °C |
| Sulfur | S | No | – | Like P and O, sulfur is generally considered as an impurity, with concentrations below 0.02%. The tendency of sulfur for segregating at the grain boundary is very high and this causes the reduction of the high-temperature ductility in alloy steels. This element is insoluble in Fe and can form deleterious low melting compounds, such as FeS. However, similarly to P, sulfur is intentionally added to a special class of stainless steels (e.g., AISI 303 or AISI 416HS) that have a S content between 0.08 and 0.35%, resulting in an improved machinability of the alloy. These steels are usually named *free machining* grades |
| Tantalum | Ta | – | Yes | Like niobium, tantalum is a strong carbide former. It has a very high affinity to carbon, in some austenitic stainless steels (i.e. AISI type 347 and AISI 348) it is used as carbide stabilizing agent |

(continued)

**Appendix A** (continued)

| Element | Symbol | γ-stabilization | Carbides formation | Main alloying effects |
|---|---|---|---|---|
| Tin | Sn | – | – | Tin is considered as a detrimental impurity, in view of its effects on steel properties. One of the deleterious consequences of its presence is the temper embrittlement and a worsening of the quality of the surface of continuously cast steels |
| Titanium | Ti | No | Yes | Titanium is considered as a very strong deoxidizer due to its high affinity to oxygen. During melting and refining, the addition of Ti has to be carried out carefully in order to avoid its oxidation and subsequent migration of its oxides into the slag. Titanium is also an effective former of nitrides and carbides, and its effects are similar to those of V and Nb. However, Ti-carbides and nitrides are more stable if compared with Nb- and V-based ones. Ti nitrides act as grain boundary stabilizers. Ti is also used as carbide stabilizing agent in austenitic stainless steels, ferritic and PH steels. Titanium improves intergranular corrosion resistance (i.e., sensitization) in austenitic stainless steels. This element is also used in Fe–Ni–Cr alloys (e.g., alloy A286) for the formation of $\gamma'$ precipitation. In maraging steels Ti forms the intermetallic hardening H-phase. Furthermore, the addition of Ti prevents boron nitrides formation, so that boron can retain its elemental form |
| Tungsten | W | No | Yes | Tungsten is an α-stabilizer and carbide-forming element. W carbides (WC and $W_2C$) are very hard and of same stoichiometry and structure of Mo-carbides. Tungsten induces secondary hardening during tempering treatment of quenched steels, limits grain growth at high temperature and, together with Mo, is the most important alloying element in tool steels. In superaustenitic stainless steels W can promote the formation of Laves phases. In addition with N, W seems to decrease formation kinetics of σ phase |
| Vanadium | V | – | Yes | Vanadium forms extremely hard carbides and nitrides. V-carbides have higher hardness than Cr-, Mo-carbides and Fe-carbides. Vanadium is used for grain refinement, resulting in strength and toughness increase. Vanadium can also dissolve into austenite thus increasing its hardenability. Vanadium addition introduces large secondary hardening effect after tempering, with a general improvement of the high-temperature hardness of the material |
| Zirconium | Zr | No | Yes | Zirconium in steel limits the austenite phase field. It is a carbide and nitride-forming element. Its effects on steel are similar to those of titanium, although it is very expansive and, therefore, not so common in steels |

There are at least two possible criteria for the classification of alloying elements in steels. The first is based on the stabilization of the austenite (γ) phase. The second one is based on their tendency to form carbides. The carbide formation requires that concentration of the carbon would locally overcome a threshold value, i.e., its solubility limit in the relevant alloy, otherwise the element would remain in solution. According to the aforementioned criteria, the information concerning the most important alloying elements, and also impurities, in steels are given in this appendix. Concentrations in wt% [1–3]

**Appendix B** Precipitates and secondary phases in stainless steels

| Phase type | Name | Symbol | Unit cell | Atoms per cell | Lattice parameters [nm] | Typical chemical composition/s | Additional information |
|---|---|---|---|---|---|---|---|
| Intermetallic | Alfa prime | α′ | bcc | – | a = 0.2884 | – | Typical of ferritic and duplex grades. Precipitation occurs in the 300–550 °C temperature range. Its presence is associated with a phenomenon called 475 °C embrittlement (with reference to the temperature of 475 °C of the Fe–Cr phase diagram). Its chemical composition features essentially Fe and Cr, the phase is richer in Cr than Fe. The *bcc* structure is coherent with that of ferrite. Its presence has negative effects on both mechanical properties and corrosion resistance. α′ has also been found in 17-4PH for particular exposure conditions to temperatures in the 400–480 °C range |
| Intermetallic | Beta | β | Ordered bcc | 2 | a = 0.2865–0.2887 | NiAl | Typical of AISI 631 (17-7PH) and 13-Mo PH martensitic stainless steel. The β intermetallic is a strengthening phase formed on purpose thanks to aging treatment. The presence of an Al amount that exceeds a minimum for precipitate formation is a fundamental requirement |

(continued)

**Appendix B** (continued)

| Phase type | Name | Symbol | Unit cell | Atoms per cell | Lattice parameters [nm] | Typical chemical composition/s | Additional information |
|---|---|---|---|---|---|---|---|
| Intermetallic | CRPs (Cu-rich precipitates) | – | fcc | – | a = 0.3026–0.3600 | – | Most known as the reinforcing phase in the martensitic AISI 630 (17-4PH) precipitation hardening stainless steel. Peak hardness in the material is obtained thanks to a combination of solution annealing and aging called H900 condition (aging at 480 °C). Other annealing/aging combinations are also possible in order to modify properties. When CRPS first form, its structure is supposed to be a coherent bcc. The structure of CRPs would probably transform according to the *bcc* → 9R → 3R → *fcc* sequence as precipitates grow after nucleation. APT analyses allowed to quantify the Cu content of CRPs resulting in an atomic percentage exceeding 50 at% |
| Intermetallic | Gamma prime | $\gamma'$ | fcc | 4 | a = 0.3565–0.3601<br>c = 0.7406 | $(Ni, Co, Fe, Cr)_3(Al, Ti)$ | Typical of: AISI 660 (A286), 631 (17-7PH). This ordered $L1_2$ intermetallic compound is also the reinforcing phase in Fe- and Ni-based superalloys. Formation occurs generally on purpose thanks to aging treatment. It forms in the alloy after aging treatment. For $\gamma'$ to form, the chemical composition of the alloy must contain sufficient amount of Al and Ti. This requirement leads to the need to be very careful in the melting and refining phases because both Al and Ti tend to oxidize, thus contributing to the slag formation and depletion of these elements in the liquid. This is particularly true for Ti, which has a very high affinity with oxygen |

(continued)

**Appendix B** (continued)

| Phase type | Name | Symbol | Unit cell | Atoms per cell | Lattice parameters [nm] | Typical chemical composition/s | Additional information |
|---|---|---|---|---|---|---|---|
| Intermetallic | G | – | fcc | 116 | a = 1.115–1.120 | $Ni_{16}Nb_6Si_7$<br>$Ni_{16}Ti_6Si_7$<br>$(Ni, Fe, Cr)_{16}(Nb, Ti)_6Si_7$ | Typical of: AISI 308, 310S, 329, Alloy 709, duplex stainless steels (when subjected to heating for prolonged time in the 250–500 °C temperature range). The G phase is a silicide that was first identified in 1956 in the A286 austenitic PH stainless steel. Its name derives from its prominent formation at GBs. It features a complex *fcc* crystalline structure. In low-C completely austenitic matrix the G phase precipitates at GBs and within grains when the material is heat treated in the 500–850 °C temperature range. Complete crystallographic characterization of the G phase in any stainless steel grade has not yet been well documented. The G phase occurs much less frequently and in smaller quantities in stainless steels than other phases such as σ, χ and Laves |
| Intermetallic | Eta | η | hex. | 8 | a = 0.5109<br>c = 0.8299 | $Ni_3Ti$ | Typical of overaged AISI 660 (A286). The formation of the η phase occurs at the expense of the γ′ phase due to prolonged exposure to temperatures between 600 and 850 °C. It has been observed that the η phase is formed from the γ′ phase by precipitation at the grain boundaries according to cellular reaction. η phase formation is associated to degradation of the creep behavior, fracture behavior (intergranular fracture), increased HE sensitivity, decreased fatigue life |

(continued)

**Appendix B** (continued)

| Phase type | Name | Symbol | Unit cell | Atoms per cell | Lattice parameters [nm] | Typical chemical composition/s | Additional information |
|---|---|---|---|---|---|---|---|
| Intermetallic | Mu | μ | Rhombohedral | 13 | a = 0.4762<br>c = 2.5015 | $(Fe, Co)_7(Mo, W)_6$<br>$(Cr, Fe)_7(Mo)_2(Cr, Fe, Mo)_4$ | Found in Fe-17Cr–14Ni–6–, Fe–15Cr–40Ni–4W–2Mo–Al–Ti ad superferritic UNS S44660 stainless steel. Its presence is generally associated with detrimental effects. This phase is typical of the Fe–Cr–Mo and Fe–Cr–W ternary systems. Evidence of its presence have been reported also for SASSs. The μ phase occurs much less frequently and in smaller quantities in stainless steels than other phases such as σ, χ and Laves |
| Intermetallic | R | – | hex. | 53 (159) | a = 1.090<br>c = 1.934 | $Fe_{22}Mo_{18}Cr_{13}$<br>$(Fe, Ni)_{10}Cr_5Mo_3Si_2$ | The R phase can be found in superaustenitic stainless steels like 654 SMO, Fe–22Cr–8Ni–3Mo duplex grade, and other duplex grades, including superduplex stainless steels. It essentially contains Mo–Fe–Cr, but it may also contain Ni and Si. It has been documented that this phase precipitates in the 550–650 °C temperature range in an initial stage of the thermal treatment leading to a degradation of the alloy through toughness reduction. It has been observed that in superduplex grades the amount of R phase is very small and tends to convert into σ phase with prolonged exposure to high temperature. The R phase occurs much less frequently and in smaller quantities in stainless steels than other phases such as σ, χ and Laves |

(continued)

**Appendix B** (continued)

| Phase type | Name | Symbol | Unit cell | Atoms per cell | Lattice parameters [nm] | Typical chemical composition/s | Additional information |
|---|---|---|---|---|---|---|---|
| Intermetallic | Sigma | σ | bct | 30 | a = 0.87–0.92<br>c = 0.4554–0.48 | $(Fe, Ni)_x(Cr, Mo)_y$ | It is probably the most studied and undesirable intermetallic phase and appears in several binary, ternary and quaternary systems such as: Fe–Cr, Fe–Mo, Fe–V, Fe–Mn, Fe–Cr–Ni, Fe–Cr–Mo, Fe–Cr–Mn and Fe–Cr–Ni–Mo. Some examples of stainless steels where this phase can be found are: AISI 304, 304L, 316, 316L, 321, 347, duplex grades and superduplex, superaustenitic grades. σ phase is a non-magnetic intermetallic compound, also able to change the fracture type from transgranular to intergranular as its volume fraction in the alloy increases. One of the proposed models for precipitation sequence implies metastable χ phase formation (initial stage) and subsequent transformation into the deleterious σ phase. More information about this phase is given in the main text. The typical temperature range of formation is quite broad and lays between 550 and 1050 °C |

(continued)

**Appendix B** (continued)

| Phase type | Name | Symbol | Unit cell | Atoms per cell | Lattice parameters [nm] | Typical chemical composition/s | Additional information |
|---|---|---|---|---|---|---|---|
| Intermetallic | Chi | $\chi$ | bcc | 58 | a = 0.881–0.895 | $Fe_{35}Ni_3Cr_{13}Ti_7$<br>$Fe_{36}Cr_{12}Mo_{10}$<br>$(Fe, Ni)_{36}Cr_{18}Mo_4$<br>$Fe_{36}Cr_{12}Mo_3Ti_7$ | $\chi$ phase can be found in several stainless steel grades, some of them are: AISI 316, 316L, 321, duplex and superduplex grades, superaustenitic grades. $\chi$ phase is classified as a tertiary compound containing Fe, Cr and Mo featuring a wide range of stoichiometry extending from $Fe_{36}Cr_{12}Mo_{10}$ to $Fe_{36}Cr_{12}Mo_3Ti_7$. $\chi$ pahse has been found to precipitate at the $\alpha/\alpha$ GBs prior to $\sigma$ phase. Its precipitation mechanism and its relatioship with $\sigma$ phase have been the subject of extensive study. More information about this phase is given in the main text. The typical temperature range of formation is between 600 °C and 900 °C |
| Intermetallic | Laves | – | hex. | 12 | a = 0.473–0.483<br>c = 0.772–0.786 | $Fe_2Mo$<br>$Fe_2Nb$<br>$Fe_2Ta$<br>$Fe_2Ti$<br>$Fe_2W$ | Laves phases has been documented to appear in several stainless steels such as: AISI 316, 316L, 321, 347, superaustenitic grades, AISI 441 and other ferritic grades. Their presence is associated to a decrease in the impact toughness, as well as a decrease in the high-temperature strength. The typical temperature range of formation is between 550 and 900 °C |

(continued)

**Appendix B** (continued)

| Phase type | Name | Symbol | Unit cell | Atoms per cell | Lattice parameters [nm] | Typical chemical composition/s | Additional information |
|---|---|---|---|---|---|---|---|
| Carbides | $M_{23}C_6$ | – | fcc | 116 | a = 1.057–1.068 | M = Cr, Fe, Mo<br>$Cr_{23}C_6$<br>$(Cr, Fe, Mo)_{23}C_6$<br>$(Cr_{16}Fe_5Mo_2)C_6$ in AISI 316 | Typical of nearly all stainless steels. This carbide is usually associated with sensitization of austenitic stainless steels after welding processes. Its precipitation occurs typically at the GBs. Its presence is generally undesirable because, in addition to deleterious effects on the corrosion resistance, it is also associated to a decrease in toughness and ductility. However, its presence has been demonstrated to be positive to some extent. In fact, it has been shown that grain boundary sliding can be hindered by the presence of $M_{23}C_6$ carbides, thus improving creep performances. The typical temperature range of formation is between 600 °C and 950 °C |

(continued)

**Appendix B** (continued)

| Phase type | Name | Symbol | Unit cell | Atoms per cell | Lattice parameters [nm] | Typical chemical composition/s | Additional information |
|---|---|---|---|---|---|---|---|
| Carbides | MC | – | Ordered fcc | 8 | a = 0.4131–0.4698 | M = Hf, Nb, Ta, Ti, V, Zr<br>HfC<br>NbC<br>TaC<br>TiC<br>VC<br>ZrC | Typical of: AISI 321, 347, 660 (A286), 630 (17-4PH) and martensitic grades containing Nb and V. In austenitic stainless steels the MC carbide forms as a consequence of a eutectic reaction. Carbides belonging to the MC type are used for stabilization of austenitic grades in order to avoid precipitation of $M_{23}C_6$ carbide, since MC carbides are more thermodynamically stable than $M_{23}C_6$. However, some of the stabilizing elements (e.g., Hf, Ti and Zr) can react with nitrogen to form MN nitrides, that are even more stable than MC carbides. N content must be therefore carefully controlled. Precipitation of MC carbides is mainly intragranular, on stacking faults and dislocations (however, precipitation at GBs can also occur). In some alloys, such as A286, a two-step aging treatment is used also to control morphology and size of MC carbides and related positive effects on mechanical properties. In Nb-containing grades the volume fraction of MC carbides increases with Nb content. As Nb is added, the overall bulk hardness increases due to the increased fraction of MC carbide. However, a Nb content above a certain threshold value can lead to excessive hardness difference between MC carbides and matrix phase, thus leading to a decrease in mechanical properties. The typical temperature range of formation is between 700 °C and the melting temperature |

(continued)

**Appendix B** (continued)

| Phase type | Name | Symbol | Unit cell | Atoms per cell | Lattice parameters [nm] | Typical chemical composition/s | Additional information |
|---|---|---|---|---|---|---|---|
| Carbides | $M_3C$ | – | – | – | – | M = Cr, Fe, Mn<br>$Cr_3C$<br>$Fe_3C$<br>$Mn_3C$ | The most known $M_3C$ carbide is cementite. Few studies have been published on this carbide type in stainless steels and there is evidence of its presence only in certain types of martensitic and superaustenitic grades. However, literature data are limited and further investigation would be helpful in clarifying this issue |
| Carbides | $M_6C$ | – | fcc | 112 | a = 1.085–1.128 | M = Cr, Fe, Mo, Nb, W, V<br>$(Cr, Fe, Mo, Nb, W, V)_6C$<br>$Fe_3Nb_3C$ | It can be found in austenitic stainless steels containing Mo, W and Nb (e.g., AISI 316, 316L), but especially Mo. It is usually found to contain more than one metallic element and this leads to stoichiometries such as $A_3B_3C$ or $A_4B_2C$. This carbide has had relatively little attention in the literature as it is usually not detected, or is detected in very small amounts in stainless steels. Given the strong dependence of its presence on Mo content, it cannot be ruled out that with the increasing study of relatively new stainless steel types such as the superaustenitics, investigations into its possible occurrence will be pursued. The typical temperature range of formation is between 700 and 950 °C |

(continued)

**Appendix B** (continued)

| Phase type | Name | Symbol | Unit cell | Atoms per cell | Lattice parameters [nm] | Typical chemical composition/s | Additional information |
|---|---|---|---|---|---|---|---|
| Carbides | $M_7C_3$ | – | Pseudo hex. | 40 | a = 1.398<br>c = 0.4523 | $(Cr, Fe)_7C_3$ | Typical of martensitic (e.g., AISI 440C) and duplex grades including superduplex and hyperduplex. It occurs in both the ternary Fe–Cr–C and quaternary Fe–Cr–Ni–C systems. It can be found in austenitic stainless steels produced according to specific cast practices (e.g., centrifugal casting for gas-reformer and pyrolysis-furnace tubes used in the petrochemical industry). Its formation is strongly influenced by the C to Cr ratio. Its presence can be also associated to secondary hardening. The typical temperature range of formation is between 950 and 1050 °C |
| Nitrides | MN | – | Ordered fcc | 8 | a = 0.4097–0.4577 | CrN<br>NbN<br>TiN<br>VN<br>ZrN | Typical of AISI 321. CrN and NbN have been reported for AISI 630 (17-4PH). MN nitrides are primary nitrides. They have same structure as MC carbides; however, they are even more stable and do not undergo dissolution during solution annealing. Given the high stability and low solubility of MN nitrides, Ti-stabilized grades are almost always subjected to their formation. On the other hand, CrN precipitation in austenitic stainless steels characterized by commercial N content usually occurs at the GB, inside the grains and at dislocations. Furthermore, N depletion in the matrix as a consequence of the precipitation of nitrides can make austenite unstable, thus leading to the possible formation of ferrite and σ phase at aging temperatures |

(continued)

**Appendix B** (continued)

| Phase type | Name | Symbol | Unit cell | Atoms per cell | Lattice parameters [nm] | Typical chemical composition/s | Additional information |
|---|---|---|---|---|---|---|---|
| Nitrides | $M_2N$ | – | hex. | 9 | a = 0.478–0.480<br>c = 0.444–0.447 | $(Cr, Fe)_2N$ | $M_2N$ is a secondary nitride that has been found in some stainless steels such as: AISI 304N, 316N, superaustenitics, duplex, superduplex and hyperduplex grades. Precipitation of $M_2N$ nitride occurs in stainless steels having high N content (0.1–0.9%). The typical temperature range of formation is between 650 and 950 °C |
| Nitrides | Z phase | – | Tetragonal | 6 | a = 0.3037–0.6784<br>c = 0.7391–7397 | CrNNb<br>(Fe, Cr)(C, N)(Nb, Mo) | Z phase forms in stainless steels containing Nb and more than 0.4 wt% N. It has been found in AISI 347N. The temperature range for Z phase precipitation has been found to be from 600 to 1200 °C. In Nb-stabilized stainless steels, in case the Z phase forms so does the NbC carbide. This latter is probably more stable than the Z phase. The Z phase forms on GB and within grains with different morphologies |
| Nitrides | Pi | π | Cubic primitive | – | a = 0.647 | $Mo_{13}Fe_7N_4$ | The π phase is a nitride that has been found in duplex grades, as well as in SASSs such as 254 SMO and 654 SMO |
| Borides | $M_2B$ | – | Orthorhombic | 48 | a = 1.4585<br>b = 0.7331<br>c = 0.4223 | $Cr_{1.04}Fe_{0.96}B$ | Typical of B-containing grades such as AISI 304 + boron. Its precipitation occurs mostly at GBs. In the moment when precipitation occurs within the grain it is often found to form at incoherent phase boundaries of primary carbides. The addition of B is associated with beneficial effects when it is in solid solution (hardenability improvement) |

(continued)

**Appendix B** (continued)

| Phase type | Name | Symbol | Unit cell | Atoms per cell | Lattice parameters [nm] | Typical chemical composition/s | Additional information |
|---|---|---|---|---|---|---|---|
| Borides | $M_3B_2$ | – | Tetragonal | 10 | a = 0.5807<br>c = 0.3142 | $FeMo_2B_2$ | Typical of B-containing grades. It is found less frequently than $M_2B$ and occurs only in steels with high Mo contents |

*ATP atomic probe tomography. GBs* grain boundaries, *HE* hydrogen embrittlement. Sulfides are deliberately excluded from this appendix [1–21]

# References


1. Gialanella, S., Malandruccolo, A.: Aerospace Alloys. Springer (2020)
2. McGuire, M. F.: Stainless Steels for Design Engineers. ASM International, Materials Park, Ohio (2008)
3. Martorano, M., et al.: Predicting delta ferrite content in stainless steel castings. ISIJ Int. **52**(6), 1054–1065 (2012)
4. Barroux, A., et al.: Pitting corrosion of 17-4PH stainless steel manufactured by laser beam melting. Corros. Sci. **169**, 108594 (2020)
5. Charles, J.: Duplex stainless steels—a review after DSS'07 held in Grado. Steel Res. Int. **79**(6), 455–465 (2008)
6. De Cicco, H., et al.: Microstructural development and creep behavior in A286 superalloy. Mater. Charact. **52**(2), 85–92 (2004)
7. De Tiedra, P., et al.: Potentiodynamic study of the influence of gamma prime and eta phases on pitting corrosion of A286 superalloy. J. Alloy. Compd. **673**, 231–236 (2016)
8. Isalgué, A., et al.: Study of the spinodal decomposition of an Fe-28Cr-2Mo-4Ni-Nb alloy by small-angle neutron scattering. J. Mater. Sci. **25**(12), 4977–4980 (1990)
9. Knyazeva, M., Pohl, M.: Duplex steels. Part I: Genesis, formation, structure. Metall. Microstruct. Anal. **2**(2), 113–121 (2013)
10. Knyazeva, M., Pohl, M.: Duplex steels. Part II: Carbides and nitrides. Metall. Microstruct. Anal. **2**(5), 343–351 (2013)
11. Kobayashi, S., Takasugi, T.: Mapping of 475 °C embrittlement in ferritic Fe-Cr-Al alloys. Scripta Mater. **63**(11), 1104–1107 (2010)
12. Mateo, A., et al.: Characterization of the intermetallic G-phase in an AISI 329 duplex stainless steel. J. Mater. Sci. **32**, 4533–4540 (1997)
13. Padilha, A.F., Rios, P.R.: Decomposition of austenite in austenitic stainless steels. ISIJ Int. **42**(4), 325–337 (2002)
14. Phillips, N.S.L., et al.: Phase transformations in cast superaustenitic stainless steels. J. Mater. Eng. Perform. **18**(9), 1285–1293 (2009)
15. Smith, D., et al.: Thermophysical properties of alloy 709. Nucl. Eng. Des. **322**, 331–335 (2017)
16. Song, Z., Pu, E.: Precipitated phases of superaustenitic stainless steel 654SMO. J. Iron. Steel Res. Int. **24**, 743–749 (2017)
17. Svoboda, M., et al.: Phase changes in superaustenitic steels after long-term annealing. Z. Met. **95**(11), 1025–1030 (2004)




A. Malandruccolo, *Superaustenitic Stainless Steels*, Engineering Materials,
https://doi.org/10.1007/978-3-031-68744-0

18. Vander Voort, G., et al.: A study of selective etching of carbides in steel. Microsc. Microanal. **10**(S02), 76–77 (2004)
19. Wang, Z., et al.: Atom probe tomographic characterization of nanoscale Cu-rich precipitates in 17-4 precipitate hardened stainless steel tempered at different temperatures. Microsc. Microanal. **23**(2), 340–349 (2017)
20. Yeli, G., et al.: Sequential nucleation of phases in a 17-4PH steel: microstructural characterization and mechanical properties. Acta Mater. **125**, 38–49 (2017)
21. Yoo, W., et al.: Study on the microstructure and mechanical properties of 17-4PH stainless steel depending on heat treatment and aging time. Solid State Phenom. **118**, 15–20 (2006)

SPRINGER NATURE

# GPSR Compliance

*The European Union's (EU) General Product Safety Regulation (GPSR) is a set of rules that requires consumer products to be safe and our obligations to ensure this.*

*If you have any concerns about our products, you can contact us on ProductSafety@springernature.com*

In case Publisher is established outside the EU, the EU authorized representative is:

Springer Nature Customer Service Center GmbH
Europaplatz 3
69115 Heidelberg, Germany

The manufacturer's authorised representative in the EU is Springer Nature Customer Service Centre GmbH, Europaplatz 3, 69115 Heidelberg, Germany. If you have any concerns regarding our products, please contact ProductSafety@springernature.com

Printed and bound by CPI Group (UK) Ltd, Croydon, CR0 4YY
01/10/2026
02225965-0008